Surface Engineering 2004—
Fundamentals and
Applications

MATERIALS RESEARCH SOCIETY
SYMPOSIUM PROCEEDINGS VOLUME 843

Surface Engineering 2004—Fundamentals and Applications

Symposium held November 30–December 2, 2004, Boston, Massachusetts, U.S.A.

EDITORS:

Soumendra N. Basu
Boston University
Boston, Massachusetts, U.S.A.

James E. Krzanowski
University of New Hampshire
Durham, New Hampshire, U.S.A.

Joerg Patscheider
EMPA
Duebendorf, Switzerland

Yury Gogotsi
Drexel University and
A.J. Drexel Nanotechnology Institute
Philadelphia, Pennsylvania, U.S.A.

Materials Research Society
Warrendale, Pennsylvania

CAMBRIDGE
UNIVERSITY PRESS

University Printing House, Cambridge CB2 8BS, United Kingdom

One Liberty Plaza, 20th Floor, New York, NY 10006, USA

477 Williamstown Road, Port Melbourne, VIC 3207, Australia

314-321, 3rd Floor, Plot 3, Splendor Forum, Jasola District Centre, New Delhi - 110025, India

79 Anson Road, #06-04/06, Singapore 079906

Cambridge University Press is part of the University of Cambridge.

It furthers the University's mission by disseminating knowledge in the pursuit of education, learning and research at the highest international levels of excellence.

www.cambridge.org
Information on this title: www.cambridge.org/9781558997912

Materials Research Society
506 Keystone Drive, Warrendale, PA 15086
http://www.mrs.org

First published 2005
First paperback edition 2013

Single article reprints from this publication are available through University Microfilms Inc., 300 North Zeeb Road, Ann Arbor, MI 48106

CODEN: MRSPDH

A catalogue record for this publication is available from the British Library

ISBN 978-1-558-99791-2 Hardback
ISBN 978-1-107-40904-0 Paperback

This symposium has been supported by the National Science Foundation under Grant Number CMS-04566508. Any opinions, findings, and conclusions or recommendations expressed in this material are those of the author(s) and do not necessarily reflect the views of the National Science Foundation.

CONTENTS

*Invited Paper

MECHANICAL PROPERTIES OF SURFACES, INTERFACES AND THIN FILMS

*Invited Paper

ION BEAM, LASER AND PLASMA MODIFICATION OF SURFACES

*Invited Paper

SURFACE INDENTATION AND PHASE TRANSFORMATIONS

SYNTHESIS AND CHARACTERIZATION OF ENGINEERED SURFACES AND COATINGS

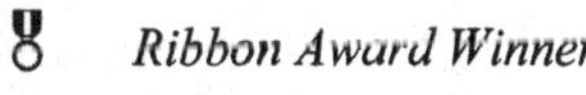

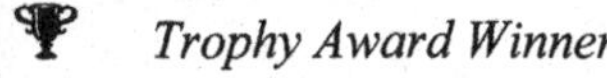

PREFACE

This volume contains papers presented at Symposium T, "Surface Engineering 2004—Fundamentals and Applications," held November 30–December 2 at the 2004 MRS Fall Meeting in Boston, Massachusetts. This symposium was the third of its kind devoted to this subject.

This symposium covered a broad range of topics related to surface engineering, surface modification and surface mechanics and presented a broad, multidisciplinary program promoting the interaction of scientists and engineers from many fields with the central goal of modifying and enhancing engineered surfaces. Topics included the effects of coating nanostructure on hardness; new developments in super-low friction coatings; control of stresses during surface modification; modification of surfaces and coatings using ion beams and lasers; and new methods for synthesizing and characterizing engineered surfaces and coatings.

The symposium featured six oral sessions, including a joint session with Symposium R, "Fundamentals of Nanoindentation and Nanotribology III," as well as a poster session. These sessions included more than 60 oral presentations and over 40 poster presentations and were attended by over three hundred participants. The symposium also featured ten invited presentations by distinguished scientists.

The symposium co-organizers would like to gratefully acknowledge Balzers AG, Renishaw Incorporated and the Surface Engineering and Material Design Program of the U.S. National Science Foundation for financial support.

Soumendra N. Basu
James E. Krzanowski
Joerg Patscheider
Yury Gogotsi

February 2005

MATERIALS RESEARCH SOCIETY SYMPOSIUM PROCEEDINGS

Volume 807— Scientific Basis for Nuclear Waste Management XXVII, V.M. Oversby, L.O. Werme, 2004, ISBN: 1-55899-752-0
Volume 808— Amorphous and Nanocrystalline Silicon Science and Technology—2004, R. Biswas, G. Ganguly, E. Schiff, R. Carius, M. Kondo, 2004, ISBN: 1-55899-758-X
Volume 809— High-Mobility Group-IV Materials and Devices, M. Caymax, E. Kasper, S. Zaima, K. Rim, P.F.P. Fichtner, 2004, ISBN: 1-55899-759-8
Volume 810— Silicon Front-End Junction Formation—Physics and Technology, P. Pichler, A. Claverie, R. Lindsay, M. Orlowski, W. Windl, 2004, ISBN: 1-55899-760-1
Volume 811— Integration of Advanced Micro- and Nanoelectronic Devices—Critical Issues and Solutions, J. Morais, D. Kumar, M. Houssa, R.K. Singh, D. Landheer, R. Ramesh, R. Wallace, S. Guha, H. Koinuma, 2004, ISBN: 1-55899-761-X
Volume 812— Materials, Technology and Reliability for Advanced Interconnects and Low-k Dielectrics—2004, R. Carter, C. Hau-Riege, G. Kloster, T-M. Lu, S. Schulz, 2004, ISBN: 1-55899-762-8
Volume 813— Hydrogen in Semiconductors, N.H. Nickel, M.D. McCluskey, S. Zhang, 2004, ISBN: 1-55899-763-6
Volume 814— Flexible Electronics 2004—Materials and Device Technology, B.R. Chalamala, B.E. Gnade, N. Fruehauf, J. Jang, 2004, ISBN: 1-55899-764-4
Volume 815— Silicon Carbide 2004—Materials, Processing and Devices, M. Dudley, P. Gouma, P.G. Neudeck, T. Kimoto, S.E. Saddow, 2004, ISBN: 1-55899-765-2
Volume 816— Advances in Chemical-Mechanical Polishing, D. Boning, J.W. Bartha, G. Shinn, I. Vos, A. Philipossian, 2004, ISBN: 1-55899-766-0
Volume 817— New Materials for Microphotonics, J.H. Shin, M. Brongersma, F. Priolo, C. Buchal, 2004, ISBN: 1-55899-767-9
Volume 818— Nanoparticles and Nanowire Building Blocks—Synthesis, Processing, Characterization and Theory, O. Glembocki, C. Hunt, C. Murray, G. Galli, 2004, ISBN: 1-55899-768-7
Volume 819— Interfacial Engineering for Optimized Properties III, C.A. Schuh, M. Kumar, V. Randle, C.B. Carter, 2004, ISBN: 1-55899-769-5
Volume 820— Nanoengineered Assemblies and Advanced Micro/Nanosystems, J.T. Borenstein, P. Grodzinski, L.P. Lee, J. Liu, Z. Wang, D. McIlroy, L. Merhari, J.B. Pendry, D.P. Taylor, 2004, ISBN: 1-55899-770-9
Volume 821— Nanoscale Materials and Modeling—Relations Among Processing, Microstructure and Mechanical Properties, P.M. Anderson, T. Foecke, A. Misra, R.E. Rudd, 2004, ISBN: 1-55899-771-7
Volume 822— Nanostructured Materials in Alternative Energy Devices, E.R. Leite, J-M. Tarascon, Y-M. Chiang, E.M. Kelder, 2004, ISBN: 1-55899-772-5
Volume 823— Biological and Bioinspired Materials and Devices, J. Aizenberg, C. Orme, W.J. Landis, R. Wang, 2004, ISBN: 1-55899-773-3
Volume 824— Scientific Basis for Nuclear Waste Management XXVIII, J.M. Hanchar, S. Stroes-Gascoyne, L. Browning, 2004, ISBN: 1-55899-774-1
Volume 825E—Semiconductor Spintronics, B. Beschoten, S. Datta, J. Kikkawa, J. Nitta, T. Schäpers, 2004, ISBN: 1-55899-753-9
Volume 826E—Proteins as Materials, V.P. Conticello, A. Chilkoti, E. Atkins, D.G. Lynn, 2004, ISBN: 1-55899-754-7
Volume 827E—Educating Tomorrow's Materials Scientists and Engineers, K.C. Chen, M.L. Falk, T.R. Finlayson, W.E. Jones Jr., L.J. Martinez-Miranda, 2004, ISBN: 1-55899-755-5
Volume 828— Semiconductor Materials for Sensing, S. Seal, M-I. Baraton, N. Murayama, C. Parrish, 2005, ISBN: 1-55899-776-8
Volume 829— Progress in Compound Semiconductor Materials IV—Electronic and Optoelectronic Applications, G.J. Brown, M.O. Manasreh, C. Gmachl, R.M. Biefeld, K. Unterrainer, 2005, ISBN: 1-55899-777-6
Volume 830— Materials and Processes for Nonvolatile Memories, A. Claverie, D. Tsoukalas, T-J. King, J. Slaughter, 2005, ISBN: 1-55899-778-4
Volume 831— GaN, AlN, InN and Their Alloys, C. Wetzel, B. Gil, M. Kuzuhara, M. Manfra, 2005, ISBN: 1-55899-779-2
Volume 832— Group-IV Semiconductor Nanostructures, L. Tsybeskov, D.J. Lockwood, C. Delerue, M. Ichikawa, 2005, ISBN: 1-55899-780-6

MATERIALS RESEARCH SOCIETY SYMPOSIUM PROCEEDINGS

Volume 833— Materials, Integration and Packaging Issues for High-Frequency Devices II, Y.S. Cho, D. Shiffler, C.A. Randall, H.A.C. Tilmans, T. Tsurumi, 2005, ISBN: 1-55899-781-4
Volume 834— Magneto-Optical Materials for Photonics and Recording, K. Ando, W. Challener, R. Gambino, M. Levy, 2005, ISBN: 1-55899-782-2
Volume 835— Solid-State Ionics—2004, P. Knauth, C. Masquelier, E. Traversa, E.D. Wachsman, 2005, ISBN: 1-55899-783-0
Volume 836— Materials for Photovoltaics, R. Gaudiana, D. Friedman, M. Durstock, A. Rockett, 2005, ISBN: 1-55899-784-9
Volume 837— Materials for Hydrogen Storage—2004, T. Vogt, R. Stumpf, M. Heben, I. Robertson, 2005, ISBN: 1-55899-785-7
Volume 838E—Scanning-Probe and Other Novel Microscopies of Local Phenomena in Nanostructured Materials, S.V. Kalinin, B. Goldberg, L.M. Eng, B.D. Huey, 2005, ISBN: 1-55899-786-5
Volume 839— Electron Microscopy of Molecular and Atom-Scale Mechanical Behavior, Chemistry and Structure, D. Martin, D.A. Muller, E. Stach, P. Midgley, 2005, ISBN: 1-55899-787-3
Volume 840— Neutron and X-Ray Scattering as Probes of Multiscale Phenomena, S.R. Bhatia, P.G. Khalifah, D. Pochan, P. Radaelli, 2005, ISBN: 1-55899-788-1
Volume 841— Fundamentals of Nanoindentation and Nanotribology III, D.F. Bahr, Y-T. Cheng, N. Huber, A.B. Mann, K.J. Wahl, 2005, ISBN: 1-55899-789-X
Volume 842— Integrative and Interdisciplinary Aspects of Intermetallics, M.J. Mills, H. Clemens, C-L. Fu, H. Inui, 2005, ISBN: 1-55899-790-3
Volume 843— Surface Engineering 2004—Fundamentals and Applications, J.E. Krzanowski, S.N. Basu, J. Patscheider, Y. Gogotsi, 2005, ISBN: 1-55899-791-1
Volume 844— Mechanical Properties of Bioinspired and Biological Materials, C. Viney, K. Katti, F-J. Ulm, C. Hellmich, 2005, ISBN: 1-55899-792-X
Volume 845— Nanoscale Materials Science in Biology and Medicine, C.T. Laurencin, E. Botchwey, 2005, ISBN: 1-55899-793-8
Volume 846— Organic and Nanocomposite Optical Materials, A. Cartwright, T.M. Cooper, S. Karna, H. Nakanishi, 2005, ISBN: 1-55899-794-6
Volume 847— Organic/Inorganic Hybrid Materials—2004, C. Sanchez, U. Schubert, R.M. Laine, Y. Chujo, 2005, ISBN: 1-55899-795-4
Volume 848— Solid-State Chemistry of Inorganic Materials V, J. Li, M. Jansen, N. Brese, M. Kanatzidis, 2005, ISBN: 1-55899-796-2
Volume 849— Kinetics-Driven Nanopatterning on Surfaces, E. Wang, E. Chason, H. Huang, G.H. Gilmer, 2005, ISBN: 1-55899-797-0
Volume 850— Ultrafast Lasers for Materials Science, M.J. Kelley, E.W. Kreutz, M. Li, A. Pique, 2005, ISBN: 1-55899-798-9
Volume 851— Materials for Space Applications, M. Chipara, D.L. Edwards, S. Phillips, R. Benson, 2005, ISBN: 1-55899-799-7
Volume 852— Materials Issues in Art and Archaeology VII, P. Vandiver, J. Mass, A. Murray, 2005, ISBN: 1-55899-800-4
Volume 853E—Fabrication and New Applications of Nanomagnetic Structures, J-P. Wang, P.J. Ryan, K. Nielsch, Z. Cheng, 2005, ISBN: 1-55899-805-5
Volume 854E—Stability of Thin Films and Nanostructures, R.P. Vinci, R. Schwaiger, A. Karim, V. Shenoy, 2005, ISBN: 1-55899-806-3
Volume 855E—Mechanically Active Materials, K.J. Van Vliet, R.D. James, P.T. Mather, W.C. Crone, 2005, ISBN: 1-55899-807-1
Volume 856E—Multicomponent Polymer Systems—Phase Behavior, Dynamics and Applications, K.I. Winey, M. Dadmun, C. Leibig, R. Oliver, 2005, ISBN: 1-55899-808-X
Volume 858E—Functional Carbon Nanotubes, D.L. Carroll, B. Weisman, S. Roth, A. Rubio, 2005, ISBN: 1-55899-810-1
Volume 859E—Modeling of Morphological Evolution at Surfaces and Interfaces, J. Evans, C. Orme, M. Asta, Z. Zhang, 2005, ISBN: 1-55899-811-X
Volume 860E—Materials Issues in Solid Freeforming, S. Jayasinghe, L. Settineri, A.R. Bhatti, B-Y. Tay, 2005, ISBN: 1-55899-812-8
Volume 861E—Communicating Materials Science—Education for the 21st Century, S. Baker, F. Goodchild, W. Crone, S. Rosevear, 2005, ISBN: 1-55899-813-6

Prior Materials Research Society Symposium Proceedings available by contacting Materials Research Society

Hard and Protective Coatings

Mater. Res. Soc. Symp. Proc. Vol. 843 © 2005 Materials Research Society T1.2

Investigations of the effect of $(Cr_{1-x},Al_x)N$ coatings' micro Structure on Impact Toughness

K. Bobzin, E. Lugscheider, O. Knotek, M. Maes *

Material Science Institute, RWTH Aachen University, Augustinerbach 4-22, D-52056 Aachen, Germany

Abstract

Originated from the tooling industry, PVD (Physical Vapor Deposition) coating development focused on increasing the wear resistance. Nowadays, a steadily increasing market is evolving by coating machine parts. The requirements that have to be met due to the needs of this new market segment focus on tribological behavior. This means, that the focus of wear resistance is shifted towards properties like coefficient of friction, wetting behavior and the response of coatings towards dynamic loads. For many tribological applications, coatings are exposed to severe alternating loads, which are usually left out in common test methods. The approach of common coating test methods are based on the static behavior of deposited coatings. The impact tester is a testing device with a novel approach to dynamic load behavior of both bulk and coated materials. In this paper, the effect of the coatings' microstructure and Young's modulus on the impact toughness was investigated. A change in microstructure was provoked by changing deposition parameters like aluminum content. In a second stage these coatings were then tested with respect to their response to high alternating loads. For this purpose both load and number of impacts were varied.

Key words: CrAlN, impact testing, tribology

* M.Maes

Email address: maes@msiww.rwth-aachen.de (K. Bobzin, E. Lugscheider, O. Knotek, M. Maes).

URL: http:/www.rwth-aachen.de/ww (K. Bobzin, E. Lugscheider, O. Knotek, M. Maes).

1 Introduction

Physical Vapor Deposition (PVD) or Plasma Enhanced Chemical Vapor Deposition (PECVD) carbon based coatings are usually related to machine parts. Chromium based coatings are mostly not considered for this tribological application, since the dry friction coefficient is overshadowed by those of carbon based coatings. However, preliminary investigations[1] show, when lubricated, the friction torque in machine parts (for instance: cylindrical roller bearings) is equal to that of a low friction carbon based coating Graded Zirconium carbide (ZrC_g).

The advantages of chromium based coatings are manifold. Although Cr_xN exceeds the hardness of a hardened steel, it cannot match the hardness of TiN or TiC [2]. Its toughness is remarkable, yet difficult to quantify. Low surface energy of the immediately formed Chromium Oxides Cr_2O_3, when exposed to oxygen, can offer good protection against adhesive wear and corrosion. The formation of protective, passivating Cr_2O_3 coatings [3,4] also leads to an improved oxidation resistance ($< 800°C$), in comparison to TiN or TiC coatings ($< 550°C$). For this reason Cr_xN coatings are used amongst others for cutting of non-ferrous heavy metals and in plastic processing [5]. Other points of interest for tribological applications are the non depending friction coefficient against steel when exposed to air's humidity. Also the higher temperature stability than carbon based coatings is of interest. The compressive residual stress within these coatings can help to prevent thermal or mechanical crack initiation [6,7]. This is particularly interesting since most tribological applications deal with high alternating loads. Former impact testings have proven Cr_xN coatings to be more resistant against impact loading and were able to achieve a much higher service lifetime [9,10]. These arguments have left us to believe, that chromium based coatings show an extraordinary potential as a tribological coating. They can offer a perfect alternative, where carbon based coatings fail to perform.

The system $(Cr_{1-x},Al_x)N$ is a novel generation of chromium based coatings and extensively investigated by [11,12,7,13] at deposition temperatures of ($> 400°C$). By adding Aluminum the Young's modulus is altered and a composite coating generated. Also the addition of aluminum effected the structure of the synthesized coatings, although it is not suggested, that this is aluminium content related. Aim of this paper was to observe the dynamic load behavior of the differently synthesized coatings.

2 Experimental Details

Since coating development of machine parts is focus of this paper, steel (100Cr6) and carbide inserts were used for deposition. All samples were mirror polished and ultrasonically cleaned in alkaline solutions of different concentrations, rinsed with deionised water and finally dried. The 100Cr6 samples were subjected to a hardening and annealing process resulting in a final hardness of 60 Rockwell C. The samples were hardened at 860 °C for 1h and tempered for 2h at 180 °C. After cleaning, the samples are immediately batched in a pre conditioned deposition chamber and all three stages (Heating, Etching and Deposition) were carried out in this same chamber. Pre conditioning of the deposition chamber comprises baking out the vacuum chamber in order to get rid of entrapped water within the chamber walls.

The coating equipment was a CemeCon CC800/9 equipped with four Targets of either Aluminum and/or Chromium see Fig.1. The first two stages of the process, heating (T_{Max}=160 °C, 1h) and ion etching (1h) ensure a degassing and a full removal of surface oxides, which both strongly effects coating adhesion. The gas flow during deposition was controlled for both argon and nitrogen with a fixed ratio of Ar:N_2 resp. 300 sccm to 60 sccm resulting in a chamber pressure of 580 mPa. To avoid annealing of the temperature sensitive substrates the sum of power rates for all cathodes never exceeded 8kW during the entire deposition process. To generate coatings of different aluminium content, cathode arrangement and local cathode power was varied between 1 and 3kW see Table 1. However the sum was kept constant at 8kW at all times.

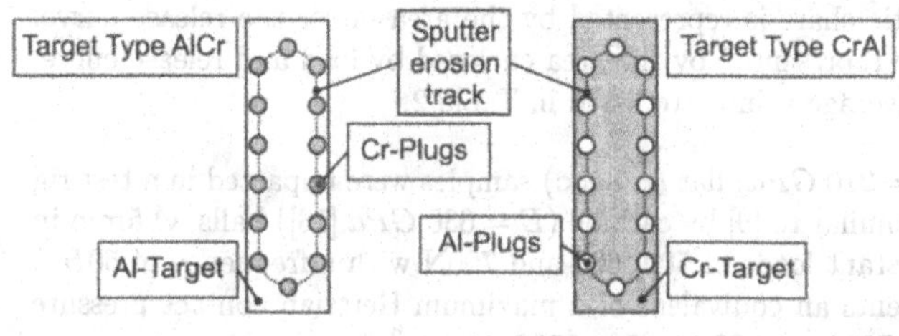

Fig. 1. Target design

Table 1
Experimental data on coating deposition

Batch	Position CrAl	Power	Position AlCr	Power
2054.CrAlN	1,2,3,4	4x2 kW	-	-
2070.CrAlN	1,3	2x1 kW	2,4	2x3 kW
2090.CrAlN	-	-	1,2,3,4	4x2 kW

The experimental values are given in ref. [14].

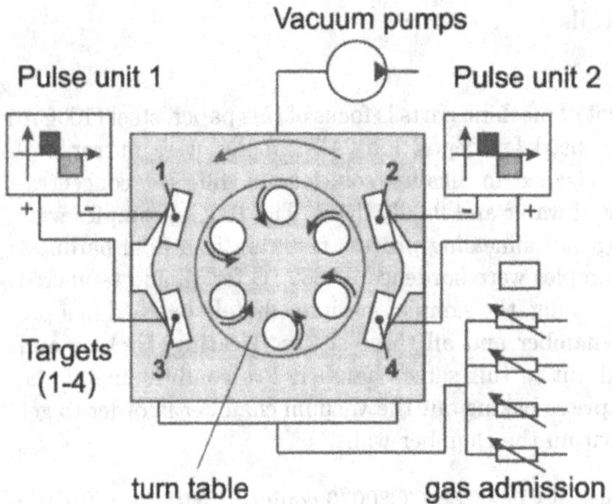

Fig. 2. Experimental setup

After deposition, the coating properties were determined with respect to coating thickness, adhesion [15], hardness and impact toughness. Nanoindentation (Equipment: MTS Systems Nanoindentation XP) was performed with a Berkovic indenter .The indenter penetrated the coated surface perpendicular at a constant load of 10 mN. Calculations of the Young's modulus are based on Oliver and Phar's equations [16,17] and Poisson ratio was kept constant at $\nu = 0.25$. Additionally both elastic and plastic deformation of the coating was calculated by numeric integration from the load/displacement curves Fig. 3. The plastic share is represented by the area under the release curve. The elastic share represented by the area enclosed by load and release curve. The calculated average values are listed in Table 2.

The 100Cr6 ($E = 210\ GPa$) flat ($r \to \infty$) samples were impacted in a test rig (Discription[8]) similar to [9] by carbide ($E = 630\ GPa$ [18]) balls Ø $5mm$ in diameter, at constant loads of 500, 600 and 700N with a frequency of $50Hz$. This load represents an equivalent of a maximum Hertzian contact pressure of respectively $-7600, \ -8000\ and\ -8500\ N/mm^2$.

$$\sigma_{Z,\ Max} = \sigma_0 = -\frac{1}{\pi} \cdot \sqrt[3]{\frac{1.5 \cdot F \cdot E^2}{r^2 \cdot (1-\nu^2)^2}} \tag{1}$$

$$\frac{1}{r} = \frac{1}{r_1} + \frac{1}{r_2} \tag{2}$$

$$E = 2 \cdot \frac{E_1 \cdot E_2}{E_1 + E_2} \tag{3}$$

Table 2
Deposition parameters and coating analysis.

Process ID	2054.CrAlN	2070.$CrAlN$	2090.$CrAlN$
$(Cr_{1-x},Al_x)N$	$x = 0.23$	$x = 0.48$	$x = 0.66$
$T_{R,\ Max}$ [°C]	120	190	120
Deposition time t_B [s]	4800	4800	8400
Deposition rate $\partial s/\partial$ t [μ m/h]	2.7	0.9	1.7
Coating thickness s_D [μ m]	3.6	1.2	3.9
Hardness Rockwell C	59 HRC	60 HRC	60 HRC
Coating hardness [GPa]	13.1	22.0	19.8
Young's modulus [GPa]	305	445	283
$W_{plastic}$[mN·nm]	541.5	389.2	421.8
$W_{elastic}$[mN·nm]	278.9	282.1	342.0
Ratio $W_{elastic}/W_{plastic}$	0.52	0.72	0.81
Adhesive strength HF [15]	HF 1	HF 1	HF 1
Critical Scratch Load L_C [N]	>90	20	70

The experimental values are given in ref. [14].

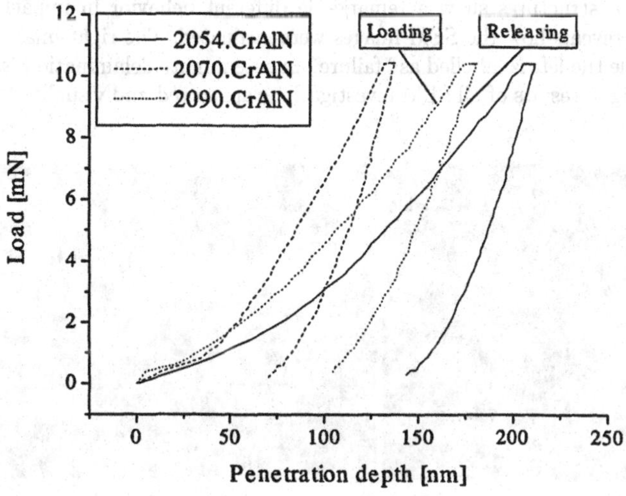

Fig. 3. Load/displacement curves CrAlN

3 Results

The results of mechanical testing are listed in Table 2. The coatings' microstructure is shown in Fig. 4. For both high and low aluminum contents (x=0.23 and x=0.66) a columnar growth of the coatings is observed. At an aluminum content of x=0.48 this columnar growth disappears almost entirely, resulting in finer grained coating structure. This coating also exhibits highest hardness, Young's modulus and deposition temperature. Additionally a widening of the columns towards the surface is observed for the coating with lowest aluminum content (x=0.23). In contrast the coating with highest aluminum content (x=0.66) shows a more even, and smaller column width.

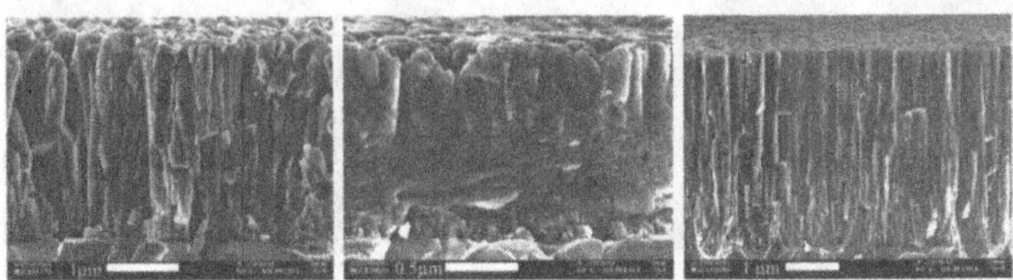

Fig. 4. SEM Fracture image 2054.CrAlN, 2070.CrAlN, 2090.CrAlN

The synthesized structures show a remarkable different behavior in impact testing. Fig. 5 reveals how the SEM images were evaluated. The right image is "intact" while the left is labelled as "failure" in case minute delamination is observed. In Fig. 6 results of all SEM investigations are sorted and visualized.

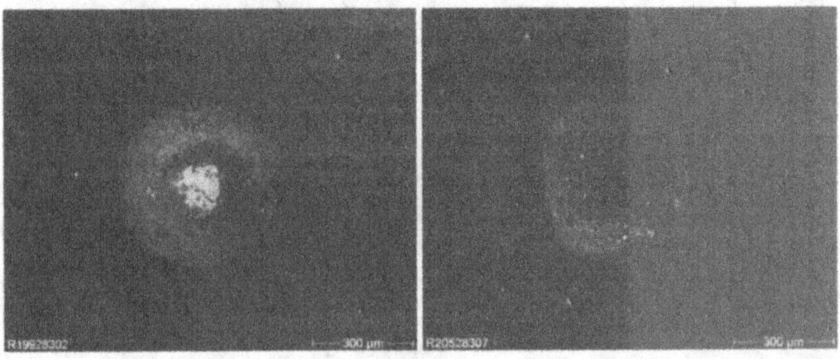

Fig. 5. SEM image 2054.CrAlN, 2090.CrAlN. Load: 600N, Impacts: 10^6

Most coatings eventually show small delaminations within the impact crater, similar to the left image Fig. 5, when either load or number of impacts is increased. Fig. 7 reveals a different behavior. The image reveals crack initiation at the border of the impact crater, whilst the center area of the impact is completely delaminated.

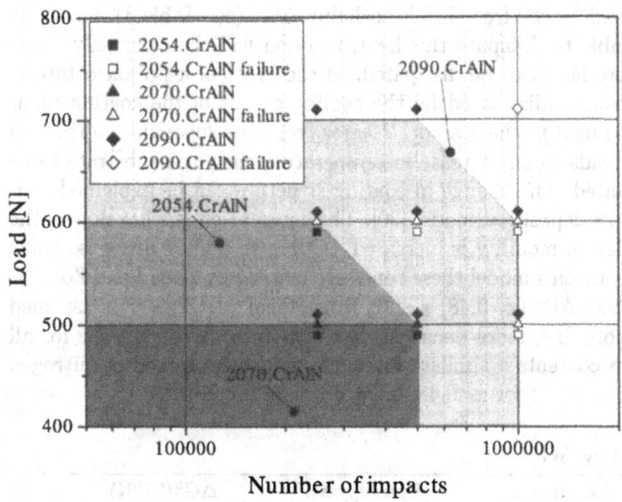

Fig. 6. Results impact testing of 2054.CrAlN, 2070.CrAlN, 2090.CrAlN

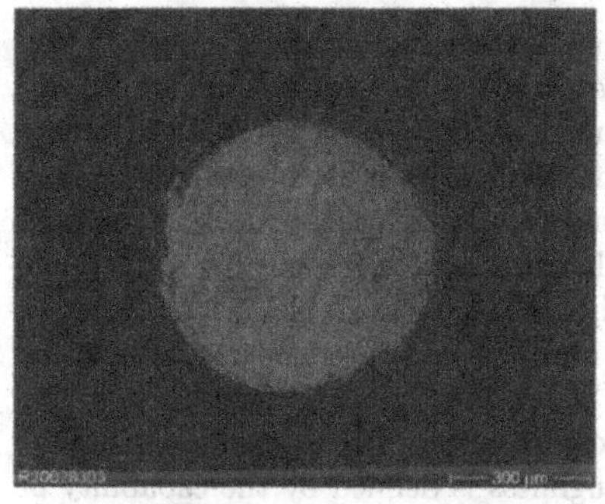

Fig. 7. SEM image 2070.CrAlN. Load: 600N, Impacts: 10^6

4 Discussion

The results show, that although all deposition parameters were kept constant a remarkable change in structure is observed for the coating with an aluminum content of x=0.48. This observation can be explained by the degree of target poisoning[19]. This is a phenomenon, which is observed when a surplus of reactive gas is present during sputtering. Consequently, during sputtering a reactive layer is formed at the targets' surface. The reactive heat will be dissipated at the cathodes. In case of reactive deposition heat generated is represented by the free Gibb's enthalpy ΔG_f° (see Table 3). Since the gas phase is unable to dissipate this heat, reaction will therefore only occur at surfaces, where heat can be dissipated. In the event of a poisoned target, the heat is eventually dissipated by the cooling circuit of the coating plant and does not add heat to the reactor. Whereas reactive deposition in the non poisoned mode leads to an increase in temperature since the substrate table is thermally isolated. The change in coating structure can be explained considering Thorton's deposition model [20]. The assumption is, that due to the reactive heat adatom mobility is increased, leading to a more dense isotropic structure. In Thornton's model these zones are marked by Zone I and Zone T. For coating 2070.CrAlN (x=0.48) a shift from ZoneI $\rightarrow$ Zone T is assumed resulting in a more fine, more dense istropic structure. By our means for all other aluminium contents a similar structure could be obtained by nitrogen flow adjustment(increase for x<0.48 and a decrease for x>0.48).

Table 3
Free Gibb's enthalpy ΔG_f°

Formula	Reaction	EntahlpyΔH°	ΔG_f°(773K)
CrN	$CrN \Leftrightarrow Cr + \frac{1}{2}N_2$	118-123[kJ/mol]	-61.6[kJ/mol]
AlN	$AlN \Leftrightarrow Al + \frac{1}{2}N_2$	289-318[kJ/mol]	-219.2[kJ/mol]

The experimental values are given in ref. [21].

With respect to impact resistance, columnar coating structures perform surprisingly well. Best performance is achieved by a fine grained columnar coating with a high aluminum content and a Young's modulus closest to the supporting substrate 100Cr6. The coating with an aluminum content of x=0.48 exhibits the worst impact resistance. It is likely to assume, that a lack of support caused by a decrease in hardness of the substrate is responsible for the low impact resistance. However no annealing effects (see Table 2) are observed and therefore no change in substrate hardness can be made responsible for a lack of support. Consequently, it is most likely, that either the columnar structure, the Young's modulus or a combination of both is responsible for the high impact toughness. Impact toughness is defined by the capability of the composite material to absorb or store impact energy without adding permanent damage to the material. In case of the columnar structure the energy induced by the impact is absorbed by friction between columns, plastic deformation or stored in elastic deformation. The columnar coating contains small intergranular spaces allowing the material to expand while being compressed. The induced tension is passed on to the substrate, without inducing tensile stresses parallel to the coating. In contrast, impact load applied to a fine granular structure will induce immediate tensile stresses within the coating causing rupture, flaking, break out etc.

The Young's modulus measured by nanoindentation is inevitably structure depending. Since the berkovich indenter penetrates the surface perpendicular and therefore in longitudinal axis of the column. Nanoindenter results between different coating structures are difficult to compare, but comparison between similar structures is, by our means, valid. Nevertheless, adding aluminum to the system CrAlN is helpful with respect to impact toughness. Due to the increase of aluminumnitride, an increase in hardness, as well as a reduction in Young's modulus is observed. FEM [1] have already shown, that bringing coating's and substrates' Young's modulus closer to one another helps to reduce stresses within the interface. Also the effect of structure refinement caused by the aluminum content helps to increase impact toughness.

5 Conclusions

The results clearly show that impact toughness was succesfully increased by adding aluminum to the $(Cr_{1-x},Al_x)N$ coatings. Additionally a columnar structure showed an improved resistance regarding impacts. Provided the load is applied perpendicular to the surface, a columnar coating structures shows best performance. The behavior of these coatings, when adding additional shear stresses to normal load has yet to be investigated. This kind of load can be achieved by tilting the substrate, whilst being impacted. Expected is, that these coatings will have difficulties to withstand these kind of loads. Up until now these coatings have already shown outstanding behavior for machine parts. They are well able to reduce fretting in gear wheels, and increase wear resistance in cylinder roller bearings. Future research will aim at combining both columnar and isotropic structures to reach even higher impact toughness. Hence increasing wear resistance of machine parts even more.

6 Acknowledgements

The authors wish to express their gratitude to the German Research Association for its financial support within the scope of the collaborative research program SFB 442 "Environmentally-compatible tribological systems".

References

[1] K. Bobzin, E. Lugscheider, M. Maes, P.W. Gold, J. Loos, M.Kuhn,Surface and Coating Technology, Vol. 188-189,Nov-Dec 2004, pp. 649-654

[2] C. Friedrich, G. Berg, E. Broszeit, C. Berger, Materialwissenschaften und Werkstofftechnik 28 (1997) pp. 59-76

[1] Finite Element Modelling

[3] O. Knotek, W. Bosch, M. Atzor, W.-D. Münz, D. Hoffmann, J. Goebel, High Temperature - High Pressures 18 (1986), pp. 435-442 .

[4] S. Hofmann, H.A. Jehn, Werkstoffe und Korrosion 41, (1990), pp. 756-760

[5] J. Vetter, E. Lugscheider, S.S. Guerreiro, Surface and Coatings Technology 98 (1998), pp. 1233-1239

[6] E. Lugscheider, K. Bobzin, Th. Hornig, M. Maes, Thin Solid Films,Vol. 420-421,Dec 2002, pp.318-323

[7] S. S. Guerreiro, PhD Thesis RWTH Aachen (1998), VDI-Verlag ISBN 3-18-353305-7

[8] E. Lugscheider, O. Knotek, C. Wolff and S. Brwulf, Surface and Coatings Technology, Volumes 116-119, September 1999, pp. 141-146

[9] K.-D. Bouzakis, N. Vidakis, A. Lontos, S. Mitsi, K. Davis, Surface and Coating Technology 133-134 (2000) pp. 489-496

[10] J.C.A. Batista, C. Godoy, A. Matthews, Surface and Coatings Technology 163-164 (2003) pp. 353-361

[11] M. Atzor, PhD Thesis RWTH Aachen (1989), Fortschritt-Berichte VDI: Reihe 5, ISBN 3-18-145605-5

[12] H.-J., Scholl, PhD Thesis RWTH Aachen (1993), Mainz-Verlag ISBN3-930085-07-0

[13] Th. Hornig, PhD Thesis RWTH Aachen (2002), Mainz-Verlag ISBN 3-89653-935-3

[14] K. Bobzin, E. Lugscheider, M. Maes, $(Cr_{1-x},Al_x)N$ PVD Niedertemperaturbeschichtung zum Verschleißschutz von Bauteilen, Proceedings GFT-Conference 2004

[15] N.N, DIN-Fachbericht 39, Charakterisierung dünner Schichten, Beuth Verlag, Berlin 1993 2004

[16] W.C. Oliver, G.M. Pharr, Journal of Materials Research, Vol. 7, No.6, June 1992, pp. 1564-1583

[17] G.M. Pharr, W.C. Oliver, MRS Bulletin, July 1992, pp. 28-33

[18] Grewe, Kolaska, Werkstoffkunde und Eigenschaften von Hartmetallen und Schneidkeramik, VDI-Z 125 Nr. 18, VDI-Verlag, 1983

[19] I. Safi, Surface and Coatings Technology, Vol. 127, Issues 2-3, 22 May 2000, Pages 203-218

[20] J.A. Thornton, J. Vac. Sci. Technol., 11, pp. 666- 670 (1974).

[21] N.N., J. Phys. Chem. Ref. Data, Vol. 14, 1985

Mater. Res. Soc. Symp. Proc. Vol. 843 © 2005 Materials Research Society T1.4

Diamondlike Carbon-Metal Nanocomposite Films for Medical Applications

R. J. Narayan[1], H. Abernathy[1], L. Riester[2], C. J. Berry[3], and R. Brigmon[3]

[1] School of Materials Science and Engineering, Georgia Institute of Technology, Atlanta, GA 30332-0245
[2] Metals and Ceramics Division, Oak Ridge National Laboratory, Oak Ridge, TN 37831
[3] Environmental Biotechnology Section, Savannah River National Laboratory, Aiken, SC 29808

Abstract

Silver and platinum were incorporated within diamondlike carbon (DLC) thin films using a multicomponent target pulsed laser deposition process. Transmission electron microscopy of the DLC-silver and DLC-platinum composite films reveals that these films self-assemble into particulate nanocomposite structures that possess a high fraction of sp^3-hybridized carbon atoms. DLC-silver-platinum films demonstrated exceptional antimicrobial properties against *Staphylococcus* bacteria.

Introduction

The introduction of implantable medical devices into the body has been shown to greatly increase the risk of infection. Infections involving artificial organs, synthetic vessels, joint replacements, or internal fixation devices usually require reoperation. Some infections are more serious than others; infected cardiac, abdominal, and extremity vascular prostheses result in amputation or death [1-2].

Bacteria form a glycocalyx, an adherent coating that forms on all foreign materials placed in vivo [3-5]. This 5-50 μm thick glycoprotein-based coating protects bacteria through a diffusion limitation process, and serves to decrease their antibiotic sensitivity by 10 to 100 times. In addition, the constituents of many alloys and polymers can inhibit both macrophage chemotaxis and phagocytosis. Finally, tissue damage caused by surgery and foreign body implantation further increases the susceptibility to infection.

The sustained delivery of silver ions into the local micro-environment of implants systemic side effects and exceeds usual systemic concentrations by several orders of magnitude [6-8]. Silver nanoparticles have been shown to possess an unsurpassed antimicrobial spectrum, with efficacy against 150 different pathogens. Silver ions bind strongly to electron donor groups on sulfur-, oxygen- or nitrogen-containing enzymes. These ions displace other cations (e.g., Ca^{2+}) important for enzyme function. In addition, nanocrystalline silver also provides broad-spectrum fungicidal action. Very low silver ion concentrations required for microbicidal activity (in the range 10^{-9} mol/l). Films containing both silver and platinum may demonstrate enhanced antimicrobial activity due to formation of a galvanic couple that accelerates silver ion release [7].

We have recently developed a variant of the conventional pulsed laser deposition process in order to incorporate antimicrobial metals during DLC film deposition. Briefly, a single multicomponent target is loaded into the pulsed laser deposition chamber. This target contains pure graphite that is covered by a piece of the desired modifying element (Figure 1). The focused laser beam sequentially ablates the graphite target component and the modifying

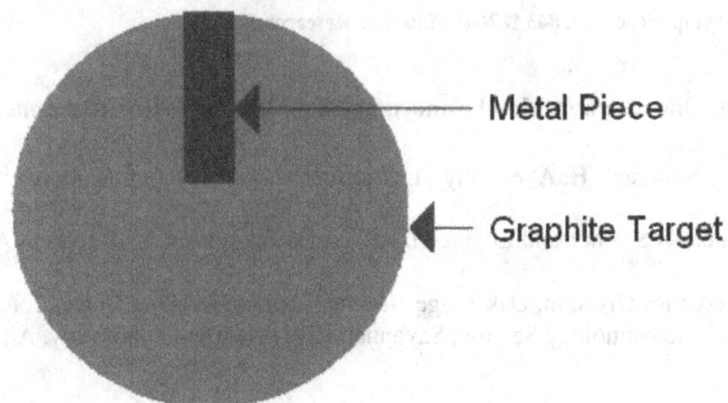

Figure 1. Schematic of target configuration used in this study.

element target component to form composite layers. The metal composition in these films can be controlled through altering the following parameters: (1) the scanning radius of the laser beam on the target surface, (2) the laser beam position, (3) the position of the circular target, (4) the size of the metal piece on the target, and (5) the laser energy density. The fraction of metal atoms incorporated into the diamondlike carbon film can be estimated using the following relation:

$$\text{fraction of metal within diamondlike carbon-metal nanocomposite film} = \alpha\, \delta\, (1-R_d)/2\, \pi\, \gamma(1-R_c) \quad [1]$$

in which α is the laser ablation ratio, γ is the laser beam scanning radius, R_c is the reflectivity of carbon, R_d is the reflectivity of the metal strip, and δ is width of the metal piece. We have prepared several diamondlike carbon-silver nanocomposite films using this modified pulsed laser deposition process. These films were examined using transmission electron microscopy, electron energy loss spectroscopy, visible Raman spectroscopy, nanoindentation, and microbial biofilm attachment testing.

Experimental

Several 1 cm x 1 cm pieces of silicon (100) were cleaned in acetone and methanol within an ultrasonic cleaner. The silicon substrates were subsequently dipped in hydrofluoric acid to remove silicon oxide, which resulted in a hydrogen-terminated surface. The Lambda Physik LPX 200 KrF excimer laser (248 nm wavelength) was used for ablation of the multicomponent target. One piece of silver and platinum were placed over the graphite target, which was rotated at 5 rpm throughout the deposition. The metal piece covered roughly one-sixth of the target surface area. The laser was operated at a frequency of 10 Hz and at a pulse duration of 25 ns. The laser energy output of 215 mJ and laser spot size of 0.043 cm^2 imparted an average energy density of ~ 5 J/cm^2 to the target. The target-substrate distance was maintained at 4.5 cm.

High resolution Z-contrast images were obtained using a JEOL 2010 F scanning transmission electron microscope equipped with field emission gun and Gatan Image Filter (GIF). In STEM-Z contrast imaging, an image is formed by collecting large-angle scattered electrons using an annular detector. The resulting contrast is proportional to atomic number squared (Z^2). Parallel electron energy loss spectroscopy was used to obtain information on carbon bonding; spectra were collected from zero loss up to 1000 eV energy loss. Micro-Raman spectra were obtained using an argon ion laser operating at 483-514 nm.

Biofilm attachment studies were done using *Staphylococcus sp* (wild type) grown in liquid media. Organisms were grown in nutrient broth (8 g Difco #0003-05-02 in 1L water, pH 6.8, sterilized) at 37 ºC for 15 minutes under 15 pounds pressure on a shaker platform. When log phase growth was obtained, one square centimeter DLC-silver/silicon, DLC-silver-platinum/silicon, and silicon chips were rinsed with ethyl alcohol, added to the media, and incubated for 24, 48 and 72 hours. Incubated materials were then rinsed three times in FA buffer (10 g of Difco #223142 in 1 liter deionized ultrafiltered water, pH = 7.2), vortexed for one minute, and then rinsed twice more in FA buffer. Samples were then heat fixed at 60º C for 12 minutes, stained with 4'-6-Diamidino-2-phenylindole dihydrochloride (DAPI) (0.7 micrograms of DAPI per ml phosphate buffer, pH 7.5) for five minutes, and rinsed with FA buffer. Stained microbial cells were counted and examined using a Zeiss Axioskop2 plus epifluorescent microscope and a Zeiss 510 Meta Laser Scanning Microscope.

Results and Discussion

Transmission electron microscopy

Z-contrast scanning transmission electron microscopy provides unique information on nanostructured composite thin films [8]. An image is formed by scanning a 2.2 Å probe across the sample. The Z-contrast signal is collected from a high angle annular detector, and the electron signal scattered through large angles (typically 75 to 150 mrad) is analyzed. Contrast is proportional to the atomic number (Z) squared. For example, the silver: carbon contrast is over 60:1 and the platinum:carbon contrast is 169:1. Dark field Z-contrast images of diamondlike carbon-silver and diamondlike carbon-platinum nanocomposite films are shown in Figure 2 (a) and Figure 2 (b), respectively. The bright regions correspond to the higher atomic number metal regions, and the dark regions correspond to the DLC matrix. The noble metals are dispersed as nearly spherical metal clusters in the diamondlike carbon matrix. The average size of these nanocrystalline particles varies between 3 and 5 nm. These images demonstrate that noble metals are segregated into nanoparticle arrays within the diamondlike carbon matrix.

Electron energy loss spectra between 280 to 310 eV were acquired. The sp^3 fraction was determined from the K edge loss spectra using an empirical technique developed by Cuomo et al. [9]. In this technique, the peak in the region from 285 to 290 eV results from excitation of electrons from the 1s ground state to the vacant π^* antibonding state. The peak in the region above 290 eV results from excitation to the higher σ^* state. The ratio of the integrated areas under these two energy windows is approximately proportional to the relative number of π and σ^* orbitals. Using this information, the atomic fraction of sp^2 bonded carbon (x) can be determined using the expression:

$$(I(\pi)/I(\sigma))_s/(I(\pi)/I(\sigma))_r=3x/(4-x) \quad [2]$$

in which $I(\pi)$ is the intensity in the range from 284 to 289 eV and $I(\sigma)$ is the integrated intensity in the range from 290 to 305 eV. The subscripts s and r refer to the ratio determined for the DLC specimen and a reference material with 100% sp^2 bonding, respectively. The sp^3 content was determined to be 63% for a diamondlike carbon film on silicon (100). The sp^3 content was determined to be 47% for a diamondlike carbon-silver nanocomposite film on Si (100). This data suggests that a moderate reduction in sp^3 content occurs in the metal alloyed films.

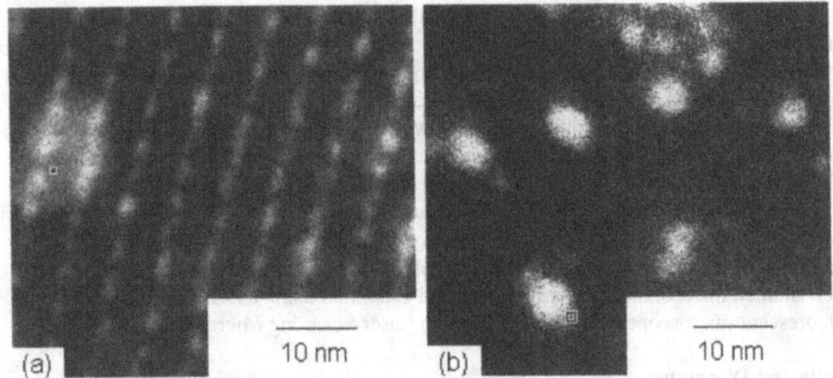

Figure 2. (a) Z-contrast image of diamondlike carbon-silver nanocomposite film. Well aligned arrays of silver nanodots were observed. (b) Z-contrast image of diamondlike carbon-platinum nanocomposite film. Metal nanodots appear in arrays within the diamondlike carbon matrix.

Raman spectroscopy

Adhesion of diamondlike carbon thin films is dependent on several factors, including film stress, film/substrate chemical bonding, and substrate topology [10-12]. Large internal compressive stresses as high as 10 GPa have been observed in diamondlike carbon thin films, regardless of the deposition process used. These stresses limit maximum film DLC thickness to 0.1–0.2 micrometers, and prevent widespread medical use. Lifshitz et al. have attributed these stresses to "subplantation" (low energy subsurface implantation) of carbon ions during diamondlike carbon film growth [13-14]. They suggest that carbon ions with energies between 10-1000 eV undergo shallow implantation to depths of 1-10 nm of during film growth. Carbon species are trapped in subsurface sites due to restricted mobility. This process leads to the development of very large internal compressive stresses.

The Raman spectra of diamondlike carbon thin films contain characteristic peaks that reflect carbon bonding and internal stress. All of the spectra show the following: (1) a broad hump centered in the 1510-1557 cm^{-1} region, which is known as the G- band, and (2) a small shoulder at 1350 cm^{-1}, which is known as the D- band. The G-band is the optically allowed E_{2g} zone center mode of crystalline graphite, and is typically observed in diamondlike carbon films. The D-band is the A_{1g} mode of graphite.

Diamondlike carbon films containing high sp^3 fractions demonstrate the following: (1) a relatively symmetrical G-band, and (2) a lesser D-band, suggesting an absence or a low amount of graphite clusters. The presence of metal atoms leads to a shift in the G-peak to lower wavenumbers, and a slight increase in the D-peak height (Figure 3). The G peak position shift can be produced by a decrease in sp^3 content, an increase in graphitic cluster size within the film, and/or a decrease in compressive stress within the film [15-16]. Silver and platinum have significantly smaller elastic moduli than diamondlike carbon, and noble metal nanoparticles may be expected to absorb stresses from the diamondlike carbon matrix.

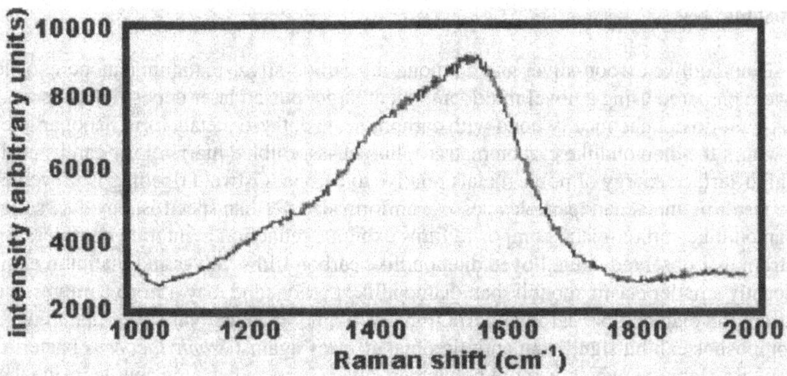

Figure 3. Visible Raman spectrum of diamondlike carbon-platinum nanocomposite film.

Biofilm Attachment Studies

The results of the biofilm attachment study showed a significant difference in the amount of microbial colonization per unit area between uncoated silicon wafers and DLC-silver and silver-platinum nanocomposite-coated silicon wafers. Unalloyed diamondlike carbon films delaminated from silicon substrates, and were unavailable for antimicrobial testing. The gram positive *Staphylococcus sp* also produced a quantifiable biofilm after 24, 48 and 72 hours. Using rich media and long incubation times, impacted biofilm formation on the DLC-metal composite films and control silicon wafers was observed (Figure 4). The DLC-silver-platinum nanocomposite films demonstrated one order of magnitude lower gram positive bacteria densities than the uncoated silicon wafers. Colonization rates were between 100 percent to half an order of magnitude lower on the DLC-silver-platinum nanocomposite films than on the uncoated silicon wafers.

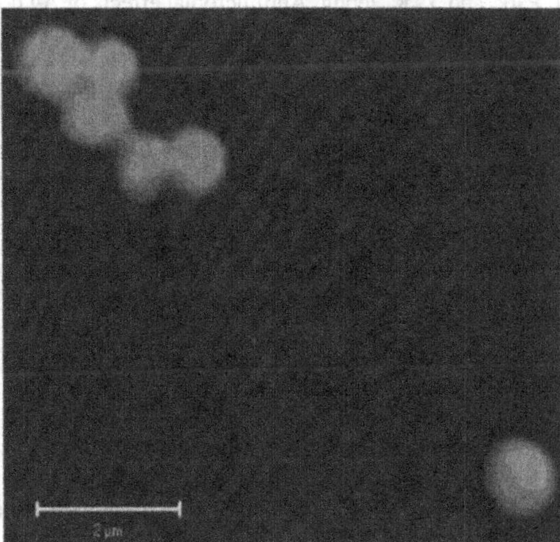

Figure 4. LSM meta system image of diamondlike carbon-silver nanocomposite film inoculated with *Staphlococcus sp*.

Conclusions

Diamondlike carbon-silver and diamondlike carbon-silver-platinum nanocomposite films were prepared using a novel multicomponent target pulsed laser deposition process. Silver and platinum do not chemically bond with carbon; instead, these metals form nanoparticle arrays within the diamondlike carbon matrix. This self-assembled morphology can be attributed to the high surface energy of noble metals relative to carbon. Ostwald ripening is prevented, and the resulting metal nanoparticles possess uniform size. Raman spectroscopy data suggest that diamondlike carbon-metal composite films exhibit a reduction in internal compressive stress from that observed in unalloyed diamondlike carbon films. Silver and platinum exhibit significantly smaller elastic moduli than diamondlike carbon, and may absorb compressive stress from the diamondlike carbon matrix. Finally, diamondlike carbon-silver-platinum nanocomposites exhibit significant antimicrobial efficacy against *Staphylococcus* bacteria. It is believed that platinum, which is more noble than silver, forms a galvanic couple inside the *in vitro* biofilm testing environment. Silver release (and, by extension, antimicrobial function) is greatly increased in these diamondlike carbon-silver-platinum nanocomposite films. The *in vitro* antimicrobial susceptibility of several bacterial and fungal pathogens to diamondlike carbon-biofunctional metal coatings is currently being assessed.

References

1. G. Printzen, *Injury-Inter. J. Care Injured* **27**, 9 (1996).
2. Z. U. Isiklar, G. C. Landon, and H. S. Tullos, *Clin. Orthop. Related Res.* **299**, 173 (1994).
3. K. Merritt, A. Gaind, and J. M. Anderson, *J. Biomed. Mater. Res.* **39**, 422 (1998).
4. C. C. Chang, K. Merritt, *J. Biomed. Mater. Res.* **26**, 197 (1992).
5. K. K. Jefferson, *FEMS Microbiology Lett.* **236**, 163 (2004).
6. R. O. Darouiche, Anti-infective efficacy of silver-coated medical prostheses, Clinical Infectious Diseases, Vol 29 (No. 6), 1999, p 1371-1377
7. D. J. Stickler, Biomaterials to prevent nosocomial infections: is silver the gold standard?, Curr. Opinion Infectious Diseases, Vol 13 (No. 4), 2000, p 389-393
8. K. S. Oh, S. H. Park, and Y. K. Jeong, Antimicrobial effects of Ag doped hydroxyapatite synthesized from co-precipitation route, Key Engineering Materials, Vol 264-268 (No. 1-3), 2004, p 2111-2114
9. D. P. Dowling, A. J. Betts, C. Pope, M. L. McConnell, R. Eloy, M. N. Arnaud, *Surface Coatings Technol.* **163**, 637 (2003).
10. S. Lopatin, S. J. Pennycook, J. Narayan, and G. Duscher, *Appl. Phys Lett.* **81**, 2728 (2002).
11. J. Bruley, D. B. Williams, J. J. Cuomo, and D. P. Pappas, *J. Microscopy (Oxford)* **180**, 22 (1995).
12. E. H. A. Dekempeneer, R. Jacobs, J. Smeets, J. Meneve, L. Eersels, B. Blanpain, J. Roos, and D. J. Oostra, *Thin Solid Films* **217**, 56 (1992).
13. K. Bewilogua, D. Dietrich, G. Holzhuter, and C. Weissmantel, *Phys. Stat. Sol. A* **71**, 56 (1982).
14. D. G. McCulloch, D. R. McKenzie, and C. M. Goringe, *Phys. Rev. B* **61**, 2349 (2000).
15. Y. Lifshitz, *Diamond Related Mater.* **5**, 388 (1996).
16. Y. Lifshitz, G. D. Lempert, E. Grossman, I. Avigal, C. Uzansaguy, R. Kalish, J. Kulik, D. Marton, and J. W. Rabalais, *Diamond Related Mater.* **4**, 318 (1995).
17. B. K. Tay and P. Zhang, *Thin Solid Films* **420**, 177 (2002).
18. V. V. Uglova, V. M. Anishchik, Y. Pauleau, A. K. Kuleshov, F. Thièry, J. Pelletier, S. N. Dub, and D. P. Rusalsky, *Vacuum* **70**, 181 (2003).

Mater. Res. Soc. Symp. Proc. Vol. 843 © 2005 Materials Research Society **T1.5**

Hard PVD Coatings for Austenitic Stainless Steel Machining: New Developments

J.L. Endrino, A. Wachter, E. Kuhnt, T. Mettler, J. Neuhaus, C. Gey
Balzers AG, SBU Tools.
Balzers, FL-9496, Liechtenstein.

ABSTRACT

The machinability of austenitic stainless steels is, in general, considered to be a difficult process. This is due in great part to the high plasticity and tendency to work-harden of the workpiece, which normally results in extreme conditions imposed on cutting edges. Additionally, austenitic stainless steels have much lower thermal conductivities in comparison to plain carbon and tool steels; this inflicts high thermal loads within the chip-tool contact zone, which can significantly increase the wear rate. The complex machining of austenitic stainless steels can be, in part, relieved by use of a hard coating with low thermal conductivity and an adequate coating surface quality. This can lead to a low friction coefficient between the parts and an improved chip evacuation process. In this study, stainless steel plates were machined using a high speed finishing process and cemented carbide tools coated with four high aluminum containing coatings namely, AlCrN, AlCrNbN, fine grained (fg) AlTiN, and nanocrystalline (nc) AlTiN. Both AlTiN and AlCrN-based coatings are known for their high oxidation resistances and the formation of aluminum oxide surface layers during oxidation [1]. The coating surface textures before and after post-deposition treatment were analyzed by means of the Abbot-Firestone ratio curves in order to study the influence of surface configurations leading to reduced tool wear. A maximum tool life of 150 m was observed for the tool coated with the nc-AlTiN coating.

INTRODUCTION

High work hardening rates even at low deformations and low thermal conductivities characterize austenitic stainless steels [2]. These two characteristics make austenitic stainless steels (γ-SS) more difficult to machine than carbon steels, low alloy steels and non-austenitic stainless steels. The high toughness and high ductility leads to the formation of long chips and the bonding of the material to the tool, which results in an increase adhesive wear. Moreover, high temperatures at the tool-chip interface result in an increased diffusion and chemical wear. In addition, the formation of build-up edges can also lead to variable machining forces and an increase in mechanical wear of the edge, which can cause tool breakage. To increase the machinability of γ-SS optimum cutting parameters such as an optimum cutting speed and feed rate can increase the tool life many times [3,4]. Due to the high tool-chip interaction, surface coatings play a major role in increasing the tool life. Numerous studies in the last two decades have shown that high aluminum containing nitride coatings (such as AlCrN's and AlTiN's) can provide better wear protection than non aluminum containing nitrides (CrN and TiN) at high temperatures due to their higher resistance to oxidation, higher hot hardness and lower thermal conductivities [5].

In our present research, we are investigating the influence of AlCrN, AlCrNbN, fg-AlTiN and nc-AlTiN in the wear mechanism and tool life of carbide end mills during finishing of AISI

316 stainless steel. Even though previous studies have shown the superior oxidation resistance and hot hardness of AlCrN based coatings compared to AlTiN's, we demonstrate that by tailoring of the surface with an adequate coating's crystallinity and surface texture longer tool lifes can be achieved.

EXPERIMENTAL DETAILS

The films used in this research were deposited using a Balzers' Rapid Coating System (RCS) deposition machine in cathodic arc ion plating mode. Customized $Al_{70}Cr_{30}$, $Al_{70}Cr_{25}Nb_5$ and $Al_{67}Ti_{33}$ in a reactive nitrogen atmosphere were used to obtain stoichiometric AlCrN, AlCrNbN, fg-AlTiN and nc-AlTiN thin films. Cemented carbide 8mm-diameter finishing tools were used as substrates. In addition, mirror-polished cemented carbide inserts were added to carry out x-ray diffraction studies. The temperature of the substrates during deposition was held at approximately 500°C for the AlCr-based coatings while for the AlTi-based coatings, the temperature was held at approximately 600°C. The deposition times were adjusted in order for the thicknesses of all the deposited coatings to be within 3.5 ± 0.2 microns.

A Siemens D500 diffractometer operated with a CuK_{α} tube and the Θ/2Θ mode was used to perform XRD analysis and identify the phases formed. Surface height characterization and the material ratio curves for the deposited coatings before and after surface post-treatment processes were obtained using a [0]Mahr Perthometer model M1 and the MarSurf XR 20 surface texture analysis software.

In the cutting tests, the wokpiece material was AISI 316 austenitic stainless steel hot rolled, quenched and plasma cut with a hardness of160HB (R_m = 550 N/mm^2). The cutting test were performed on a Mikron VCP 600 milling machine with a cutting speed of 120 m/min, a feed rate of 0.05 mm, a depth of cut of 10 mm and a radial depth of cut of 0.5 mm. Also, a LEO 1530 scanning electron microscope (SEM) was used to carry out observations from the flanks of the tools after 7 m running-in life.

RESULTS

XRD Patterns

Two AlCrN-based coatings and two AlTiN coatings were deposited onto the cemented carbide cutting tools. Figure 1(a) shows the XRD patterns for AlCrNbN and AlCrN coatings. These films present a cubic B1 (NaCl) structure, and their peaks are located close to those corresponding to cubic AlN peaks [JCDPS Card No. 25-1495]. Nonetheless, these two coatings show unequal textures, while AlCrN has a (111) preferential orientation, AlCrNbN has a (200) to (111) diffraction ratio of 4.2. This change in orientation by incorporating Nb into the CrAlN structure could in part be driven by the lower solubility of the aluminum nitride cubic phase in NbN compared to the excellent solubility of c-AlN in CrN [6,7]. This leads to a critical cubic/hexagonal Al ratio composition and the consequent change in texture. The XRD patterns for fg-AlTiN and nc-AlTiN are shown in figure 1(b). Both coatings showed a cubic AlN structure, which can also be expected in high aluminum containing $Ti_{1-x}Al_xN$ layers for x<0.7 [8,9]. Despite the fact that both coatings showed a reduced grain size and a (200) texture, effective crystallite size measurements indicate that the nc-AlTiN coating has crystallites around two thirds the size of those ones in the fg-AlTiN coating.

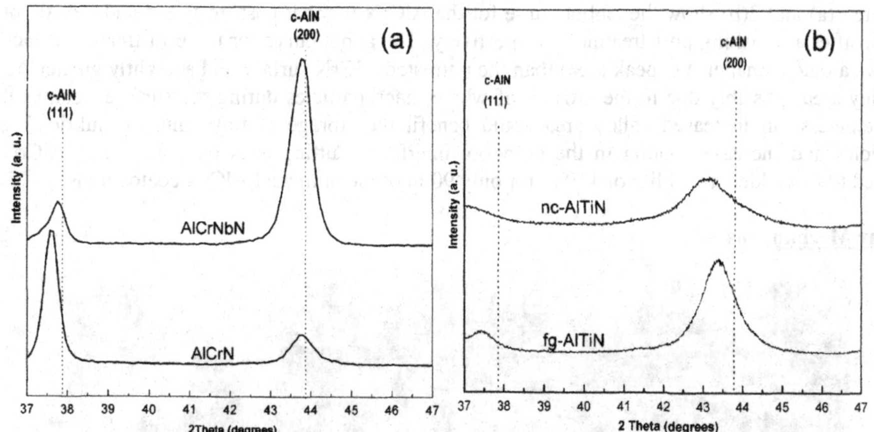

Figure 1. X-ray diffraction patterns for high aluminum PVD coatings. (a) AlCrNbN compared to AlCrN. (b) Nano-crystalline AlTiN compared to fine-grained AlTiN.

Post-treatment Surface Textures

The formation of surface coating macroparticles is relatively common in cathodic arc evaporation processes. The presence of these macroparticles in the surface has a negative effect in the machining of stainless steels. The use of common post-treatment polishing processes can reduce the density of macroparticles in the coating's surface, which leads to an increase in cutting performance [10]. The roughness texture of coated tools can be accurately characterized by means of material ratio curves (Abbot-Firestone curves) [11,12] .

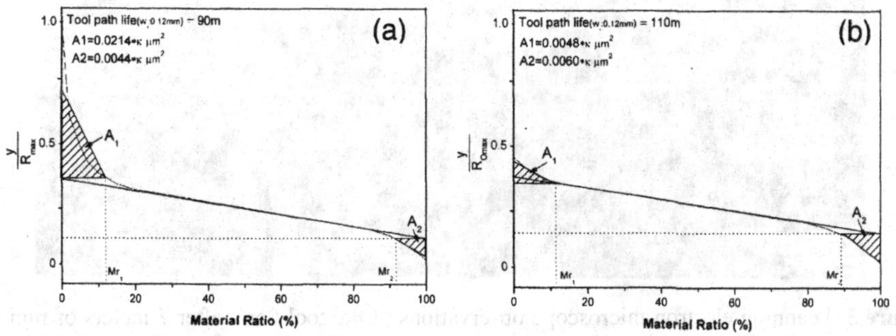

Figure 2. Characterization of surface textures by the material ratio curves for AlCrN. (a) Before surface treatment. (b) After surface treatment.

Figure 2(a) and 2(b) show the Abbot curve for the AlCrN without post treatment and the Abbot curve after a polishing post-treatment, respectively. The abbot curve for the post-treated surface shows a much smaller A1 (peak area) than the untreated AlCrN surface and a slightly greater A2 (valley area) possibly due to the erosion of whole macroparticles during the surface treatment. Nonetheless, an increased valley area could benefit the storage of tiny emulsion lubricating particles and increase cooling in the chip/tool interface. Cutting tests on post-treated AlCrN coated tools yielded a tool life of 110 m for only 90 m of the untreated AlCrN coated tools.

Wear Mechanisms

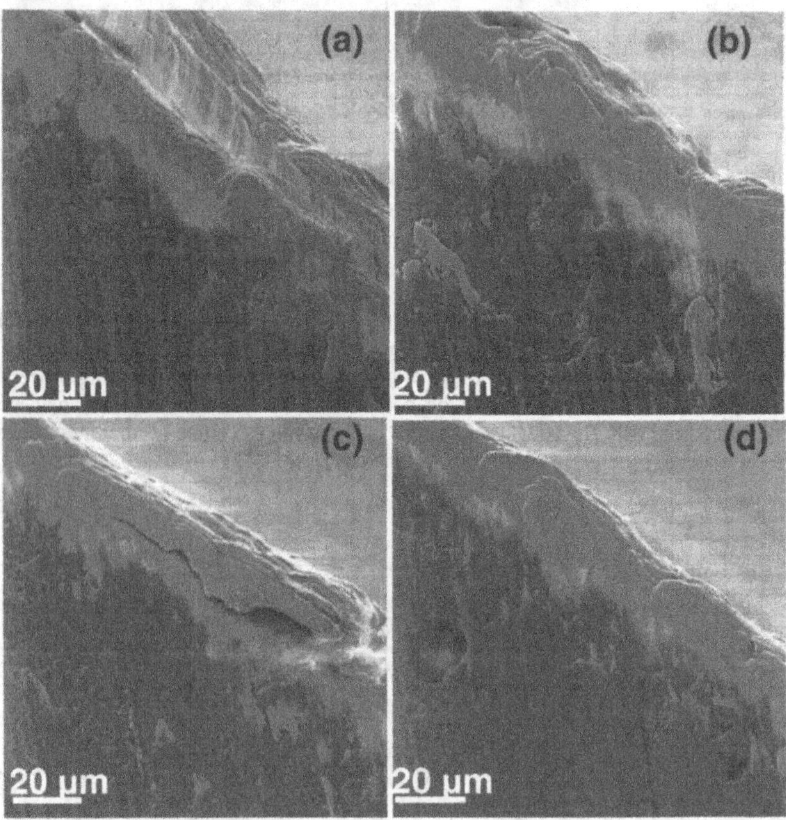

Figure 3. Scanning electron microscope observations of the tool flank after 7 meters of running life machining AISI 316 austenitic stainless steel. (a) AlCrN. (b) AlCrNbN. (c) Fine grained AlTiN. (d) Nano-crystalline AlTiN.

The wear behavior of the coated cemented carbide tools was studied by means of scanning electron microscopy. SEM observations after a running-in phase of 7 meters for the four coatings are shown in figure 3(a)-(d). In all the cases, a thick build-up of γ-ss can be observed at the cutting edge. Figures 3 (a) and (c) indicate that the build-up edges grow and break periodically as a result of the high shear forces at the edge during the machining process. In addition and between the build-up zone and the surface coating, a light color area corresponding to the exposed WC/Co substrate is observed. In comparison to the AlCrN coated tool, the AlCrNbN coating (figure 3(b)) appears to have a thinner average build-up edge although a much higher adhesion of the workpiece material into the flank. In the case of the nanocrystalline AlTiN (figure 3(d)), it appears to have much lower adhesion of the workpiece to the flank and a more stable state of than in the case of the fg- AlTiN coated tool (figure 3(c)).

Cutting Tests

Machining tests results for the AlCrN-based coated tools are shown in figure 4(a). Although no significant differences in wear could be spotted in the initial running-in stage, the niobium containing coating had a tool life 20 meter longer than that of the pure AlCrN coated tool, which could indicate a benefit of (200) surface texturing of AlCrN in protecting the cemented carbide substrate in the critical wear stage. Figure 4(b) shows the flank wear with respect to tool life for fg-AlTiN and nc-AlTiN coated tools. In this case, the fine-grained AlTiN seem to be more beneficial in initial stages of wear. However, the performance of the nc-AlTiN coated tool was remarkably better than that of the fg-AlTiN coated tool in the later stages of the wear. This could be due in part to the smaller crystalline size of the surface coating and its higher grain boundary density, which may favor the more rapid diffusion of aluminum to the surface [13]. This results in a fast formation of a thin alumina surface protective layer decreasing the thermal conductivity of the surface of the tool and minimizing the stickiness of the stainless steel to the tool's surface.

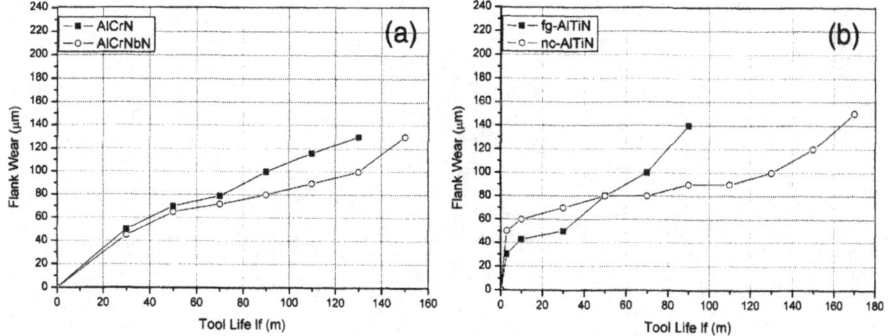

Figure 4. Cutting test results of coated cemented carbide end mills during machining of AISI 316 austenitic stainless steel. (a) AlCrN and AlCrNbN. (b) Fine-grained AlTiN and nanocrystalline AlTiN.

SUMMARY

In this study, an investigation of the effects of coating selection and surface post treatments in the machining of austenitic stainless steels has been performed. As surface post-treatments always resulted in longer tool lifes, quantification by means of the material ratio curves indicated not only a decrease in the "peak area" but an increase in the "valley area" as well. The selection of the hard PVD coating also seems to have a strong influence in tool life. The nanocrystalline structured AlTiN coating outperformed the fine grained AlTiN coating in the latter stages of wear and had almost twice its tool life. In AlCrN-based coatings, the AlCrNbN coating presented a (200) texture and performed significantly better than the AlCrN coating with a (111) texture.

REFERENCES

1. M. Kawate, A. Kimura, T. Suzuki, *Surf. Coat. Technol.*, v165, pp.163-167 (2003).
2. C. Koepfer, "Making Stainless More Machinable", *Modern Machine Shop Magazine*, July (2003).
3. T-R Lin, *J. Mater. Procc. Tecnol.*, v. 127, pp. 8-16, (2002).
4. I. Korkut, M. Kasap, I. Ciftci, U. Seker, *Materials & Design*, v.25 pp. 303-305 (2004).
5. M. Arndt, T. Kacsich, *Surf. Coat. Technol.*, v. 163-164 pp.674-680 (2003).
6. A. Sugishima, H. Kajioka, Y. Makino, Surf. Coat. Technol., v. 97, pp. 590-594 (1997).
7. Y. Makino, ISIJ International, v38, pp. 925-934 (1998).
8. M. Arndt, T. Kacsich, *Surf. Coat. Technol.*, v. 163-164 pp.674-680 (2003).
9. T. Ikeda, H. Satoh, *Thin Solid Films*, v. 195, pp. 99-110 (1991).
10. K.-D. Bouzakis, N.Michailidis, S. Hadjiyiannis, K. Efstathiou, E. Pavlidou, G. Erkens, S. Rambadt, I. Wirth, *Surf. Coat. Technol.*, v. 146-147, pp. 443-450 (2001).
11. R. S. Sayles, *Tribology International*, v.34, pp. 299-305 (2001).
12. ISO13565-2:1996, Geometrical Product Specification (GPS) – Surface texture: Profile method; Surfaces having stratified functional properties.
13. G.S. Fox-Rabinovich, G.C. Weatherly, A. I. Dodonov, A.I. Kovalev, L.S. Shuster, S.C.Veldhuis, G.K. Dosbaeva, D.L. Wainstein, M.S. Migranov, Surf. Coat. Technol., v. 177-178, pp. 800-811 (2004).

Mater. Res. Soc. Symp. Proc. Vol. 843 © 2005 Materials Research Society  T1.6

Nanocomposite TiC/a-C coatings: structure and properties

Jeff Th.M. De Hosson [1], Yutao Pei [1], Damiano Galvan [1] and Albano Cavaleiro [2]

[1] Dept. of Applied Physics, Materials Science Centre and the Netherlands Institute for Metals Research, University of Groningen, Nijenborgh 4, 9747 AG Groningen, The Netherlands.
[2] Departamento de Engenharia Mecanica, FCTUC - Universidade de Coimbra Pinhal de Marrocos, 3030 Coimbra, Portugal

ABSTRACT

This contribution deals with fundamental and applied concepts in nano-structured coatings, in particular focusing on the characterization with high-resolution (transmission) electron microscopy. Both unbalanced and balanced magnetron sputtering systems were used to deposit nc-TiC/a-C nanocomposite coatings with hydrogenated or hydrogen-free amorphous carbon (a-C) matrix, respectively. The contents of Ti and C in the coatings have been varied over a wide range (7~45 at.% Ti) by changing the flow rate of acetylene gas or the locations of substrates relative to the center of C/TiC targets. Different levels of bias voltage and deposition pressure were used to control the nanostructure. The nanocomposite coatings exhibit hardness of 5~35 GPa, hardness/E-modulus ratio up to 0.15, wear rate of 4.8×10^{-17} m³/Nm, coefficient of friction of 0.04 under dry sliding and strong self-lubrication effects. The nanostructure and elemental distribution in the coatings have been characterized with cross-sectional and planar high-resolution transmission electron microscopy (HRTEM) and energy filtered TEM. The influences of the volume fraction and size distribution of nano-crystallites TiC (nc-TiC) on the coating properties were examined.

INTRODUCTION

Nanocomposite coatings consisting of TiC nanoparticles embedded in an amorphous carbon (a-C) matrix are able to combine high fracture toughness and wear resistance with low friction coefficient. Coatings currently employed for such applications are based on a hydrogenated a-C matrix. However, if a hydrogen-free a-C matrix could be made, improvement in thermal stability of the nanocomposite coating is anticipated, together with a decrease in the coefficient of friction (CoF) of sliding in a humid environment. By controlling the size and volume fraction of nanocrystalline phases in the a-C matrix and consequently the separation width of the amorphous matrix among the nanocrystallites, the properties of the nanocomposite coatings can be tailored, e.g. to make a balance between hardness and elastic modulus to permit close match to the elastic modulus of substrates [1], and particularly to obtain high toughness that is crucial for applications under high loading contact and surface fatigue [2, 3, 4]. In addition to these characteristics, amorphous carbon based nanocomposite coatings exhibit not only excellent wear resistance but also low friction due to self-lubrication effects, which make them attractive for environmental reasons because liquid lubricants can be omitted.

This paper focuses on the deposition and characterization of TiC/a-C nanocomposite coatings. Detailed mechanical examinations, including nanoindentations, scratch and tribo-tests, have been performed. Cross-sectional and planar TEM observations and energy filtered TEM are employed to characterize the nanostructure, elements distribution in the coatings and chemistry of the carbon matrix, respectively. The influences of the volume fraction and size distribution of nanocrystalline TiC (nc-TiC) on the coating properties were examined.

EXPERIMENTAL

Hydrogen-free nc-TiC/a-C nanocomposite coatings were deposited by co-sputtering graphite and TiC or Ti targets in a balanced magnetron sputter with a combination of RF and DC power supply. The initial pressure of the deposition chamber before sputtering was $4\sim5\times10^{-6}$ mbar and the deposition pressure of argon was 5×10^{-3} mbar or 1×10^{-2} mbar.

Hydrogenated nc-TiC/a-C:H coatings were deposited with closed-field unbalanced magnetron sputtering in an argon/acetylene atmosphere in a Hauzer HTC-1000 coating system, which was configured of two Cr targets and two Ti targets opposite to each other. The detailed set-up of the coating system has been documented elsewhere [5]. The Cr targets were used to create an intermediate layer between the nc-TiC/a-C:H coating and the substrate material. The flow rate of acetylene and substrate bias varied in the range of 80-125 sccm and 0-150 V, respectively, to obtain different C/Ti contents and nanostructures in the coatings. The substrates used for each coating were ø30×5 mm discs of hardened M2 steel, 304 stainless steel and ø100 mm Si wafer for the measurements of residual stress by monitoring the curvature change.

Tribo-tests were performed on a CSM tribometer with a ball-on-disc configuration at 0.1m/s sliding speed and 2N or 5N normal load. A CSM Revetest scratch tester was used for measuring the interfacial adhesion strength with a diamond stylus of 200 μm radius and 0.01 m/min scratch speed. A fully calibrated MTS Nano Indenter XP was employed to measure the hardness and Young's modulus of the coatings with a Berkovich indenter and the fracture toughness with a cube-corner indenter according to Lawn-Evans-Marshall's approach [6,7].

The surface morphology and fractured cross sections of the coatings were examined using a scanning electron microscope, SEM (Philips FEG-XL30s). The investigation of the nanostructures was carried out in a high-resolution transmission electron microscope (HRTEM) (JEOL 4000 EX/II, operated at 400 kV) and an analytical TEM (JEOL 2010F-FEG, operated at 200 kV). Electron probe microanalysis (EPMA) with a Cameca SX-50 equipment was used to determine the chemical composition of the coatings.

RESULTS AND DISCUSSIONS

1. nc-TiC/a-C nanocomposite coatings

A special configuration of substrates/targets was used to obtain a large composition gradient in a single batch of deposition. As sketched in Fig.1a, nine pieces of substrates were aligned in a row, from the center of graphite target to the center of TiC target or Ti target, and stayed stationary during deposition of the top coatings. Depending on the relative locations of substrates to the two targets, the composition of deposited coatings may continuously vary from Ti-doped DLC to pure TiC if graphite and TiC targets are used, see Fig.1b. This method allows a quick search for the desired composition and associated deposition parameters. It was intended to use Ti as one of the two targets for the deposition of Ti intermediate layer and for the supplement of Ti elements for the top nanocomposite coating. Due to the low sputtering yields of graphite target, however, it was rather difficult to optimize the microstructure of both the intermediate layer and the top coating simultaneously. Eventually a compromise of using TiC and C targets, instead of Ti and C targets, was necessary to obtain the desired C contents in the coatings and the intermediate layer was thus TiC.

Table 1 summarizes the deposition parameters used and the corresponding coating performances. In general, an increase in substrate bias results in a significant increase in the hardness of nanocomposite coatings. Under the same deposition conditions (i.e. in the same batch), both the hardness and Young's modulus of the coatings increase with Ti content. However, the increasing amplitude of the latter is larger than that of the former, which leads to a decrease in H/E ratio with Ti content, as shown in Fig.1c. Substrate bias and deposition pressure may also have an influence on the H/E ratio by altering the intensity of ion bombardment that governs the sizes/distribution of TiC nanocrystallites and the residual stress in the coatings, but not so pronounced as effects exerted onto the individual H and E parameters.

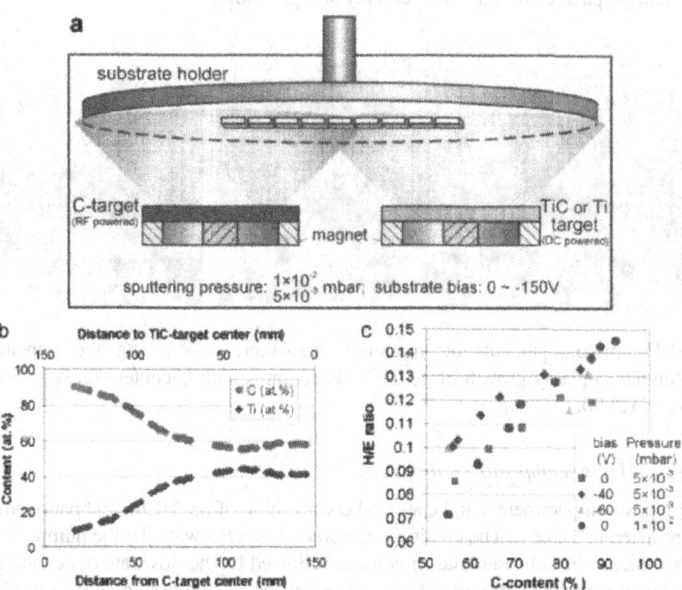

Fig. 1. (a) Sketch of setup for deposition of nc-TiC/a-C coatings; (b) C/Ti content gradient achieved in a batch deposited at -40V substrate bias and 5×10^{-3} mbar pressure and (c) H/E ratio vs. carbon content under various deposition conditions.

Table 1. Deposition parameters and properties of nc-TiC/a-C coatings deposited with TiC and C targets

Deposition pressure	Substrate bias (V)			
	0	-40	-60	-80*
5×10^{-3} mbar	Ti-cont.: 11.9~49 at.% Hardness: 10~16 GPa H/E ratio: 0.121~0.086	Ti-cont.: 10~45 at.% Hardness: 14~27.5 GPa H/E ratio: 0.143~0.100	Ti-cont.: 12~45 at.% Hardness: 20~37 GPa H/E ratio: 0.131~0. 099	Ti-cont.: 12~75 at.% Hardness: 21~35 GPa H/E ratio: 0.117~0.082
1×10^{-2} mbar	Ti content: 6~45 at.% Hardness: 6.9~9.5 GPa H/E ratio: 0.145~0.072	-	Ti content: 8~45 at.% Hardness: 11~14.2 GPa H/E ratio: 0.118~0.080	-

* with Ti and C targets

A high H/E ratio can be attained, as shown in Fig.1c. Although for a long time hardness has been regarded as a primary material property affecting wear resistance, the elastic strain to failure, which is related to the H/E ratio, is a more suitable parameter for predicting wear resistance [8]. Within a linear-elastic approach, this is understandable according to the relations that the yield stress of contact is proportional to (H^3/E^2) and the equation $G_c=\pi a\sigma_c^2/E$. It indicates that the fracture toughness of coatings defined by the so-called 'critical strain energy release rate' G_c would be improved by both a low Young's modulus and a high critical stress (σ_c) for failure hardness.

Associated to the change in properties is the evolution in microstructure of the nanocomposite coatings. SEM micrographs of the fractured cross sections presented in Fig. 2 clearly show the transition from glassy to columnar growth and also enhanced surface roughness of the coatings as Ti-content increases. As shall be discussed later the columnar boundaries are potential weak points of the coatings where growth defects occur and cracks can initiate and propagate through under contact of high loading.

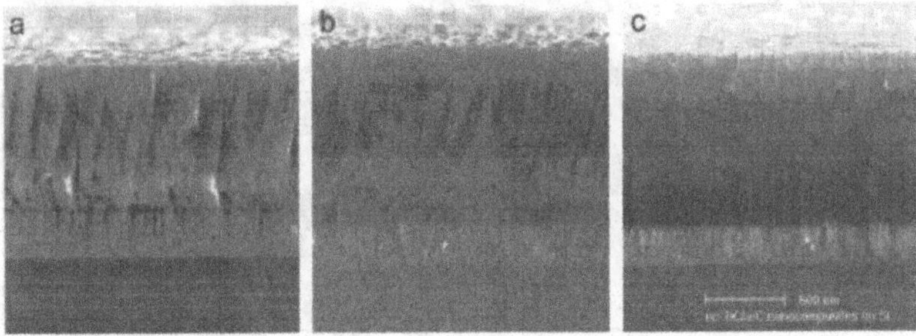

Fig. 2. SEM micrographs of the fractured cross-sections showing the transition from columnar to glassy growth of nc-TiC/a-C coatings with C content (at.%) of (a) 66, (b) 83 and (c) 90.

2. nc-TiC/a-C:H nanocomposite coatings

The deposition parameters and chemical composition of nc-TiC/a-C:H nanocomposite coatings are listed in Table 2. The coatings are named in such a way that the numbers before the letter V indicate the substrate bias in voltage, followed by the flow rate of acetylene in standard cubic centimeters per minute (sccm). Besides the change from columnar to glassy growth with increasing the C content, similar to that observed in nc-TiC/a-C coatings, nc-TiC/a-C:H nanocomposite coatings also exhibit such a transition with a increase in substrate bias voltage, as shown in Fig. 3. The columnar microstructure gets slightly weaker at 60V bias compared with that at 0V (floating bias) and is hardly traceable at 100V so that the fractured cross section looks featureless. Increasing the substrate bias further to 150V, the coatings become harder and are fully free of columnar growth and consequently, the fractured cross sections appear flat and show only some contrast of exposed nanoparticles. Cross sectional TEM observations show a transition of growth mechanism and the columnar structures are restrained at higher substrate bias and C-content.

Table 2. Deposition parameters and composition of nc-TiC/a-C:H nanocomposite coatings

Coating No.	Bias [-V]	C_2H_2 [sccm]	Composition [at.%] C	Ti	O	Ti/C ratio	TiC volume fraction [%]*
0V110	floating	110	71.34	13.64	15.02	0.191	27.7
60V110	60	110	80.21	16.33	3.46	0.204	29.5
100V110	100	110	81.02	17.84	1.14	0.220	31.7
150V110	150	110	80.30	18.51	1.19	0.231	33.1
100V80	100	80	66.60	31.75	1.65	0.477	60.1
100V125	100	125	87.19	11.85	0.96	0.136	19.7

* providing 2.0 at.% saturation limit of Ti in C [9].

High resolution TEM presented in Fig. 4 clearly reveals that nanocrystalline TiC particles are homogeneously embedded in the a-C:H matrix and an increase in C content leads to the formation of finer TiC nano-particles, e.g. 5 nm vs. 2 nm in diameter observed in the two coatings 100V80 and 100V110 consisting of 66.6 at.% and 81.0 at.% C, respectively. Indexing the lattice fringes and also the circular selected area diffraction patterns confirms the formation of TiC nanocrystalline phases. According to calculations based on the Ti-content of the nanocomposite coatings, the volume fraction of TiC nanocrystallites in all the coatings can be estimated as listed in Table 2, assuming the density of the TiC phase and amorphous carbon are 4.9 and 2.0 g/cm³, respectively. The results are consistent with the estimate obtained from a simple model of hard spheres in an fcc arrangement with corresponding separation of a-C:H matrix among TiC nanocrystallites of different sizes observed by HR-TEM.

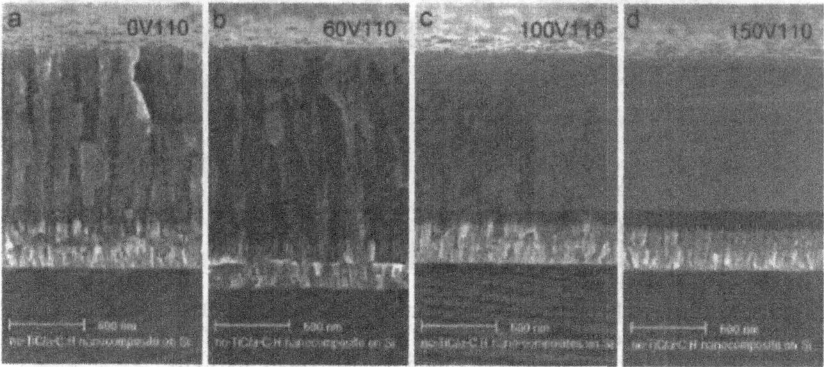

Fig. 3. SEM micrographs showing the fractured cross-sections of TiC/a-C:H nanocomposite coatings deposited at 110sccm flow rate of acetylene gas and different substrate bias: (a) floating, (b) -60V, (c) -100V and (d) -150V.

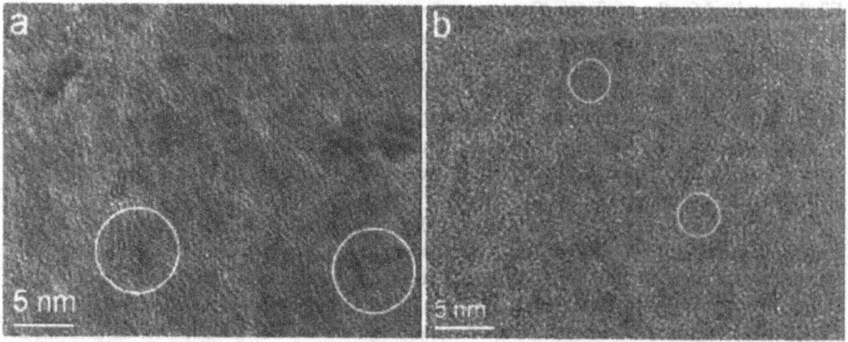

Fig. 4. HRTEM micrographs showing TiC nanocrystallites embedded in a-C:H matrix in nc-TiC/a-C:H nanocomposite: (a) coating 100V80 with 60.1 vol.% TiC nanoparticles and (b) coating 100V110 with 31.7 vol.% TiC nanoparticles.

Table 3. Mechanical properties of nc-TiC/a-C:H nanocomposite coatings

Coating No.	Residual stress (MPa)	H (GPa)	E (GPa)	H/E	Fracture toughness (MPa $m^{½}$)
0V110	-649	5.5	61.3	0.091	-
60V110	-352	11.8	99.8	0.118	23.66
100V110	-761	15.6	136.6	0.114	36.15
150V110	-1595	19.8	168.3	0.118	49.94
100V80	-1184	20.0	229.4	0.087	32.21
100V125	-1154	15.8	128.5	0.123	55.78

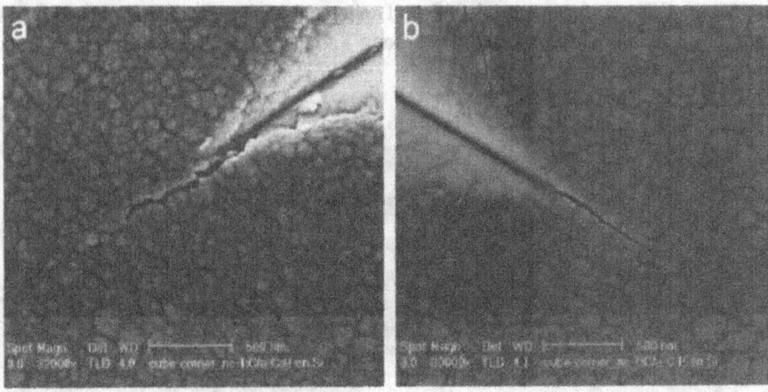

Fig.5 SEM micrographs of nanoindentation impressions made at 190mN normal load: (a) radial crack meanderingly propagated through column boundaries in the coating 100V80; (b) straight radial crack formed in the coating 100V110

The mechanical properties of the nc-TiC/a-C:H coatings are listed in Table 3. Increase of the substrate bias voltage leads to an obvious increase in both the hardness and elastic modulus of the nanocomposite coatings. However, the H/E ratio stays almost constant when the substrate bias is above 60 V. On the other hand, a higher flow rate of acetylene gas i.e. a higher C-content results in a markedly higher H/E ratios, which is analogues to the results in Fig.1c obtained for the nc-TiC/a-C nanocomposite coatings.

Nanoindentation with a cube-corner tip have revealed a significant increase in the fracture toughness of the coatings with increasing C content or substrate bias, referring to Table 3. The coatings with glassy microstructure exhibit substantial toughening effects. SEM observations on indentation indents demonstrate that the radial cracks propagate readily through the columnar boundaries and leave a meandering crack path in coating 100V80, as seen in Fig. 5a. In contrast, the radial cracks are straight in coating 100V110 where columnar structures are restrained (Fig. 5b).

In this sense, one may conclude that an ideal coating should be free of columnar structures, but that of course depends also on the sp^2/sp^3 ratio of the inter-columnar a-C. The toughening of the nanocomposite coatings has been achieved actually on two different scales, namely by restraining the formation of columns on a micro-scale and through manipulating the nanostructure on a nano-scale. The columnar boundaries (CBs) are detrimental in terms of micro-crack initiation and propagation under loading contact, as seen in Fig. 5a and 9a, and can be fully prevented by increasing substrate bias and/or carbon content. Regarding the toughening mechanism on a nano-scale, localized plastic deformation in the a-C:H matrix due to a large amount of nano-defects may be delocalized and lead to the spread of slip bands around TiC nanoparticles under high loading contact, once the a-C matrix separation among the TiC nanograins is wide enough. Here we consider the a-C:H as a system with a density distribution caused by localized defects having either severe shear or hydrostatic stress field components. The hydrostatic stress field component, representing free volume, can be annealed out but localized defects with shear stress components cannot. The later trigger the formation of shear bands leading to brittle failure that can be turned into more ductile behavior provided the TiC particles become of the same size as the width of the shear bands. For that reason 100V125 exhibits a better fracture toughness (see Table 3).

3. Tribological behavior of nanocomposite coatings

Ball-on-disk tribo-tests have been systematically performed on the nanocomposite coatings under loading conditions of maximum 1.5 GPa Hertzian pressure with Ø6 mm balls. The friction coefficient of the coatings 0V110 and 100V80 is above 0.2, i.e. much higher than that of the remaining four coatings. Among these four coatings, the coating 100V110 shows the lowest friction coefficient of about 0.047 against 100Cr6 bearing steel balls in ambient air, as shown in Fig. 6a. Against different counterparts, i.e. sapphire, alumina and bearing steel balls, only slight differences in the friction coefficient are observed (Fig. 6b).

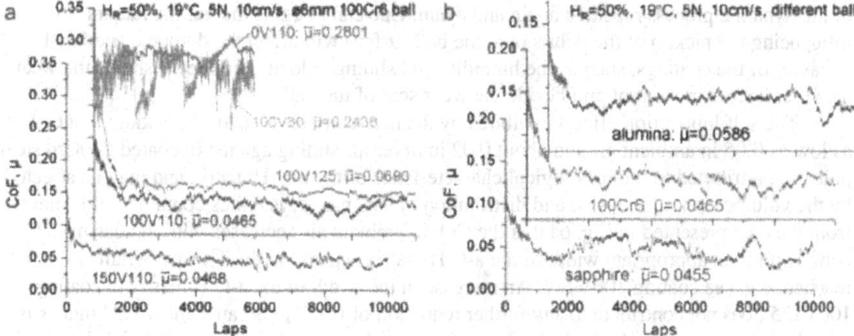

Fig. 6. (a) Friction graphs of nc-TiC/a-C:H nanocomposite coatings under the same sliding wear conditions and (b) friction graphs of the coating 100V110 sliding against different balls in ambient air (5N normal load, 10cm/s sliding velocity and 50% humidity).

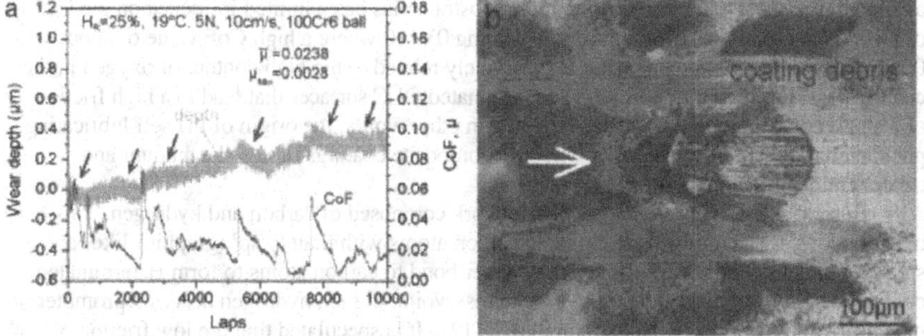

Fig. 7. (a) Dynamic frictional behavior of coating 100V110 sliding in air of 25% relative humidity and (b) wear scar of 100Cr6 steel ball (an arrow indicating the sliding direction of the coating in contact).

The most interesting results are the self-lubrication effects observed on the coatings except 0V110 and 100V80. During the course of transferring films onto the ball surface from the wear track of the coating, the friction coefficient continuously decreases until a stationary state is reached where the steady-state coefficient of friction (CoF) at a lower value is achieved. To prove that the self-lubrication is induced by the formation of transfer films, the wear depth was in-situ monitored with a resolution of 0.02μm by a RVDT sensor during the tribo-tests. As marked by the arrows in Fig. 7a, segments with a negative slope were observed in the depth vs laps graph and indicated a significant growth of the transfer film on the ball surface, rather than a real change in the depth of the wear track on the coatings. Correspondently, a substantial decrease in the CoF was detected followed by a long tail in the CoF graph. Once the transfer film stopped growing, the CoF could not decrease further and started to fluctuate. Because the transfer film covered the ball surface in contact, the wear rate of the coating diminished at that moment and resulted in less debris formed. As a result, the transfer film became thinner with sliding until it fully broke down, which led to a sudden jump of the CoF.

Sliding at higher a CoF would generate more debris from the wear track, which in turn provided the necessary materials for the growth of new transfer film. Thereafter, a new cycle of the dynamic process repeated again and again. One can imagine that all the factors influencing the pickup of the debris onto the ball surface will affect the dynamic frictional behavior of the coatings, such as the humidity and sliding velocity. Figure 7b shows the wear debris collected in front of and beside the wear scar of the ball.

The self lubrication effects exhibited by the nanocomposite coatings, leading to a CoF as low as 0.05 in ambient air and about 0.02 in dryer air sliding against uncoated 100Cr6 steel balls, are attributed to the tribological characteristics of the a-C:H matrix and may be affected by the volumetric fraction, size and distribution of TiC nanocrystallites contained. It is clear from the data presented in Fig. 6a that the CoF in ambient air decreases with increasing C content until an appropriate width of the a-C:H matrix separating the TiC nanograins is reached, e.g. see coating 100V110. An increase in the width of the a-C:H matrix in coating 100V125 does not contribute to any further reduction of friction, because the a-C:H matrix is already wide enough to shield the TiC nanoparticles, i.e. dominating the interfacial shearing ability between the coating surface and the transfer films formed on the surface of counterpart ball. If the a-C(:H) matrix cannot efficiently shield TiC particles in the transfer films as in the cases of coating 100V80 with a high volume fraction of (rather big) TiC nanocrystallites the TiC nanoparticles may cause considerable damage to the hydrophobic contact surfaces or to the transfer films formed. This may explain the fluctuations in the friction graph of 100V80 as shown in Fig. 6a. On the other hand, the substrate bias has a limited influence on the CoF of the nanocomposite coatings, except in coating 0V110 where a high CoF value of about 0.28 was recorded in ambient air. It is most likely related to the high content of oxygen in the coating 0V110, which may form oxygen terminated DLC surfaces that lead to a high friction or a small contact angle with water [10, 11]. In other words, the origin of the self lubricating phenomena observed on the TiC/a-C:H nanocomposite coatings lies on the amount and tribochemical property of the a-C:H matrix.

Basically, a-C:H is an amorphous network composed of carbon and hydrogen. This network consists of strongly cross-linked carbon atoms with mainly sp^2 (graphite-like) and sp^3 (diamond-like) bonds. Hydrogen may either bond to carbon atoms to form H-terminated carbon bonds or stay unbonded in hydrogen reservoirs. In fact, hydrogen acts as a promoter or stabilizer of the sp^3-bonded carbon phase [12]. It is speculated that the low friction of most carbon films is largely due to the fact that these materials are chemically inert and consequently they exert very little adhesive force during sliding against other materials. The major friction-controlling mechanism observed is attributed to the formation of a transfer film on the surface of the counterpart, which permits easy shear within the interfacial materials and protects the counterpart against wear [13]. However, the shearing ability strongly depends on the nature of the surrounding atmosphere present in the contact.

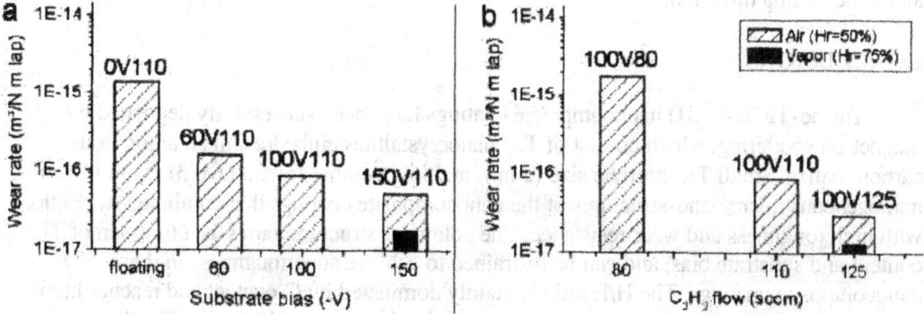

Fig. 8. Influence of the substrate bias (a) and the flow rate of acetylene on the wear rate of nc-TiC/a-C:H coatings under dry sliding against 100Cr6 ball in ambient air or humid air.

Efforts were made to measure the wear rate of different coatings after tribo-tests. With a confocal microscope, eight images of 3D profiles were captured at different positions on a wear track for reliable average. A software code was written in MatLab and used to calculate the volumetric loss (V) of a wear track and finally the wear rate (W_r) of the coatings defined as volume of wear per unit track length, per Newton of normal load and per lap. The wear rate of all the coatings is plotted in Fig. 8. It is clear that the wear rate of the nanocomposite coatings significantly decreases with increasing hardness and especially with increasing H/E ratio. Changing the counterpart has not an obvious influence on the wear rate. In contrast, the atmosphere dramatically affects the wear properties, including both the friction coefficient and the wear rate. The lowest wear rate of 1.31×10^{-17} m³/(N m lap) is achieved on the coating 100V125 in 75% humidity, with a CoF of 0.092.

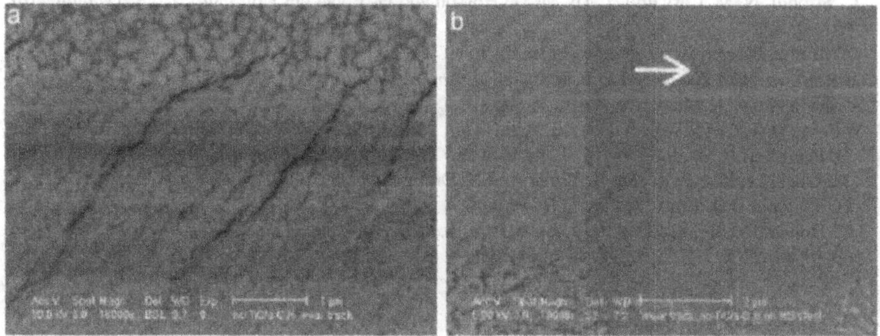

Fig. 9. SEM micrographs of the wear tracks: (a) micro-cracks formed along the columnar boundaries in the coating 100V80 and (b) smooth and featureless wear track on the coating 150V110 under the same wear conditions (5N load, 10cm/s sliding velocity, T=20°C, 50% relative humidity and against ø6mm 100Cr6 ball). An arrow indicates the moving direction of the counterparts.

SEM observations on the wear tracks reveal that micro-cracks initiate at and readily propagate through the columnar boundaries in the coatings 0V110, 60V110 and 100V80 under ball sliding contact, as shown in Fig. 9a. In contrast, no cracks are observed in the wear tracks on the coatings 100V110, 150V110 and 100V125 where columnar structures are restrained, regardless of the different normal loads applied and tested in air or in humid air. As demonstrated in Fig. 9b, the surface of the wear tracks is featureless and rather smooth along the sliding direction.

CONCLUSIONS

The nc-TiC/a-C(:H) nanocomposite coatings have been successfully deposited by magnetron sputtering, which consist of TiC nanocrystallites embedded in an amorphous carbon matrix. Small TiC particle size (2 nm) and low volume fraction (≈0.3) result in very homogeneous micro/nano-structures of the nanocomposite coatings that combine low friction with high toughness and wear resistance. The columnar structure varies as a function of Ti content and substrate bias, and can be restrained to achieve supertoughness in these nanocomposite coatings. The H/E ratio is mainly dominated by C-content and reaches high values above 0.1 at high C-contents, which result in high wear resistance. Self-lubrication can be achieved at high volumetric fraction of amorphous carbon matrix together with homogeneously distributed TiC nanocrystallites.

ACKNOWLEDGMENTS

The authors acknowledge financial support from the Netherlands Institute for Metals Research (NIMR) and the Foundation for Fundamental Research on Matter (FOM-Utrecht).

REFERENCES

[1] W.J. Meng and B.A. Gillispie, *J. Appl Phys.*, 84 (1998), 4314

[2] A.A. Voevodin and J.S. Zabinski, *Thin Solid Films*, 370 (2000), 223

[3] Y.T. Pei, D. Galvan, J.Th.M. De Hosson *Mater. Sci. Forum*, 475-479 (2005), 3655-3660

[4] Y.T. Pei, D. Galvan, J.Th.M. De Hosson, *Acta Materialia*, 2004, submitted for publication.

[5] C. Strondl, N. M. Carvalho, J. Th. M. De Hosson and G. J. van der Kolk, *Surf. Coat. Technol.*, 162 (2003), 288

[6] G.M. Phar, *Mater. Sci. Eng.* A253 (1998), 151

[7] B.R. Lawn, A.G. Evans and D.B. Marshall, *J. Am. Ceram. Soc.* 63 (1980), 574

[8] A. Leyland and A. Matthews, *Wear*, 26 (2000), 1

[9] W.J. Meng, R.C. Tittsworth and L.E. Rehn, *Thin Solid Films*, 377-378 (2000), 222

[10] D. Paulmier, H. Zaidi, H. Nevy, T. Le Huu and T. Matthia, *Surf. Caot. Technol.*, 62(1993), 570

[11] M. Grischke, K. Bewilogua, K. Trojan and H. Dimigen, *Surf. Coat. Technol.*, 74-75 (1995), 739

[12] H. Tsai and D.B. Bogy, *J. Vac. Sci. Technol.* A 5(1987), 3287

[13] C. Donnet, M. Belin, J.C. Augé, J.M. Martin, A. Grill and V. Patel, *Surf. Coat. Technol.*, 68-69(1994), 626

Mater. Res. Soc. Symp. Proc. Vol. 843 © 2005 Materials Research Society T1.9

Effect of Surface Geometry on Stress Generation in Thermal Barrier Coatings During Plasma Spray Deposition

Guosheng Ye and Soumendra Basu
Department of Manufacturing Engineering
Boston University, Brookline, MA 02246, U.S.A.

ABSTRACT

A fully coupled thermo-mechanical finite element model was used to study the buildup of stresses during splat solidification, and to understand the effect of deposition conditions on crack formation during plasma spray deposition. Through the simulation, the locations and magnitudes of maximum stresses were identified, where crack formation would presumably initiate. The model showed that the stresses scaled with the temperature difference between the superheated splat and the substrate. The simulation further showed that the stresses scale with the three geometric parameters, and two independent geometric ratios were defined; ζ (defined as t/λ) and ψ (defined as A/λ). *2D* maps of maximum S_{11} and S_{22} under different combinations of ζ and ψ were constructed. The mappings showed that only roughness features on the scale of splat thickness were important in providing locations of maximum stress concentration.

INTRODUCTION

The need for cleaner and more efficient operation mandates the use of increasingly higher temperatures in advanced gas turbines. In order to protect the metallic components in the hot sections of these turbines, thermal barrier coatings (TBCs) are used. These TBCs can maintain a significant temperature drop across their thickness due to their low thermal conductivities. This allows the metallic components, typically made of high temperature superalloys, to be exposed to lower temperatures, leading to an increase in their service lifetimes.

Typically, in land-based gas turbines, air plasma sprayed TBC top-coats are deposited over a vacuum plasma sprayed bond coat. The plasma spray deposition deposition process involves introducing ceramic powders, typically yttria stabilized zirconia (YSZ), into the high temperature plasma plume. The powders melt in the plume and the molten droplets are accelerated towards the substrate. On impingement, the molten droplets spread on the substrate forming rapidly quenched splats. Typically, several successive splats are deposited over a given area in one pass, and the coating thickness is built up using several passes over the substrate.

The rapid quenching associated with the plasma spray process combined with the brittle nature of the ceramic top-coat typically leads to a high density of microcracks. Typically, plasma sprayed TBCs contain horizontal cracks at splat boundaries and vertical cracks that extend through the splat thickness. Under appropriate conditions, longer vertical cracks that extend through the thickness of the coatings (termed as 'segmentation cracks') can be formed [1]. These cracks increase coating compliance and extend their lifetimes by resisting TBC spallation. Horizontal cracks reduce the thermal conductivity of the coatings in the direction of heat transfer, making it a more effective thermal barrier. However, it has been shown that horizontal cracks close to the top coat/bond coat interface link together after long term high temperature exposures, especially in the presence of bond coat oxidation [2]. Eventually, these linked microcracks form a critical sized flaw that leads to TBC spallation. Thus, an ideal TBC crack microstructure would consist of a high density of horizontal cracks away from the interface for reduced thermal conductivity, a low density horizontal cracks close to the interface to avoid coating spallation, and a regular distribution of vertical segmentation cracks for increased coating compliance. Such a complex engineering of the coating crack microstructure necessitates a thorough understanding of the underlying physics of crack formation during the deposition process. This paper addresses some aspects of understanding this process/microstructure mapping by thermo-mechanical modeling of splat solidification.

There have been a number of experimental studies on the effects of processing parameters on the resulting crack microstructure [3-6]. Bengtsson [3] showed that both horizontal and vertical crack densities decrease with increasing substrate temperature. Sampath et al. [7] studied the morphology of single splats on polished substrates and found that above a critical substrate temperature the splats were disk shaped while below the critical substrate temperature, the splats were fragmented. Sampath [8] and Jiang [9] attributed the splashing to the inability of the adsorbed gases on the surface of the cold substrate to escape on the arrival of the molten splat. McPherson claimed [10] that in order to get reproducible crack microstructures, it was more effective to control the state of the molten particles by controlling the average particle temperature, T_p, and average particle velocity, V_p. Moreau [11] demonstrated that changing T_p and V_p changed the distribution of microcracks, reinforcing the idea that controlling the molten particle and substrate states is a viable strategy for crack microstructure engineering in TBCs.

THERMO-MECHANICAL MODELING OF SPLAT SOLIDIFICATION

Model setup

To better understand the interaction of various processing parameters, a finite element model was setup to study the evolution of stresses in the coatings. As a first cut, the magnitude of the appropriate maximum coating stresses was taken to be an indicator of the extent of horizontal and vertical microcracks in the coating.

As discussed, TBCs are built up by stacking many splats over one another. The simulation of a single splat incorporates the effects of all process variables. As discussed, the goal of crack microstructural engineering is to control the particle and substrate states, i.e., T_p, V_p and T_{sub}. In addition, there are three geometric parameters of interest; the splat thickness, t; and the surface roughness which includes the amplitude, A and the wavelength λ (see Fig. 1). It should be noted that the splat thickness is dependent on the particle diameter (D) and particle state (T_p, V_p) as described by Madejski [12], and can be explicitly stated as:

$$t = K_1 (D)^{3/5} (V_p)^{-2/5} \exp (K_2/T_p),$$

where K_1 and K_2 are experimentally determined parameters.

In this study, the stress evolution in a single splat was simulated as a fully coupled thermal-mechanical system using the commercial code ABAQUS [13]. In the simulation, a superheated splat is made to cool on a substrate by conductive heat transfer. The model accounts for the latent heat of fusion of the splat. The objective of this modeling effort was three fold; to study the effect of different variables such as superheat, substrate temperature, splat thickness and surface roughness features on stress evolution; to locate regions of maximum stresses in the coatings and the evolution of such stresses with time; and to understand the key parameters that most affect the coating crack microstructure.

Location and nature of maximum stresses

Microstructural observations in the SEM showed that cracks mostly occur at the boundaries between splats and at the boundaries between columnar grains inside splats. Thus, the stresses in the directions parallel and normal to the splat interface are critical for crack formation. After a crack forms at the location of the maximum stresses, the stresses inside the splat will be released and it will be more difficult for cracks to form at other locations within the splat. Thus, the identification of the locations of maximum stresses was one of the objectives of the numerical simulations.

As shown in Figure 1, the normal stress component parallel to the splat interface is designated as S_{11}, while the normal stress component parallel to the splat thickness is S_{22}. The related shear stress, designated as S_{12}, changes along with S_{11} and S_{22} with the position along the sinusoidal interface, as shown in Figure 1. Figure 2 shows the distribution of three stress components through the thickness of the splat at the roughness valley after 10 ms, which is a rough estimate of the time elapsed when the next molten particle impinges on a previously solidified splat in the same pass. It is evident that the magnitude of S_{12} is much smaller than S_{11} and S_{22}, Also, S_{11} and S_{22} lead to mode I fracture, while S_{12} leads to mode II fracture. Since, mode I fracture is much more likely than mode II fracture for comparable stresses, only S_{11} and S_{22} components of the stresses are considered. Effectively, S_{22} is responsible for horizontal cracks and S_{11} is responsible for vertical cracks. The simulation also shows that the locations of maximum tensile S_{11} and S_{22} are at the roughness valley and occur at the end of the simulation

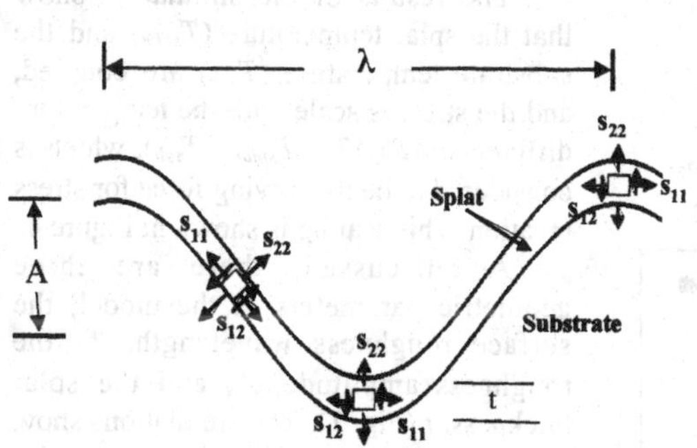

Figure 1. Direction of stress components as a function of location within the splat.

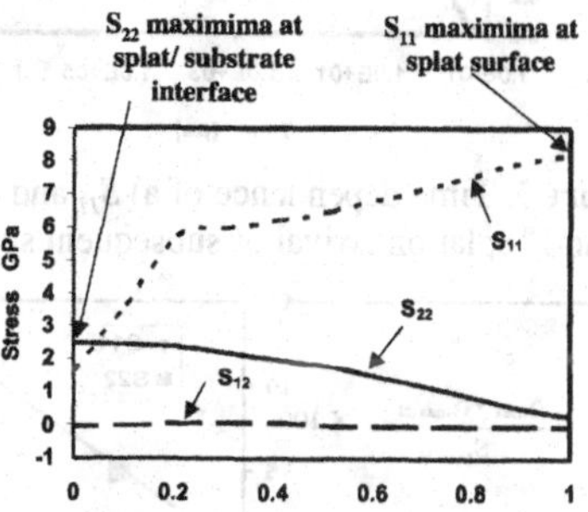

Figure 2. The amplitude of stress components at the roughness valley along the splat thickness after 10 ms.

time of 10 ms, when the splat and substrate have essentially equilibrated by cooling down together. Also, as seen in Figure 2, the location of maximum S_{11} is at the top of the splat and maximum S_{22} is at the bottom of the splat, both in the roughness valley. An interesting feature in Figure 2 is the abrupt change in the slope of the stress distribution (especially for S_{11}) at a location within the splat. Although the reason for this is currently under study, it is speculated that the location of the position of the slope change is related to the diffusion distance of heat during a period of the splat solidification time.

Simulations were also run for multiple splats to study how the maximum stresses in a given splat changes when subsequent splats solidify above it. Figure 3 shows how the maximum value of S_{11} and S_{22} in the 1st splat evolves with time, as three more subsequent splats solidify above it in the same torch pass. The figure shows that the majority of the stresses occur during the solidification of the splat itself, and although subsequent splats change the state of the stress in the already solidified splat below, the change is relatively small. This further justifies the single splat model, since it captures most of the stresses that will remain in the splat, even when more splats solidify over it. Although the actual stress distribution is dependent on the splat thickness, the trends are similar and the conclusions inferred above are independent of the actual splat thickness.

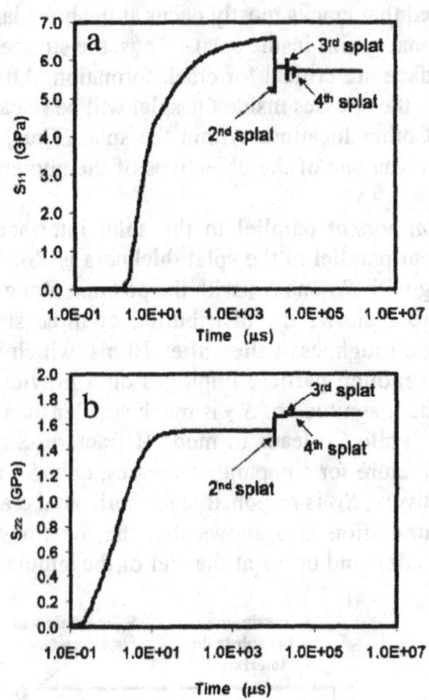

Figure 3. Time dependence of a) S_{11} and b) S_{22} in the 1st splat on arrival of subsequent splats.

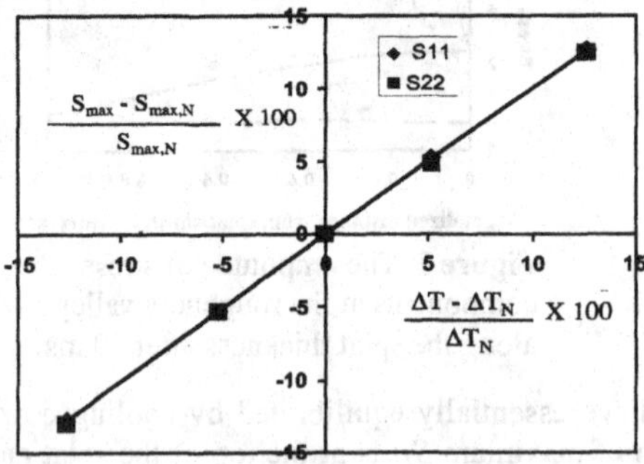

Figure 4. Plot of percentage change (from a nominal condition, N) in maximum stresses (S_{11} and S_{22}) versus the percentage change in ΔT.

Effect of processing parameters

The results of the simulation show that the splat temperature (T_{splat}) and the substrate temperature (T_{sub}) are coupled, and the stresses scale with the temperature difference ΔT ($\Delta T = T_{splat} - T_{sub}$), which is considered to be the driving force for stress creation. This scaling is shown in Figure 4.

As discussed, there are three geometric parameters in the model, the surface roughness wavelength, λ, the roughness amplitude, A, and the splat thickness, t (Fig. 1). The simulations show that the surface roughness parameters, A and λ, affect S_{22} much more than S_{11}. In fact, S_{11} is zero for solidification on a flat surface. It was also found that the stresses scaled with the three geometric parameters, t, A and λ, i.e., if the ratio of all three were kept constant, the stresses were constant. Thus, two independent geometric parameters were defined; $\psi = A/\lambda$, and $\zeta = t/\lambda$. The effect of ψ and ζ on maximum S_{11}

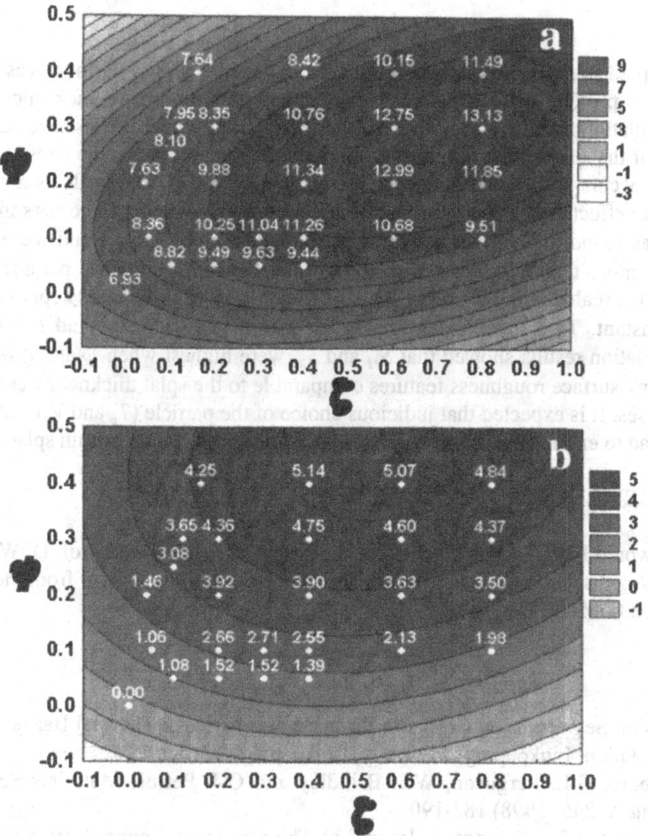

Figure 5. Contour plots of maximum a) S_{11} and b) S_{22} as a function of ψ and ζ. The numbers indicate the maximum stresses in GPa.

and S_{22} were studied over a wide range of ψ and ζ, and the results are shown in Figure 5. The figure shows that both S_{11} and S_{22} have a maxima at finite values of ψ and ζ, ($\psi \sim 0.3$ and $\zeta \sim 0.7$ for maximum S_{11} and $\psi \sim 0.5$ and $\zeta \sim 0.6$ for maximum S_{22}. This is a very interesting result since it suggests that stresses are most significant when the geometric parameters are of the same order of magnitude as the splat thickness.

A scan of the surface of a plasma sprayed coating shows features at various length scales. For large (few 100 µm) surface roughness wavelengths which are the length scale of the splat diameter and small (~100 nm) surface roughness wavelengths which is of the length scale of the cell size within a splat; λ is too large and too small respectively, compared to t (a typical splat thickness is ~ 1.5 µm) to contribute significantly to the stresses. Only wavelengths of the order of a few microns are in the correct length scale to contribute significantly to stress generation. When characterizing the surface roughness of the substrate, roughness features on this scale need to be examined carefully, since they provide locations of maximum stress concentration.

CONCLUSIONS

A fully coupled thermo-mechanical finite element model was setup and was run for a single splat case, to provide more insight into understanding the process/crack microstructure mapping. The simulation results showed that maximum S_{11}, the stress that causes vertical cracks, occurs at the top of the splat in the roughness valley; while maximum S_{22}, the stress that causes horizontal cracks, occurs at the splat bottom in the roughness valley. The analysis revealed that the key temperature effect was, $\Delta T = T_{splat}-T_{sub}$, which provides the driving force for solidification and cooling. It was found that surface roughness parameters, amplitude, A and wavelength, λ, affected S_{22} much more than S_{11}. It was also found that the three geometric parameters, splat thickness, t, λ, and A scaled with the stresses, i.e., if the ratio of all three were kept constant, the stresses were constant. Two independent geometric parameters $\psi = A/\lambda$ and $\zeta = t/\lambda$, were defined. The simulation results showed that S_{11} and S_{22} were highest when ψ and ζ were of the order of 1, implying surface roughness features comparable to the splat thickness were locations of maximum stresses. It is expected that judicious choice of the particle (T_p and V_p) and substrate (T_{sub}) states can lead to engineered crack structures based on stress control within splats.

ACKNOWLEDGEMENTS

The authors would like to acknowledge C. Cui, Prof. M. Gevelber and Prof. D. Wroblewski for many stimulating discussions. This research was funded in-part by a grant from the National Science Foundation (DMI-9713957).

REFERENCES

1. P. Bengtsson, Segmentation Cracks in Plasma Sprayed Thick Thermal Barrier Coatings. PhD Dissertation, Linkoeping University, Linkoeping, Sweden 1997.
2. A.M. Freborg, B.L. Ferguson, W.J. Brindley and G.J. Petrus, Materials Science and Engineering A, 245 (1998) 182-190.
3. P. Bengtsson and T. Johanneson, Journal of Thermal Spray Technology, 4 (1995) 245-251.
4. M. Friis, C. Persson C and J. Wigren , Surface and Coatings Technology, 141 (2001) 115-127.
5. J. Llavsky, A.J. Allen, G.G. Long, S. Krueger, C.C. Brendt and H. Herman, Journal of the American Ceramic Society, 80 (1997) 733-742.
6. S. Malmberg and J. Heberlein, Journal of Thermal Spray Technology, 2 (1993) 339.
7. S. Sampath, X.Y. Jiang, J. Matejicek, A.C. Leger and A. Vardelle, Materials Science and Engineering A, 272 (1999) 181-188.
8. S. Sampath and X. Jiang, Materials Science and Engineering A, 304-306 (2001) 144-150.
9. X. Jiang, Y. Wan, H. Herman and S. Sampath, Thin Solid Films, 385 [2] (2001) 132-141.
10. R. McPherson, Thin Solid Films, 82 [12] (1981) 297-310.
11. M. Prystay, P. Guogeon and C. Moreau, Thermal Spray: Practical Solutions for Engineering Problems. C.C. Brendt, editor. ASM International, Materials Park, OH. (1996) 517.
12. J. Madejski, Int. J. Heat Mass Transfer, 19 (1976) 1009.
13. ABACUS is licensed by Abacus Inc., Pawtucket, RI.

Tribological and Wear Resistant Surfaces and Coatings

Mater. Res. Soc. Symp. Proc. Vol. 843 © 2005 Materials Research Society T2.10

Subcritical CO_2 Assisted Polymer Surface Engineering at Low Temperatures

Yong Yang and Ly James Lee
Department of Chemical and Biomolecular Engineering, The Ohio State University, Columbus, OH 43210, U.S.A.

ABSTRACT

Polymer-based Micro/Nano Electro Mechanical Systems (MEMS/NEMS) have attracted a great deal of interest from industries and academia. The common polymer processing methods involve either organic solvents or temperatures above the glass transition temperature (T_g), which is undesirable, particularly for biomedical applications. On the basis of different properties near polymer surfaces from those in the bulk, we introduce subcritical fluids (particularly carbon dioxide, CO_2) into polymer surfaces to manipulate the polymer properties at the nanoscale so that we can achieve low temperature surface engineering. In this study, polymer surface dynamics under CO_2 were addressed using atomic force microscopy (AFM) and neutron reflectivity (NR). Monodispersed nanoparticles were deposited onto the smooth polymer surface and then embedded into the surface by annealing the sample at the pre-specified temperatures and CO_2 pressures. The embedding of nanoparticles in the proximity of the surface was measured using AFM, and thus the surface T_g profile could be determined. It was revealed that there is a rubbery layer of up to a hundred nanometers thick at the surface where the T_g is lower than that in the bulk and CO_2 dramatically reduced the surface T_g. NR studies also show that CO_2 can enhance chain mobility at the polymer surfaces below the polymer bulk T_g. These results indicated that even low concentrated CO_2 greatly could enhance polymer chain mobility below the T_g of the CO_2–plasticized polymers. The thickness of the rubbery layer can be controlled by tuning either temperatures, or CO_2 pressures, or both, which makes it possible to engineer polymer surfaces at low temperatures. Guided by the CO_2 enhanced polymer surface dynamics, we developed a novel CO_2 bonding technique to succeed in low temperature bonding of polymers at the micro/nanoscales. This CO_2 bonding technique has been applied to seal polymeric nanofluidic biochips and construct well-defined three-dimensional (3D) biodegradable polymeric tissue scaffolds.

INTRODUCTION

The demand for high-precision polymer-based miniature devices (e.g., biomedical devices) has been growing rapidly in recent years. When polymers are processed at the nanoscale, viscosity and surface tension become major limiting factors. Common polymer processing techniques involve either organic solvents or a temperature beyond the T_g. Although solvents can be used to lower the viscosity, they are detrimental to many biomolecules and surface wetting may still lead to processing difficulties. In addition, they tend to deform microstructures [1].

Recent research [2-7] of polymer thin films revealed that properties (e.g., T_g) near the polymer surface are different from those in the bulk. The competition between the polymer-free surface and the polymer-substrate interactions determines the thickness-dependent T_g shift [8]. When the polymer-free surface interaction dominates, the polymer shows a T_g depression near the surface, typically less than 100 nm [2-7]. Based on the T_g depression near the polymer surface, interfacial bonding of polymers at temperatures below their bulk T_gs has been achieved [9,10]. However, these studies showed that the bond strength developed very slowly below T_g, e.g., 0.08 MPa after 4 hours at 62 °C for polystyrene (PS), making it unsuitable for practical applications. In addition, the processing of polymers with CO_2 has developed to become a promising new technology due to its safe, low-cost and biologically benign nature. It is well known that CO_2 can depress the T_g, viscosity, and interfacial tension of polymers. Our previous studies show that CO_2 can enhance interfacial bonding of microstructures at a low temperature of 35°C compared with the bulk value of 48.3°C [11]. The bond strength of poly(DL-lactide-co-glycolide) (PLGA) approached 1 MPa with a bonding time of 30 min at 35°C and 0.79 MPa CO_2 pressure. It is necessary to explore the effect of CO_2 on the surface T_g.

Although a number of studies have been published on the surface T_g of polymers, little has known on how CO_2 affects the surface T_g. Generally, these methods of determining the surface T_g can be classified into two categories based on sample configurations. In the first category, polymer thin films were spin-coated on substrates such as silicon (Si) wafers or glass plates. The T_gs of these thin films can be measured via ellipsometry [2,3]. In the second one, freestanding polymer thin films with a thickness less than 100 nm were prepared and studied via methods including Brillouin light scattering [4], positron annihilation spectroscopy [12], and local thermal analysis [13]. Recently, a new approach was proposed to investigate the surface T_g of polymers using AFM [14,15]. Gold nanoparticles are placed onto a smooth polymer surface and the sample is annealed at a pre-specified temperature; the apparent height of the nanoparticles embedded onto the surface is measured using AFM. From AFM data, the surface layer where the T_g is less than the bulk T_g can be determined.

In this study, we applied both the gold nanoparticle/AFM approach and neutron reflectivity to evaluate the effect of CO_2 on surface dynamics of polymers. In light of the studies we demonstrated the capability of CO_2 assisted polymer surface engineering at low temperatures by achieving low-temperature bonding of polymers at the nanoscale.

EXPERIMENTAL

Materials

The materials used include PLGA and PS. PLGA (Alkermes Medisorb® 5050 DL High IV) and PS (STYRON 685D, M_n = 120,000) were donated by Alkermes, Inc. and Dow Chemical, respectively. These materials were used without further purification. For NR studies, both hydrogenated PS (hPS; M_n = 214,000, M_w/M_n = 1.03) and deuterated PS (dPS; M_n = 178,000, M_w/M_n = 1.10) were supplied by Polymer Source Inc. Unconjugated gold colloid with a particle size of 20 nm (20.2 ± 8%) was supplied by Ted Pella, Inc. The carbon dioxide used was a bone dry grade supplied by Matheson at a purity of 99.8%.

Measurements of surface T_g under CO_2

The PLGA films were prepared by spin-coating 2.5 wt.% PLGA acetonitrile solution on the polished side of Si wafers pre-treated with detergent, acetone, methanol, and 18 MΩ•cm water. PLGA solution was filtered through a 0.1 μm inorganic membrane filter (Whatman®, Anotop 25). The spinning speed was selected to produce films of about 1 μm thick. The films were placed in a cleanroom environment at room temperature for 24 hours to evaporate the solvent, then annealed at 100°C in a vacuum oven for another 24 hours. A drop of 50-fold diluted gold colloid solution was placed on the polymer film surface and the water was allowed to evaporate in a refrigerator at 4°C overnight. The sample was then annealed with and without CO_2 in a pressure vessel at a pre-specified temperature. An ISCO 500C high pressure syringe pump was used to deliver and control the pressure of CO_2 in the vessel. After 2 hours annealing, the system was quenched with ice and then taken out for AFM measurement right away.

The topological investigation of these systems was carried out using a NanoScope IIIa AFM (Digital Instruments) in the tapping mode. For each sample, the mean apparent height of the nanoparticles and the standard deviation were calculated over 50 nanoparticles.

Measurements of chain mobility under CO_2

Bilayer samples on Si substrate (hPS/dPS/Si) were prepared by using floating technique [16]. First, 2" wafers were cleaned by immersion in piranha solution (sulfuric acid: hydrogen peroxide=1:1) for 15 minutes and then 10 seconds dip in buffer oxide etchant. Secondly, the dPS toluene solution was spin-coated onto the wafer to form a thin film with the thickness of 738 Å, determined by spectroscopic-reflectometry (Nanospec 1000, Nanometrics Inc.). Independently, an approximately 800 Å thick hPS film was spin-coated onto a 3" wafer, which was cleaned in the same manner and then had an 1150 Å dry silicon oxide layer thermally grown on it at 1050°C for 2.7 h. All the films were annealed at 150°C for 12 h in vacuum to remove the residual solvent and relax the residual stress imposed by spin coating. The perimeter of the hPS film was scored with a blade, and then the film was floated off onto the surface of 7:1 buffer oxide etchant. Finally, the hPS film was rinsed with 18 MΩ•cm water and picked up onto the dPS film. Again, the residual water was removed by annealing the bilayer samples at 65°C in vacuum for 5 h.

The bilayer samples were annealed at 70°C in vacuum or under 1.38 MPa CO_2 pressure. NR measurements were carried out using the unpolarized neutron reflectometer (POSY II) at the Intense Pulsed Neutron Source, Argonne National Laboratory.

CO_2 assisted polymer bonding at the nanoscale

Nano-sized channels with a spectrum of widths from 100 to 600 nm were patterned on a negative tone resist AZPN114 coated on a Si wafer using electron beam lithography (EBL). The polymethylsiloxane (PDMS) mold with the inverse pattern was obtained via casting and curing PDMS on the nano-sized channels at room temperature for 24 hours. A PS film (~50 μm, prepared via compression molding) was preheated to 220°C for 1 minute, and a pressure of 0.1 MPa was used to emboss the PDMS mold into the molten PS film. After 30s, ice water was used to cool down the system below 40°C, and the PS layer with nanochannels was peeled off from the PDMS mold. Another PS film (~2 μm) was formed as the lid by spin-coating 5 wt.% PS toluene solution on a pre-cleaned glass slide. The film with the slide was floated onto a pool of deionized water and the film was captured on a cured PDMS plate. The film with the PDMS plate was then dried under vacuum for 24 h at 90 °C to remove residual water.

The PS lid on the PDMS plate was aligned onto the patterned PS layer so that a part of the nanochannels was not covered by the lid. The aligned sample was placed between two glass slides and a contact pressure was applied on the sample. The whole assembly was placed in the pressure vessel. After saturation with CO_2 at a pre-specified temperature and pressure for a period of time, the pressure was slowly released and the samples were taken out from the vessel. The cross-section profiles of bonded layers were measured using a Hitachi S-3000H scanning electron microscope (SEM).

RESULTS AND DISCUSSION

The temperature and CO_2 pressure dependences of the mean apparent height of the gold nanoparticles on the PLGA film surface were systematically investigated as shown in Figure 1A. At the condition of 15°C and 0.31 MPa CO_2 pressure, the detected particle height of nearly 20 nm on polymer surface indicated that the particles still stayed on the top of the PLGA surface, demonstrating the typical behavior of a glassy system. At 35°C, the AFM measurement showed that, in vacuum, the mean apparent height was 12.6 ± 1.3 nm. Considering the gold nanoparticle size of 20.2 ± 1.6 nm, the particles were embedded into the PLGA surface 7.6 nm. This implies that, underneath the PLGA surface there is a rubbery layer 7.6 nm thick with a T_g less than 35°C, which is far below the bulk T_g of PLGA (48.3°C) [15]. It was notable that when CO_2 was introduced, the particles were embedded deeper into the PLGA surface, indicating that the chain mobility of PLGA was enhanced and the thickness of the rubbery surface layer increased. For instance, compared with that in vacuum, the thickness of the rubbery surface layer of PLGA increased by 9.1 nm in the presence of 0.79 MPa CO_2 at 35°C. By re-arranging the data we could plot the surface T_g gradient of PLGA under CO_2 (Figure 1B). It was found that T_g at the polymer-air surface was substantially lower than the bulk T_g and increased toward the bulk value with the depth from the surface. Subcritical CO_2 could greatly depress T_g near the surface.

Understanding the dynamics near polymer surface under CO_2 is essential to establish guideline for CO_2-assisted polymer surface engineering. NR is a high-resolution technique (~1 nm) capable of penetrating up to several hundred angstroms, thus is highly suitable for studying chain diffusion at the polymer surface. Unfortunately, there is no deuterated PLGA available. In this study, PS was taken as the model material. In the bilayer configuration, the thickness of dPS is 738 Å, four times larger than its radius of gyration (R_g) of 168.9 Å and the PS-substrate interaction is negligible. Figure 2 shows that, in vacuum PS chains could diffuse across the interface at 70.0°C, i.e. 30.2°C below their bulk T_g, and then reach a "dead" bulk region which is still in the glassy state [16]. When low-pressure CO_2 was introduced polymer chains diffused faster and reached a deeper depth. Addition of CO_2 increases the free volume among the polymer segments, thus leading to larger T_g depression, higher chain mobility at the polymer surface, and thicker rubbery surface layer.

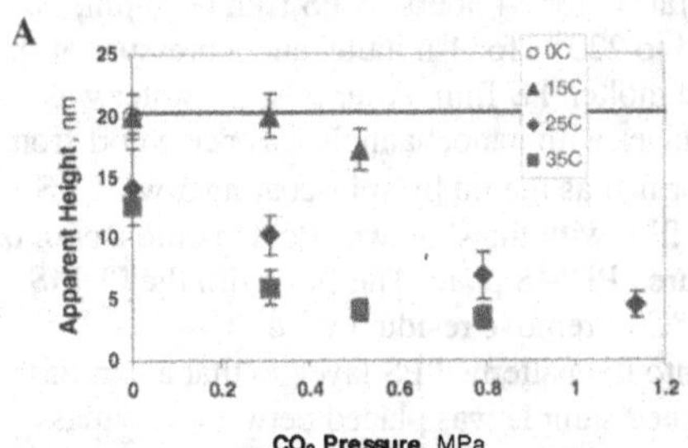

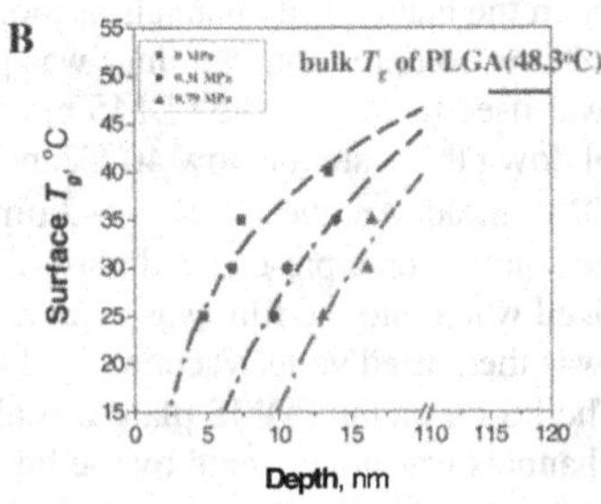

Figure 1. (**A**) Temperature and CO_2 pressure dependence of the mean apparent height of the gold nanoparticles on the PLGA surface. The horizontal line represents the nanoparticle size, 20.2 ± 1.6 nm (provided by the manufacturer). (**B**) Surface T_g gradient of PLGA.

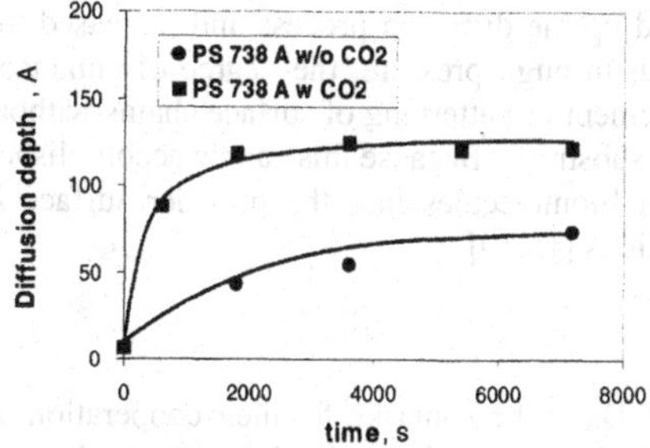

Figure 2. Time evolution of diffusion depth of bilayers. Diffusion conditions: Dots, 70°C in vacuum; squares, 70°C and 1.38 MPa CO_2 pressure.

Guided by the CO_2 enhanced polymer surface dynamics, we explored the capability of CO_2 bonding for polymer nanostructures. At the nanoscale, the surface roughness is comparable to the nano-sized features. It is necessary to have the intimate contact between the lid and the nanochannels. In our study, the PS lid was first transferred onto a PDMS plate because the PDMS plate could be pressed to intimately contact with the top surface of the nanochannels due to its elastomeric nature. A contact pressure of 0.28 MPa was required to ensure good interfacial wetting without deforming the nanostructure. Figure 3 shows the SEM micrographs of bonded PS nanochannels at 70°C and 1.38 MPa CO_2 pressure. Compared with PS nanochannels before bonding, it can be seen that the nanostructures, including surface roughness, are well preserved.

We have applied this CO_2 bonding technique to seal polymeric micro/nanofluidic chips and to construct 3D PLGA tissue scaffolds with well-defined structure [11,17]. Although there are other methods to bond micro-/nanoscale polymer features, we believe subcritical fluid bonding is the most affordable and probably the only biologically benign and sterile method, in particular for simultaneous assembly of a large number of micro-/nanostructures.

CONCLUSIONS

We applied both nanoparticles/AFM approach and neutron reflectivity to investigate surface dynamics of polymers under CO_2. The results show the existence of a rubbery surface layer of

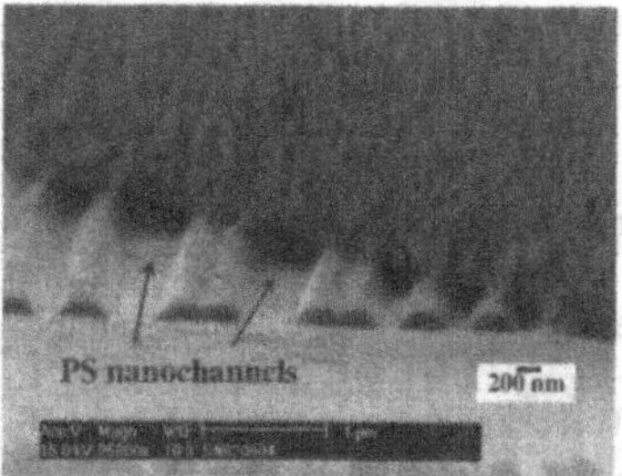

Figure 3. PS nanochannels sealed at 70°C, 1.38 MPa CO_2 pressure with 0.28 MPa contact pressure for 2 h.

polymers. Presence of subcritical CO_2 both speeded up the diffusion process and increased the "mobile" region. By controlling the amount of CO_2 through pressure, the degree of enhanced chain mobility can be controlled to result in entanglement or patterning of surface chains without altering the underlying mechanical properties of the substrate. Because this can be accomplished at near ambient temperatures, the impregnation of biomolecules into the polymer surface is possible under biologically benign and sterile conditions [18,19].

ACKNOWLEDGMENTS

We would like to thank Mr. Dehua Liu and Dr. David L. Tomasko for their cooperation on AFM measurements of surface T_g of PLGA, Dr. Mark Cheng, Mr. Richard J. Goyette Jr., and Ms. Jiong Shen for their support during neutron reflectivity studies, and Dr. Wu Lu for providing the EBL nano-sized pattern. We would also like to thank Alkermes Inc. and Dow Chemical for material donations.

REFERENCES

1. K. R. King, C. Wang, J. P. Vacanti, and J. T. Borenstein, *Mater. Res. Soc. Symp. Proc.* **729**, 3 (2002).
2. J. L. Keddie, R. A. L. Jones, and R. A. Cory, *Europhys. Lett.* **27**, 59 (1994).
3. J. L. Keddie and R. A. L. Jones, *Isr. J. Chem.* **35**, 21 (1995).
4. J. A. Forrest, K. Dalnoki-Veress, J. R. Stevens, and J. R. Dutcher, *Phys. Rev. Lett.* **77**, 4108 (1996).
5. F. Xie, H. F. Zhang, F. K. Lee, B. Du, O. K. C. Tsui, Y. Yokoe, K. Tanaka, A. Takahara, T. Kajiyama, and T. He, *Macromolecules* **35**, 1491 (2002).
6. K. L. Ngai, A. K. Rizos, and D. J. Plazek, *J. Non-Cryst. Solids* **235-237**, 435 (1998).
7. A. M. Mayes, *Macromolecules* **27**, 3114 (1994).
8. D. T. Hsu, H.-K. Kim, F. G. Shi, B. Zhao, M. R. Brongo, P. Schilling, and S.-Q. Wang, *Proc. SPIE-Int. Soc. Opt. Eng.* **3883**, 60 (1999).
9. Y. M. Boiko and R. E. Prud'homme, *Macromolecules* **30**, 3708 (1997); **31**, 6620 (1998).
10. X. Zhang, S. Tasaka, and N. Inagaki, *J. Polym. Sci., Part B: Polym. Phys.* **38**, 654 (2000).
11. Y. Yang, C. Zeng, and L. J. Lee, *Adv. Mater.* **16**, 560 (2004).
12. G. B. DeMaggio, W. E. Frieze, D. W. Gidley, M. Zhu, H. A. Hristov, and A. F. Yee, *Phys. Rev. Lett.* **78**, 1524 (1997).
13. D. S. Fryer, P. F. Nealey, and J. J. De Pablo, *Macromolecules* **33**, 6439 (2000).
14. V. M. Rudoy, O. V. Dement'eva, I. V. Yaminskii, V. M. Sukhov, M. E. Kartseva, and V. A. Ogarev, *Colloid Journal* (Translation of Kolloidnyi Zhurnal) **64**, 746 (2002).
15. J. H. Teichroeb and J. A. Forrest, *Phys. Rev. Lett.* **91**, 016104/1 (2003).
16. D. Kawaguchi, K. Tanaka, T. Kajiyama, A. Takahara, and S. Tasaki, *Macromolecules* **36**, 1235 (2003).
17. Y. Yang, S. Basu, L. J. Lee, S.-T. Yang, and D. L. Tomasko, *Biomaterials* (online available on Sep. 15, 2004).
18. T. L. Sproule, J. A. Lee, H. Li, J. J. Lannutti, and D. L. Tomasko, *J. Supercritical Fluids* **28**, 241 (2004).
19. A. K. Dillow, F. Dehghani, J. S. Hrkach, N. R. Foster, and R. Langer, *Proc. Natl. Acad. Sci. U. S. A.* **96**, 10344 (1999).

Mater. Res. Soc. Symp. Proc. Vol. 843 © 2005 Materials Research Society T2.7

X-Ray Studies of Near-Frictionless Carbon Films.

Nikhil J Mehta[1, 5], Sujoy Roy[3], Jacqueline Anne Johnson[1], John Woodford[1], Alexander Zinovev[4], Zahirul Islam[2], Ali Erdemir[1], Sunil Sinha[3], George Fenske[1], Barton Prorok[5]
[1] Energy Technology, Argonne National Laboratory, Argonne, IL 60439, U.S.A.
[2] Advanced Photon Source, Argonne National Laboratory, Argonne, IL 60439, U.S.A.
[3] Physics, University of California San Diego, San Diego, CA 92093, U.S.A.
[4] Materials Science, Argonne National Laboratory, Argonne, IL 60439, U.S.A.
[5] Materials Engineering, Auburn University, Auburn, AL 36849, U.S.A.

ABSTRACT

Carbon-based coatings exhibit many attractive properties that make them good candidates for a wide range of engineering applications. Tribological studies of the films have revealed a close correlation between the chemistry of the hydrocarbon source gases and the coefficients of friction and wear rates of the diamond-like carbon films. Those films grown in source gases with higher hydrogen-to-carbon ratios had much lower coefficients of friction and wear rates than did films derived from source gases with lower hydrogen-to-carbon ratios. The mechanism for this low friction is as yet not properly understood. Ongoing structural characterization of the films at Argonne National Laboratory is gradually revealing this mechanism. Recent studies have included x-ray photoelectron spectroscopy (XPS), near edge x-ray absorption fine structure (NEXAFS) and x-ray reflectivity (XRR). XPS showed ~10% oxygen at the surface, which was largely removed after a 1 minute sputter; NEXAFS showed a high sp^2:sp^3 ratio implying a highly graphitic material; and XRR has given a comprehensive depth profile, with three layers of increasing density as the substrate was approached. The paper discusses the results and correlation with previous friction measurements.

INTRODUCTION

Diamond-like carbon (DLC), a metastable form of amorphous carbon, is a widely studied thin film due to its combination of high strength, hardness, excellent tribological properties, chemical inertness, and IR transparency [1, 2]. Carbon readily hybridizes in a number of different forms: sp^3, sp^2 and sp, each with different physical and chemical properties. Hence, tailoring the properties of these films to the desired requirements may be accomplished by controlling growth mechanisms and hydrogen content to give different sp^1:sp^2:sp^3 ratios. At Argonne National Laboratory films were grown via radio frequency plasma-enhanced chemical vapor deposition (RF PECVD) using CH_4/CD_4 and H_2/D_2 precursor gases. Argonne's DLC, near-frictionless carbon (NFC) films with high plasma hydrogen content have exhibited extremely low μ in inert environments [2]. To get an insight into the behavior of these films, the present study characterizes them using the following x-ray techniques: near-edge x-ray absorption fine structure (NEXAFS), x-ray photoelectron spectroscopy (XPS) and x-ray reflectivity (XRR).

NFC films grown with different precursor gas (H_2/D_2 and CH_4/CD_4) compositions are shown in Table 1.

Table 1. Composition of various NFC films analyzed.

Films Analyzed (%H_2 content in source gas during deposition)		Films Analyzed (%D_2 content in the source gas during deposition)	
NFC 7 (0)	NFC 8 (25)	D- NFC 7 (0)	D- NFC 8 (25)
NFC 2 (50)	NFC 6 (75)	D- NFC 2 (50)	D- NFC 6 (75)

EXPERIMENTAL DETAILS

All of the films studied were grown on Si single-crystal wafer pieces in a modified Perkin-Elmer 2400 RF PECVD system. All substrates were sputter cleaned in an argon gas discharge plasma using RF at power levels between 900- 1100 W prior to thin film growth. Deposition of NFC films was then carried out using appropriate precursor gas mixtures. Prior to deposition, the chamber was pumped to a base pressure of ~ $3 \cdot 10^{-7}$ Torr using a turbomolecular pump. The deposition pressure ranged between 20 – 24 mTorr, depending on the type of film grown. The critical parameter during deposition is the RF self-bias to the substrate; RF power was held so as to maintain a value of -500V in order to obtain the optimal film structure. Plasma chemistry during film growth was monitored using a Dycor Dymaxion mass spectrometer. Further details of growth are discussed in references [2] and [3].

NEXAFS measurements were conducted at the 10m toroidal grating monochromator (TGM) beam line at the synchrotron radiation center (SRC) at the University of Wisconsin-Madison. The measured total electron yield (TEY) intensity, $\boldsymbol{I}$, was normalized with respect to the baseline curve, $\boldsymbol{I_0}$, obtained by measuring absorption from an atomically clean Si wafer sample. The energy step size was 0.1 eV and the collecting time of 2 sec was averaged from 3 readings for each channel. As no angular dependence was expected for the film studied due to their amorphous structure, the data were collected with incident photons normal to the sample surface. The sp^2 content was derived by comparison with highly oriented pyrolytic graphite (HOPG) probed with photons incident at 54.3^0 (magic angle) to suppress the effect due to x-ray polarization [4,5]. NEXAFS C K-edge and O K-edge spectra were obtained in the energy range of 280-315 eV and 520-550 eV in the TEY mode.

An XPS study on NFC films was done with a home-made system at the Materials Science Division, Argonne National Laboratory using an Mg***K***α (1253.6 eV) source with a hemispherical energy analyzer (HEA) in fixed absolute resolution (FAT) mode. The detail of spectra analyzed is in Table 2. The spectrometer calibration was done using a gold XPS emission line (Au$4f_{7/2}$ with binding energy 84 eV). For depth profiling of samples, the surfaces were sputtered with an Ar^+ ion beam with a total output current 30 $\mu A\ cm^{-2}$.

Table 2. Information for XPS data collection

Analysis Type	Mode	Energy Resolution	Spatial resolution (elliptical)	Energy Step
Survey spectra	FAT 44	1.1 eV	4 x 3 mm	1.0 eV
Detail core level peaks	FAT 22	0.8 eV	4 x 3 mm	0.2 eV

XRR experiments were carried out at the Advanced Photon Source at Argonne on beam line 4ID-D and 6ID-B with beam line energy of 8 KeV. The data for all films were scanned from 0.001 to 0.295 in q_z ($Å^{-1}$). The data fitting was done by a recursive least square fitting procedure using Parrat formalism [6].

RESULTS

The NEXAFS study of the C K-edge spectra for both hydrogenated and deuterated NFC films are shown in Figures 1 and 2. All spectra show pre-edge resonance at around 285 eV, which is attributed to transitions from C 1s orbitals to unoccupied π^* orbitals of sp^2 (C = C) sites. This peak also includes the contribution of sp (C ≡ C) sites if they are present [4]. In all NFC films only broad spectra appear from 288-310 eV. The amorphous nature of NFC films induces spread and overlapping of the C 1s → σ^* transitions at sp, sp^2 and sp^3 sites [4,5].

XPS studies were done on hydrogenated NFCs and results are given in Table 3. As-is films had surface adsorbates such as N, Cl, F, and O and after 1 min sputtering by Ar^+ ions, C and a negligible amount of O are detected.

XRR uses the concept of total external reflection, as the refractive index of solids for x-rays is slightly less than unity. Recursive formalization of Fresnel equations were used to simulate the experimental results of overall reflection, and film density was calculated from those results [1, 6-8]. NFC films were studied to reveal surface roughness, thickness and density. The experimental results were fitted computationally over the entire reflectivity curve. The fit for NFC6 is shown in Figure 3. The results obtained show variation of density with depth fitted with a 3-layer model as shown in Table 4.

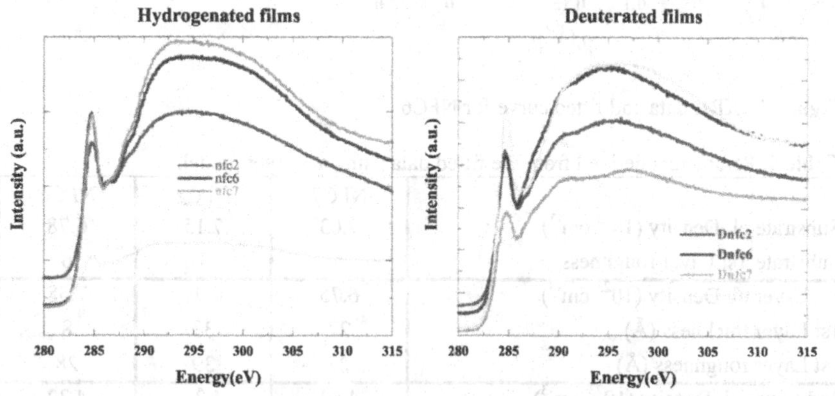

Figure 1. C K-edge for NFC films

Figure 2. C K-edge for d-NFC films

Table 3. Elemental composition for NFC films using XPS

Element	As-is (%)			After 1 min sputter (%)		
	NFC7	NFC2	NFC6	NFC7	NFC2	NFC6
Carbon	87.1	86.1	88.1	99.6	98.5	98.3
Oxygen	11.7	11.0	11.9	0.4	1.5	1.7
Chlorine		0.4				
Fluorine		2.5				
Nitrogen	1.2					
Total	100.0	100.0	100.0	100.0	100.0	100.0

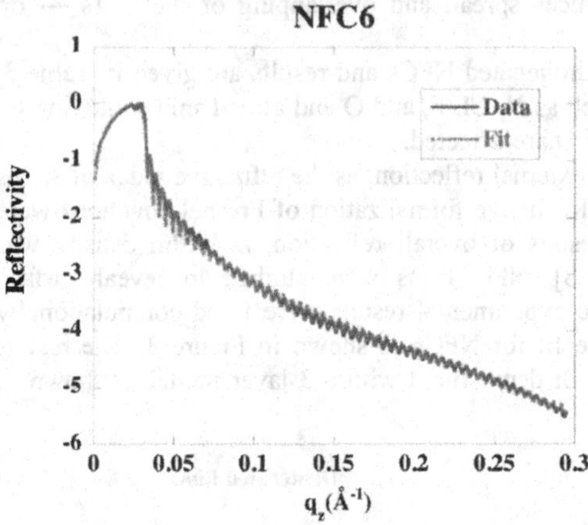

Figure 3. XRR data and fitted curve for NFC6

Table 4. Parameters derived from the fitted data using a 3-layer model.

	NFC7	NFC2	NFC6
Substrate el. Density (10^{23} cm^{-3})	7.03	7.15	6.78
Substrate/1st Layer roughness	6	7	6
1st Layer el. Density (10^{23} cm^{-3})	6.75	6.91	7.58
1st Layer thickness (Å)	27	33	8
1st Layer roughness (Å)	27	39	28
2nd Layer el. Density (10^{23} cm^{-3})	4.63	4.23	4.32
2nd Layer thickness (Å)	2342	3644	1231
2nd Layer roughness (Å)	3	7	1
3rd Layer el. Density (10^{23} cm^{-3})	4.12	3.87	4.04
3rd Layer thickness (Å)	28	27	27
3rd Layer roughness (Å)	4	9	4

DISCUSSION

NFC films were studied by NEXAFS to quantify the sp^2:sp^3 ratio; the calculation is shown in references [4] and [5]. The % of sp^2 in the films is shown in Figure 4. The films on average have 72 ± 6% sp^2 bonding with no visible trend. They high sp^2 content resonates with highly graphitic behavior, in particular, low friction coefficients.

XRR results show that the film density varies with depth, breaking into three fairly distinct layers. A thin, dense layer is found close to the substrate, followed by a less dense bulk layer and an even lower density top layer. The top surface layer is ~30 Å and may be responsible for initial high coefficient of friction values shown during bench scale measurements at Argonne. From the fit, the roughness of the first layer is greater than the thickness. Roughness here signifies the electron density profile decreasing or increasing continuously into the next layer or some intermixing of layers and is not physical like a smeared step function in conventional terms. Previous neutron reflectivity results [9] revealed a two-layer model with an inverse in the density profile. The x-ray data is thought to be more accurate due to the extended Q-range and the greater penetration depth of x-rays.

XPS gives us surface chemical composition and is primarily used to find different adsorbates on the film. As-received films showed elemental traces of F, Cl, N and a comparatively larger amount of O; they are most likely physiadsorbed, as they were not present during film growth. The majority of the oxygen and all the other adsorbates are gone after a 1 minute sputter, which erases ~25 Å. The remaining oxygen could be physiadsorbed or chemically bonded to the film. This correlates well with the XRR measurements and previous neutron reflectivity (NR) measurements, which show a surface layer of ~30 Å and may be due to all the adsorbed materials.

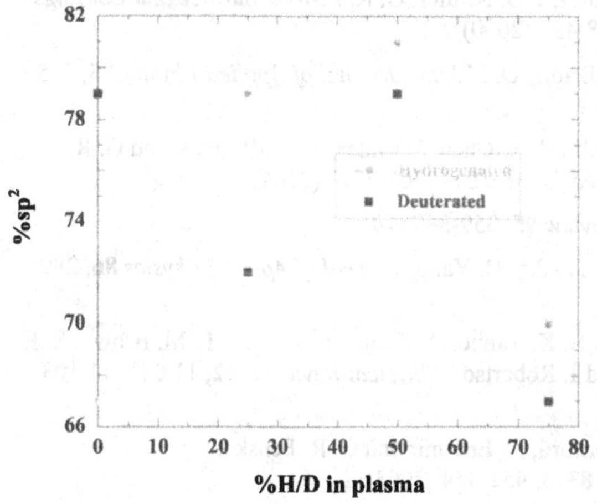

Figure 4 - Percent sp^2 content in NFC and D-NFC films based on H/D content in the plasma.

CONCLUSION

Examination of NFC films using a range of x-ray techniques gave some insight into the films. NEXAFS studies did not reveal any trend of sp^2:sp^3 ratio with source gas composition, regardless of whether the film is hydrogenated or deuterated; all films have high and nearly equal sp^2 content; the high graphitic nature of the films is likely to contribute to the low friction coefficients that these films exhibit. XPS shows surface elemental composition and XRR showed a decrease in density from the bottom to the top over the three layers above the substrate. From XRR and previous neutron reflectivity results, it was shown that we have a surface layer of ~30 Å. Sputtering for 1 minute during XPS measurements removes approximately this thickness of material. Friction tests on the films show that initial friction is high, then after a "wear-in" period reduces to a very low steady-state value, which correlates well with the existence of a surface layer exhibiting different properties. Consequently, these measurements have revealed films of a highly graphitic nature and an oxygen containing surface layer.

Further XRR and NR experiments have been done in inert atmospheres in an effort to reduce the incidence of moisture and other elements being physiadsobed on the surface, the results are presently being analyzed.

REFERENCES

[1] J. Robertson, *Materials Science and Engineering R* **37**, 129-281 (2002)

[2] A. Erdemir, *Tribology International* **37**, 577-583 (2004)

[3] A. Erdermir, O. L. Eryilmaz, I. B. Nilufer, G. R. Fenske, *Surface and Coatings Technology* **133-134**, 448-454 (2000)

[4] C. Lenardi, P. Piseri, V. Briois, O. Milani, *Journal of Applied Physics* **85**, 7159-7167 (2000)

[5] J. A. Johnson, J. B. Woodford, X. Chen, J. Anderson, A. Erdemir and G. R. Fenske, *Journal of Applied Physics* **95**, 7765-7771 (2004)

[6] L.G. Parratt, *Physical Review* **95**, 359-369 (1954)

[7] Q. Zhang, S. F. Yoon, R. J. Ahn, H. Yang, *Journal of Applied Physics* **86**, 289-296 (1999)

[8] A. C. Ferrari, A. Libassi. B. K. Tanner, V. Stolojan, J. Yuan, L. M. Brown, S. E. Rodil, B. Kleinsorge and J. Robertson, *Physical Review B* **62**, 11 089- 11 103 (2000)

[9] J. A. Johnson, J. B. Woodford, A. Erdemir and G. R. Fenske. *Applied Physics Letters,* **83**, 3, 452-454 (2003)

Mater. Res. Soc. Symp. Proc. Vol. 843 © 2005 Materials Research Society T2.8

Synthesis and Characterization of Nanocrystalline Diamond and Its Biomedical Application

Zhenqing Xu[1, 2], Arun Kumar[2], Ashok Kumar[1, 2], Arun Sikder[2]
[1]Department of Mechanical Engineering, University of South Florida, Tampa, FL, 33620, USA
[2]Nanomaterials and Nanomanufacturing Research Center, University of South Florida, Tampa, FL, 33620, USA

ABSTRACT

Diamond is known as the material that has excellent mechanical, electrical and chemical properties. Diamond is also an ideal interface that is compatible with microelectronics process and biological environments to work as a biosensor platform with excellent selectivity and stability. In our study, nanocrystalline diamond (NCD) films were grown on Si substrates by the microwave plasma enhanced chemical vapor deposition (MPECVD) method. Parameters such as gas composition, temperature and pressure are investigated to get the best film quality. Scanning electron microscopy (SEM) and Raman spectroscopy were used to characterize the NCD films. Then the NCD films were treated by hydrogen plasma in the CVD chamber to obtain the hydrogen terminated surface. This hydrogenated NCD film is ready for bio-modification and can work as the platform of the biosensors. Fourier Transform Infrared Spectroscopy (FTIR) and X-ray Photoelectron Spectroscopy (XPS) were employed to confirm the surface hydrogenation.

INTRODUCTION

In recent years, there has been much interest in the diamond films deposited using the chemical vapor deposition (CVD) technique due to the unique properties and potential applications in various areas. However, the synthesis of microcrystalline diamond films (MCD) results in large grain size and a very rough surface. Presently, the studies of nanocrystalline diamond films have been widely focused [1-4]. Compared with the microcrystalline diamond films, nanodiamond has a great deal of outstanding properties including a smooth surface, low coefficient of friction, ultrahigh nuclei density and negative electron affinity [5-7]. These properties show nanocrystalline diamond could be used in a wide range of applications including microelectrical mechanical systems, abrasive coatings, field emission electrodes and biosensors [8-10].

There are several methods to synthesize nanocrystalline diamond films. Gruen et al has demonstrated that NCD thin films can be deposited from $Ar/H_2/CH_4$ as reacting gas at different concentration of Ar. It was observed that the structures of the films changed from microcrystalline to nanocrystalline with the increase of Ar gas [11]. Other than using Ar, it has been reported that smooth NCD films were produced using a high methane concentration in a hydrogen environment [12]. Also, in order to avoid the surface damaging by the ex situ (scratching or ultrasonic) treatments, the NCD films are obtained directly on a mirror polished substrate using biased enhanced growth [2] by the MPECVD method.

In this paper, nanocrystalline diamond films were deposited using an MPECVD system. In order to study the parameters' effect during the synthesis, different gas combinations, different pressures and methane concentration was employed to grow the films. The properties of the films were characterized and compared to optimize the system. Surface morphology and other properties were then characterized by scanning electron microscopy and Raman spectroscopy. In order to work as a biosensor platform, NCD films were treated by hydrogen plasma in CVD chamber. This procedure preferentially etches any graphitic carbon and leaves the diamond surface terminated with C-H bonds [13]. This hydrogenated surface is ready for bio-modification and can be used as the biosensor platform.

EXPERIMENTAL

(100) silicon samples (1 inch by 1 inch) were used as the substrates. Surface pretreatment was employed before the deposition to enhance the nucleation density of diamond. 0 to 0.5 μm diamond powders were used to polish the sample surface by hand scratching and the substrates were ultrasonically cleaned in ethanol for 10 min. The NCD films were deposited in a 1 KW 2.45-GHz ASTeX system. A mixture gas of Argon, hydrogen and methane gas was used as a source gas. The first set of samples was deposited using $Ar/H_2/CH_4$ gas combination. Total gas was kept at 200 sccm and different concentrations of Ar (50%, 70% and 95%) were selected. Growth time was around 3 hours and chamber pressure was kept at 100 torr. Based on the results, several sets of samples were grown with different growth parameters. A high percentage of methane (up to 5 %) and different pressure was selected to investigate the parameter effect on the NCD growth to optimize the system. Nitrogen up to 20% was added to the reacting gas to dope the NCD films. Electrical properties were measure by the four-point-probe method. The NCD films were exposed to hydrogen plasma (30 torr) for 1 hour at 700 ºC to generate the hydrogen terminated surface. The diamond films were examined with SEM and Raman spectroscopy to investigate the film quality. The hydrogenated surface was characterized by XPS and FTIR.

RESULTS AND DISCUSSION

Synthesis and characterization of NCD films

SEM observations of the first set of samples indicate that the surface morphology is significantly affected by adding Ar to the reacting gas. Fig.1 shows SEM micrographs of NCD grown with different Ar concentration. At 50 vol% Ar, shown in Fig.1a, a few small crystals are grown on the surface with a lot of big diamond grains. The number of density of the small crystals increases and the large diamond crystallites start to disappear at 70 vol% Ar (Fig.1b). At 95 vol% Ar (Fig.1c), well-faced (111) and (100) diamond crystallites of micro size no longer exit, the surface structure of the diamond film changes form microcrystalline to nanocrystalline, the crystal size is less than 100 nm. These SEM micrographs demonstrate that the microstructure of the diamond films changes from microcrystalline to nanocrystalline by increasing the volume percentages of Ar in reacting gases. It is believed that a secondary nucleation with C_2 dominates the nucleation process in the high Ar composition range and results in the formation of nanocrystalline diamond [11].

Figure 2 shows the Raman scattering spectrum of NCD film deposited using 95 vol% Ar. The Raman spectrum of the sample exhibits four features near 1135, 1332 and 1480 cm^{-1}. The band near 1332 cm^{-1} is well known as the diamond peak. The broad band ranging from 1450 to 1550 cm^{-1} is related to amorphous carbon. The band near 1135 cm^{-1} is thought to the presence of hydrogen at the grain boundaries of nanocrystalline diamond [14].

Figure 1. SEM images of NCD films grown with different Ar: (a) 50% (b) 70% (c) 95%.

From the previous experiments it was demonstrated that diamond film structure changes to nanocrystalline by adding Ar to the reacting gas. The 95 vol% Ar was fixed to optimize the other growth parameters. Different methane concentration and pressure were used to investigate the effect of growth parameters on NCD deposition. Fig.3 shows the SEM micrographs of NCD films grown at 120 torr, 80 torr. Another sample was also deposited at 60 torr but there is no continuous film growth in this condition, which indicates the insufficient nucleation during the deposition. The ASTeX system was not able to maintain the microwave plasma when pressure was higher than 120 torr. It is seen that NCD films were successfully grown on the substrate between 80 torr to 120 torr in our system and the film grown at 120 torr has smallest crystals (less than 20 nm) and smooth surface. There are two keys to fabricate the nanocrystalline diamond films using CVD; one is to increase the nucleation density during the initial stages of the deposition, and the other is to control the crystal grain size at the nanometer scale. It is believed that high pressure within a certain range plays an important role in the NCD film growth due to the increase of the nucleation density, the increase of temperature inside the chamber and the higher mobility of carbon atoms. Different methane concentrations (2%, 3% and 5%) were also employed to synthesize the NCD films. However there is no significant morphology change of NCD films suggesting that the methane concentration (less than 5%) does not affect the formation of NCD.

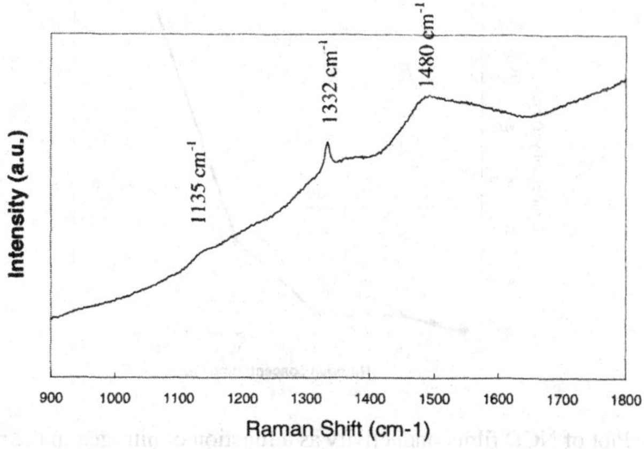

Figure 2. Raman spectrum of the diamond film deposited using 95 vol% Ar.

Figure 3. NCD films grown at different pressure: (a) 120 torr (b) 80 torr.

Nitrogen doped conductive NCD

Microcrystalline diamond has limitations in electronic devices due to it insufficient conductivity. It is known that the incorporation of nitrogen gas during the deposition could introduce the morphology and electronic change of the films to make it highly conductive [15]. Up to 20 vol% of Nitrogen gas was incorporated into the reactants to dope the NCD films and the conductivity was measured by the four-point-probe at room temperature. The conductivity vs. the nitrogen doping percentage is plotted in Fig.4. It is seen that the increase of the conductivity is dramatic, increasing from 1.16 Ω^{-1} cm^{-1} (for 1% N_2) to 126.54 Ω^{-1} cm^{-1} (for 20% N_2), which represents an increase by roughly three orders of magnitude over undoped NCD films. It is suggested that the incorporation of impurities between diamond grain boundaries (GB) makes the NCD film more conductive.

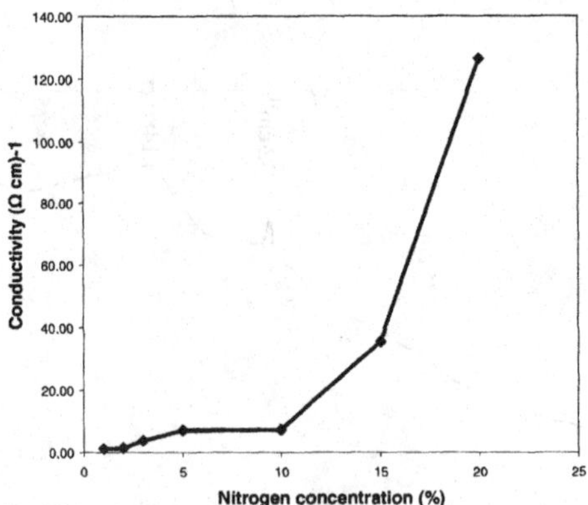

Figure 4. Plot of NCD film conductivity as a function of nitrogen in the reacting gas.

Hydrogen-terminated surface for biomedical applications

The hydrogenation of diamond surface can be accomplished by treating samples with microwave hydrogen plasma, which etches off the remnants of graphitic material from the sample and H-terminates the surface. In our study, after the diamond deposition, the sample was activated by the hydrogen plasma for about 1 hr with hydrogen purged at 300 sccm through the microwave vacuum chamber with pressure of about 30 torr. The film was characterized by XPS before and after the hydrogenation. Fig. 5 showed that the plasma hydrogenated surface exhibits a surface C 1s component shift by 0.6 eV towards a higher binding energy (from 284.5 to 285.1 eV), which is attributed to coadsorbed hydrocarbons [16]. The hydrogen-terminated surface was also investigated by the FTIR measurement. The peaks between 2800 and 3000 cm^{-1} are associated with the C-H bonding on the diamond surface [17]. These results confirmed that the diamond surface is hydrogen terminated and ready to be functionalized chemically and biologically to work as the platform for biosensors.

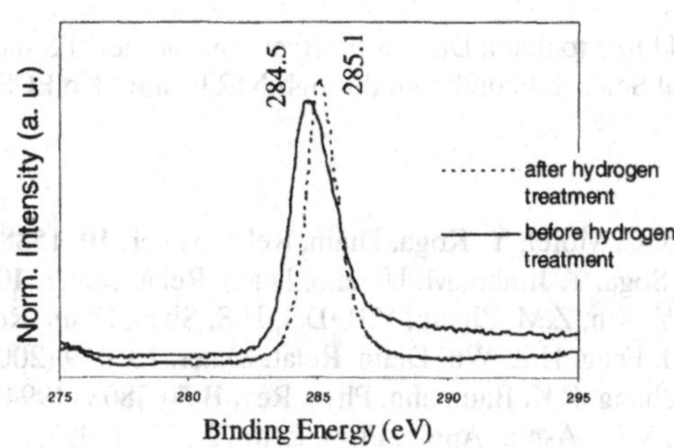

Figure 5. C 1s peak of nanocrystalline diamond film before and after hydrogen treatment.

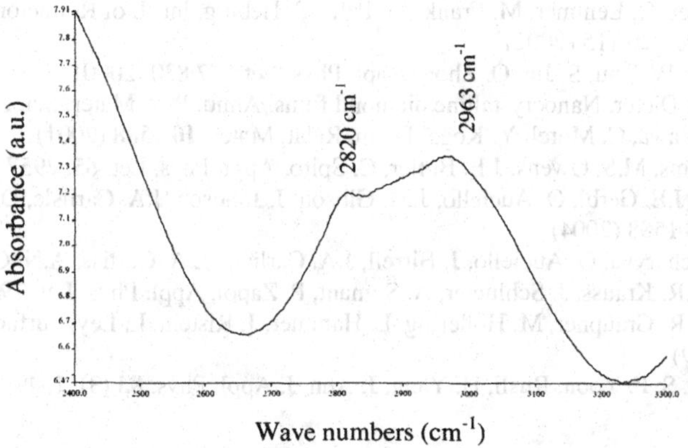

Figure 6. Infrared absorption spectrum of NCD film after hydrogenation.

CONCLUSION

Overall, nanocrystalline diamond films were synthesized by the MPECVD method. It was confirmed that by adding Ar to the reacting gas the surface morphology of the diamond films changes from microcrystalline to nanocrystalline. The SEM and Raman results suggested that Ar is important to produce NCD films due to the secondary formation of nuclei. It was observed that with the pressure increase (from 80 torr to 120 torr), the size of the diamond grains decreases to less than 20 nm. The methane concentration seems to have not much influence on the NCD growth. Up to 20 % nitrogen was used to dope the NCD and conductivity increases dramatically to 126.54 Ω^{-1} cm^{-1}. The hydrogen terminated surface was obtained by a hydrogen plasma treatment for 1 hr and the hydrogenated surface was characterized by XPS and FTIR. This hydrogen-terminated nanocrystalline diamond film is ready for bio-modification and is an excellent platform for biosensors.

ACKNOWLEDGMENT

The authors would like to thank Dr. Martin Beerbom for the XPS measurement. This work is supported by National Science Foundation through NIRT Grant No ECS 0404137.

REFERENCES

1. H. Yoshikawa, C. Morel, Y. Koga, Diam. Relat. Mater. **10**, 1588 (2001)
2. T. Sharda, T. Soga, T. Jimbo, M. Umeno, Diam. Relat. Mater. **10**, 1592 (2001)
3. T. Wang, H.W. Xin, Z.M. Zhang, Y.B. Dai, H.S. Shen, Diam. Relat. Mater. **13,** 6 (2004)
4. D.J. Qiu, A.M. Feng, H.Z. Wu, Diam. Relat. Mater. **13**, 419 (2004)
5. J. Weide, Z. Zhang, P.K. Baumann, Phys. Rev. **B 50** 5803 (1994)
6. I.L. Krainsky, V.M. Asnin, Apply. Phys. Lett. **72** 2574 (1998)
7. J.L. Davidson, W.P. Kang, A. Wisitsora-At, Diam. Relat. Mater. **12** 429 (2003)
8. A. Aleksov, M. Kubovic, M. Kasu, P. Schmid, D. Grobe, S. Ertl, M. Schreck, B. Strizker, E. Kohn, Diam. Relat. Mater. **13** 233 (2004)
9. F. Deuerler, O. Lemmer, M. Frank, M. Pohl, C. Hebing, Int. J. of Refractory Mater. and Hard Mater. **20** 115 (2002)
10. C. Bower, W. Zhu, S. Jin, O. Zhou, Appl. Phys. Lett. **77** 830 (2000)
11. M. Gruen Dieter, Nanocrystalline diamond films, Annu. Rev. Mater. Sci. **29** 211 (1999)
12. H. Yoshikawa, C. Morel, Y. Koga, Diam. Relat. Mater. **10** 1588 (2001)
13. B.D. Thoms, M.S. Owens, J.E. Butler, C. Spiro, Appl. Phys. Let. **65**, 2957 (1994)
14. J. Birrell, J.E. Gerbi, O. Auciello, J.M. Gibson, J. Johnson, J.A. Carlisle, Diam. Relat. Mater. **14** 1588 (2004)
15. S. Bhattacharyya, O. Auciello, J. Birrell, J.A. Carlisle, L.A. Curtiss, A.N. Goyette, D.M. Gruen, A.R. Krauss, J. Schlueter, A. Sumant, P. Zapol, Appl. Phys. Let. **79**, 1441 (2001)
16. F. Maier, R. Graupner, M. Hollering, L. Hammer, J. Ristein, L. Ley, Surface Sci. **443**, 177 (1999)
17. Q. Zhang, S. F. Yoon, Rusli, H. Yang, J. Ahn, J. Appl. Phys. **83** (3), 1349 (1998)

Mater. Res. Soc. Symp. Proc. Vol. 843 © 2005 Materials Research Society T3.40

Tribological Aspects of AlN-TiN Thin Composite Films

C. Waters, G. Young, S. Yarmolenko, X. Wang, J. Sankar,
North Carolina Agricultural and Technical State University,
Greensboro, NC 27411, USA

Abstract: Physical properties, and the friction and wear are important issues in small-scale applications, it is therefore essential that the materials used have good micromechanical and tribological properties. The adhesion, fracture toughness and wear properties of AlN-TiN thin composite films is being investigated in this study. The multilayered structures are generated using Pulsed Laser Deposition (PLD). The durability and functionality of thin films is subject to the adhesion between the coating and the underlying substrate in addition to it's resistance to cracking. The magnitude of the critical load during a scratch test is related to the adhesion of the substrate to the coating. In this test a Berkovich indenter is used for measurements and is drawn across the surface of a coating under an increasing load. The magnitude of the critical load will be studied for various films from a monolayer TiN film to different AlN-TiN films and those relative results compared to their facture toughness and their wear properties. Despite the adhesion, the critical load depends on several other parameters including the friction coefficient. The critical characteristic load is shown to depend on the number of layers and the relative AlN-TiN thickness. The fracture toughness showed a weak dependence on the layer characteristics.

Introduction: The scratch hardness test has been used to supply a simple and quick means of assessing the adherence of thin, hard and wear resistant coatings to substrates. In the test's simplest form, the diamond indenter is drawn across the coated sample at increasing values of normal load P until the coating is observed to become detached or fractured at some critical load see Figure 1. [1]. Higher scratch critical load means the material can bear larger scratch loads and normal stresses, which is desirable for tribomaterials and protective materials for thin film structures and devices such as nano- and micro-electromechanical systems and high density magnetic storage media.

Experimental Analysis: The (pulsed laser deposition) PLD system at the North Carolina A&T State University consists of a KrF laser, and two PLD chambers. The KrF laser (λ=248 nm, τ=25 ns) was used for ablation of polycrystalline, stoichiometric AlN and TiN (99.99 purity) onto the Si(100) substrates. The laser induced evaporated species were deposited onto substrates kept at a distance of ~6–7 cm from the target surface in an ambient background gas pressure at 5 mtorr. The deposition rate and the film or layer thickness were controlled by total deposition time [2].

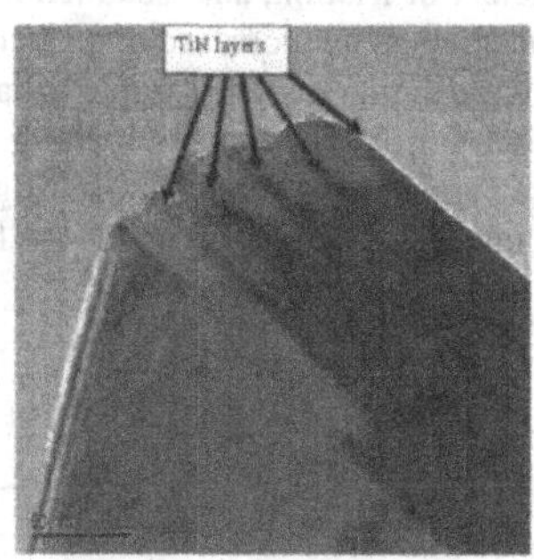

Figure 1. TEM image of 10 layer AlN-TiN film on Si(100)

All of the films were between 200 and 325 nm in total thickness. The individual layer thickness varied depending on the deposition conditions. Figure 1 displays a TEM image of a 10 layer TiN-AlN film. Although each target (TiN and AlN) was targeted for the same number of pulses the AlN layers deposit at a 3:1 rate compared with TiN. The tribological response of the films, were investigated with nanoscratch tests performed using a Nano Indenter II mechanical properties microprobe. A diamond Berkovich tip was used for the scratches. Five milli-newton (5 m N) ramped load scratch tests were conducted to investigate the films scratch resistance, coefficient of friction and critical load to failure. A scratch range of 160 μm was used for each test with the scratch speed of 1 μm/s and a ramped loading up to 160mN. As scratch behavior is known to vary with diamond tip orientation [3], all tests were conducted with the indenter oriented in the face forward direction. The in situ friction force was also obtained by measuring the lateral deflection of the scratch tip holder using a capacitance proximity probe. The coefficient of friction during scratching was calculated by dividing the friction force by the normal load.

After scratch testing, atomic force microscopy (AFM) and scanning electronic microscopy (SEM) were used to examine the scratch track morphology, and to measure the scratch section profile. Figure 2 is a AFM scan over the entire length of a scratch on a AlN-TiN layered film deposited on the Si(100) substrate.

Vickers hardness tests were run from load of 5 grams up to 300 grams. Increasing load tests were used [2] to allow fracture toughness analysis. Finally, the samples were examined by means of the SEM and Atomic Force Microscope and observations were made.

Results and Discussion: The critical positions indicated in the figures are characterized by an abrupt increase in the in situ coefficient of friction, and a sudden increase of post-scratch displacement. The increased displacement is directly associated with significant damage to the sample [4] as illustrated by AFM surface profile in Fig. 2. The ramped loading scratch causes a mixed plowing and cutting scratch (with ridge formation) evident in the AFM images. The pile-up and ridge formation is evident in the macroscopic upper part of Figure 2.

Figure 2. Composite scratch image each box 20x20mm 100mm between box scans.

Diagrammatic representation of a section through a scratch test on a coated substrate is displayed in Figure 3. The corresponding applied normal load is called the critical load, meaning the minimum load for the scratch damage with material removal, film fracture, interface delamination.

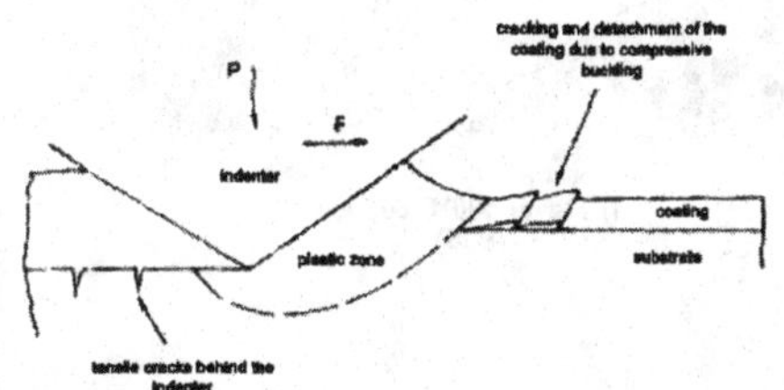

Figure 3. Schematic of coated substrate scratch behavior. The coating may buckle or spall ahead of the plastic zone associated with the generation of any ploughed groove. If there are significant adhesive forces between the indenter and the upper surface of the coating, tensile stress's generated to the rear of the indenter can lead to the formation of a series of half-moon or semicircular surface cracks, shown here [3].

This failure criterion could be delamination ahead of the indenter or chipping at the edges of the track. This critical distinction is often somewhat artificial because there is a spectrum of failure loads which are observed if repeated tests are made under nominally identical conditions [3]. This is seen in figures 4, 5 and 6. Sometimes there is an initial sharp increase in the coefficient of friction at the start of a test followed by a rapid decay down to a relatively constant value. This sharp increase is associated with the initiation of contact between the indenter and the sample surface and is not associated with critical damage of the films. For the actual applications such as protective coatings for recording devices or other thin film materials, much of the catastrophic damage occurs from isolated incidences of high loading and or three-body abrasion involving high contact stresses [4, 5]. Higher scratch critical load means the material can bear larger scratch loads and normal stresses, which is attractive for tribomaterials and protective materials for thin film structures and devices.

A critical load is not evident in the monolayer titanium films, (Fig. 5), which indicates that the scratch load bearing capacity for these films is improved by the presence of multilayers. The visual observation of the scratches verifies the friction data in figure 5. The surface roughness was measured by the AFM during the scratch test analysis. Values are given in Figure 7. The pure titanium films display a rougher surface.

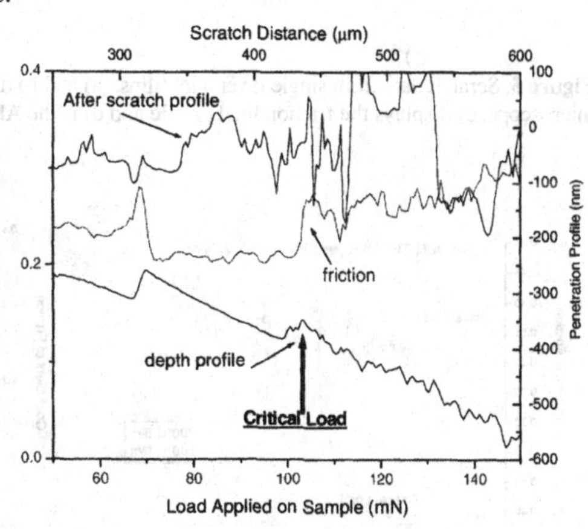

Figure 4. Scratch test data from an AlN TiN layered film, displaying the friction, load and profile data for one scratch.

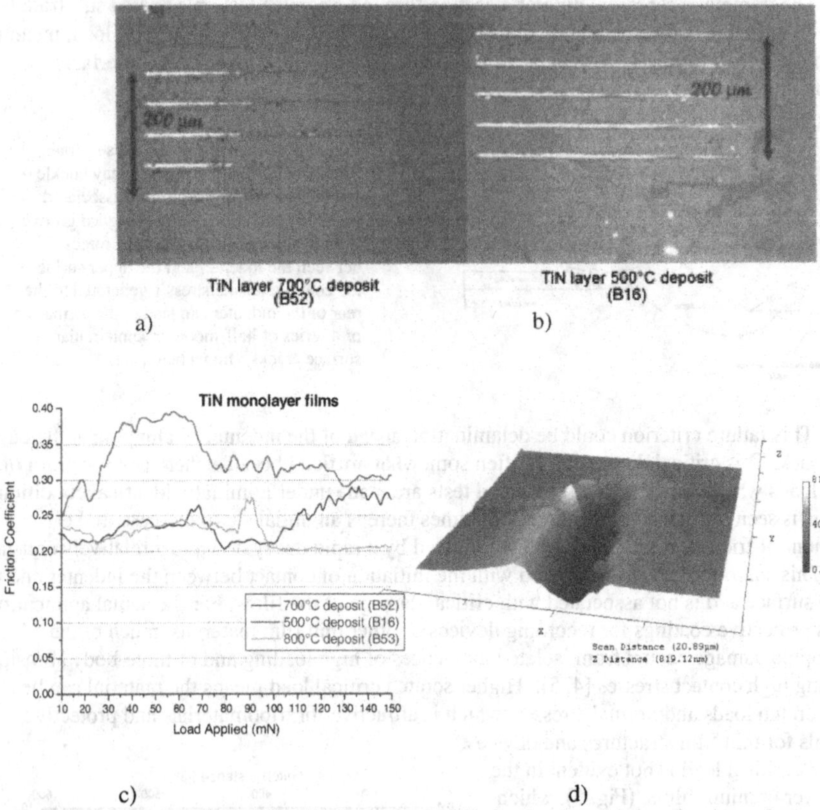

TiN layer 700°C deposit
(B52)
a)
TiN layer 500°C deposit
(B16)
b)
200 μm
200 μm
TiN monolayer films
Friction Coefficient
Load Applied (mN)
700°C deposit (B52)
500°C deposit (B16)
800°C deposit (B53)
c)
819.12nm
409.56nm
0.00nm
Scan Distance (20.89μm)
Z Distance (819.12nm)
d)

Figure 5. Scratch data from single layer TiN films. a) and b) display the after scratch surface image from an optical microscope, c) displays the friction load profile and d) is the AFM scratch depth 3-D image.

AlN-TiN 10 layer film
Critical Load
Coefficient of Friction
700°C film
500°C film
Critical Load
Load Applied (mN)
AlN-TiN 100 layer film
500°C Deposit
Coefficient of Friction
700°C Deposit
Critical Load
Load Applied on the Sample (mN)

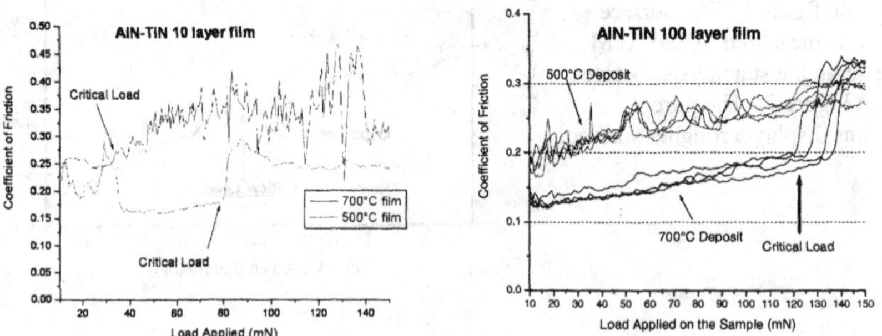

Figure 6. Scratch test data from 10 layer and 100 layer films. The 100 layer film may be said to have the highest critical load ~120mN.

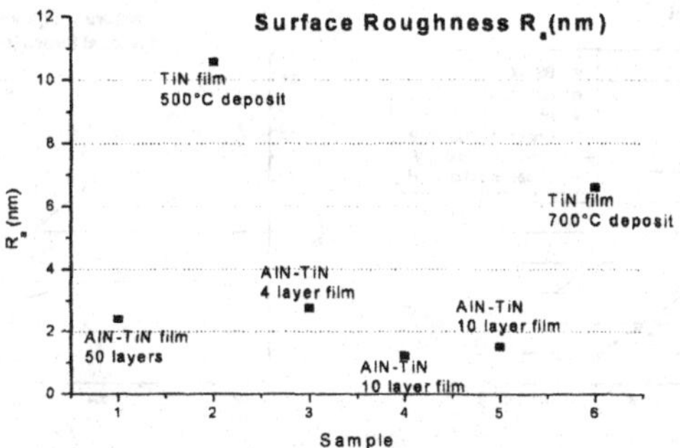

Figure 7. Surface roughness measurements taken from the AFM of film surfaces.

The critical load for fracture from the scratch test was widely varied therefore created difficulty in statistically determining for all sets of films. The berkovich indentor alone was found to create challenges during the scratch test and additional acoustic emission work is needed to accurately characterize the critical load. The 100 layered AlN-TiN film the lowest fracture toughness was consistent with its lower coefficient of friction. The fracture toughness was found according to Cook's method [6]. Employing the equation, $Z = 1.12\sqrt{\pi}\dfrac{d/c}{(3\pi/8)+(\pi/8)(d/c)^2} = 1.26$ and $x_r = x\left(\dfrac{E}{H}\right)^{1/2}$. $K_{Ic} = x_r\dfrac{P}{c^{3/2}} + Z\sigma_r c^{1/2}$ is plotted on a linear scale as shown in Figure 8 with the y-intercept determined as the K_{Ic} value and the slope the film residual stress value. During the K_{Ic} calculations there is a considerable dependence of the results on the method of determining the crack length [6].

Conclusions: In this work, thin submicron TiN and AlN-TiN films that were deposited via pulsed laser ablation were analyzed using ramped load nanoscratch and Vickers indent crack length fracture mechanics. The nano-tribological and fracture mechanics studies are offered. The critical load for fracture was too varied to accurately determine for all samples but for the 100 layered film the lowest fracture toughness was consistent with the lower coefficient of friction. Detailed AFM observations of the wear tracks indicate noteworthy plastic deformation consistent with a mixed plowing and cutting scratch with considerable ridge pile-up. The surface roughness values were generally lower for the layered films. More work is needed to more accurately quantify the films.

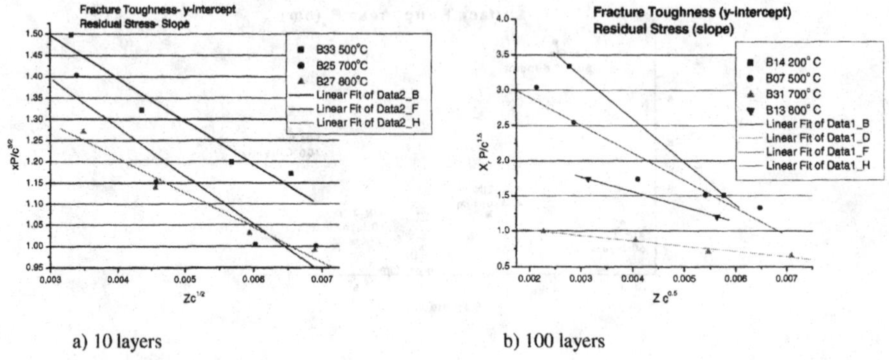

a) 10 layers b) 100 layers

Figure 8. Fracture Toughness data derived from Vickers indent tests and crack length determination. The 100 layer film displays higher fracture toughness.

References:

[1] J.A. Williams, *Tribology International*, Vol. 29, No. 8, (1996) 675-695.

[2] C.Waters, D. Kumar, S. Yarmolenko, Z. Xu and J. Sankar, X. Wang, MRS presentation, Fall 2003.

[3] S.D. McAdams, T.Y. Tsui, W.C. Oliver, G.M. Pharr, "Effects of interlayers on the scratch adhesion performance of ultra-thin films of copper and gold on silicon substrates", Thin Films: Stresses and Mechanical Properties V, vol. 356, Materials Research Society, Pittsburgh, PA, 1995, pp. 809–814.

[4] B. Bhushan, B.K. Gupta, M.H. Azarian, Wear 181–183 (1995) 743–758.

[5] X. Li, B. Bhushan, Thin Solid Films 398–399 (2001) 313–319.

[6] R.F. Cook, D.J. Morris, J Thurn, ***Mat.Res.Soc.Symp.Proc***., vol. 795, U4.1, (Fall 2003) 1.

Mater. Res. Soc. Symp. Proc. Vol. 843 © 2005 Materials Research Society T3.7

Characterization of nanocrystalline surface layer induced by shot peening and effect on their fatigue strength.

Hideo Mano[1,2], Kondo Satoru[1], Akihito Matsumuro[2], Toru Imura[3]
[1] TOGO SEISAKUSYO CORPORATION, Aichi-gun, Aichi, Japan
[2] Nagoya university, Nagoya-shi, Aichi, Japan
[3] Aichi institute of technology, Toyota-shi, Aichi, Japan

Abstract

The shot peening process is known to produce a hard layer, known as the white layer" on the surface of coil springs. However, little is known about the fatigue properties of this white-layer.

In this study, coil springs with a white-layer were manufactured. The surface of these springs was then examined using micro Vickers hardness, FE-SEM etc. to test fatigue strength of the springs.

From the results obtained, a microstructure of the white-layer with grain size of 50-100 nm was observed, with a Vickers hardness rating of 8-10 GPa.

Tow category springs were manufactured utilizing a double-peening process. These springs had the same residual stress destruction and surface roughness. Only one difference was observed: one spring had a nanocrystalline layer on the surface, while the other did not. The results of the fatigue test realized an increase in the fatigue life of the nanocrystalline surface layer by 9%.

Introduction

Shot peening process of normal condition given compressive residual stress on material surface without phase transformation. But the shot peening process is known to produce a hard layer, known as the white-layer on the surface of carbon-steel, under intensify peening conditions; intensify peening velocity and using high hardness shot media. This white-layer have high hardness than matrix phase. Therefore white-later had been thought composing martensite-phase. Recently, It was suggested white-layer is nanocrystalline phase, composed nano-ferrite [1-5].

In this study, coil springs with nanocrystalline surface layer were manufactured. Moreover fatigue examination carried out with the coil springs.

Experimental

Test springs were made from oil-tempered wire; its chemical composition is shown in Table 1. It was compressive coil spring. Table 2 shows dimensions of test springs, and photograph of accomplished test spring is shown Figure 1.

Table 1 Chemical composition of wire (mass%).

C	Si	Mn	P	S	Cu	Cr
0.57	1.43	0.7	0.02	0.006	0.01	0.68

Table 2 Dimensions of test springs.

Wire diameter (mm)	φ3.2
Mean diameter of coil (mm)	20.0
Active number of coil	4.0
Total number of coil	6.0
Free height (mm)	47.0
Spring rate (N/mm)	32.0

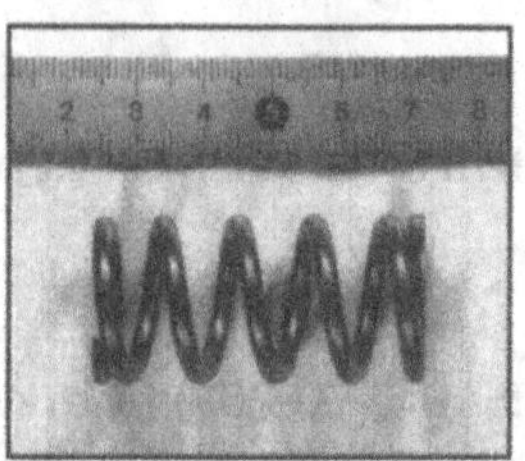

Figure1 Photograph of manufactured spring.

Test springs manufactured same process until end grinding. But, shot peening condition are different. Manufacturing process of springs is shown in Figure 2. Two kinds of springs with which shot peening conditions are different were produced. Manufactured two kinds of springs are spring A and Spring B. Spring A is a spring with a surface nanocrystalline layer, and spring B is a spring without the nanocrystalline layer. Shot peening condition is shown in Table 3. The shot peening condition of spring A were made into two-step peening. The projection material of the 1st step shot is a 0.25 mm diameter steel cut wire, and projection speed is about 100 m/s, by using air-type peening machine. The nanocrystalline layer is made by this peening process. And second step peening used steel cut wire of 0.6 mm diameter, projection speed is about 73 m/s. Spring B is one-step shot peening, projection material is 0.6 mm diameter, and projection speed is 74 m/s, by using impeller-type peening machine. The shot conditions of spring B are the same as the 2nd step shot of a spring A.

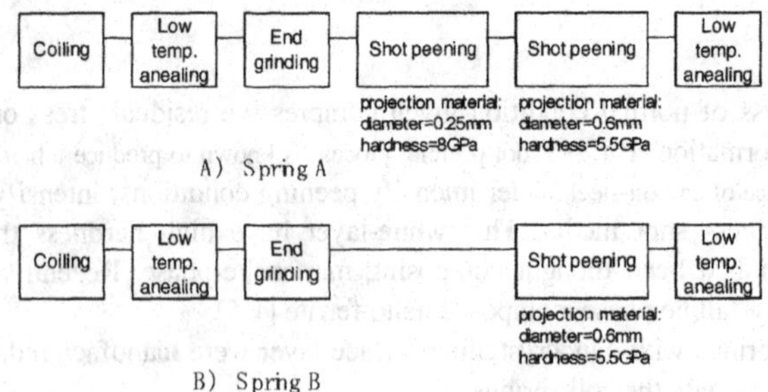

Figure 2 Manufacturing process of test spring.

Table 3 Shot peening conditions.

	Peening step	Shot media size (mm)	Shot media Hardness (GPa)	Peening time (sec.)
spring A	first	0.25	8	3000
	second	0.6	5.5	1800
spring B	-	0.6	5.5	1800

Manufactured test springs were measured surface roughness and residual stress distribution. these were observed by optical microscope and Field-Emission-Scanning-Electron-Microscope (FE-SEM).

And fatigue test carried out spring fatigue test machine at frequency of 1800 cycle/min until 5x10^7 cycles. Mean stress of fatigue test was 600 MPa. Stress amplitude was 510 MPa and 550 MPa. All of testing stress, Number of specimen were 8 springs. The spring fatigue test machine give repetitive excursion to test springs by biased cam. Test springs locate radially around bias cam. This test method is used to inspect performance of compressive coil springs, close to the actual working condition of valve springs.

Result and Discussion

Study of surface of manufactured springs

Cross-sectional optical microscope image of the surface section on manufactured springs were shown in Figure 3. these specimen etched by 3% natal. Difference of microstructure was observed these springs. Microstructure of spring A, based on Figure 3, consist of two phase, matrix phase and surface white-layer. Microstructure of spring B consist of matrix single phase.

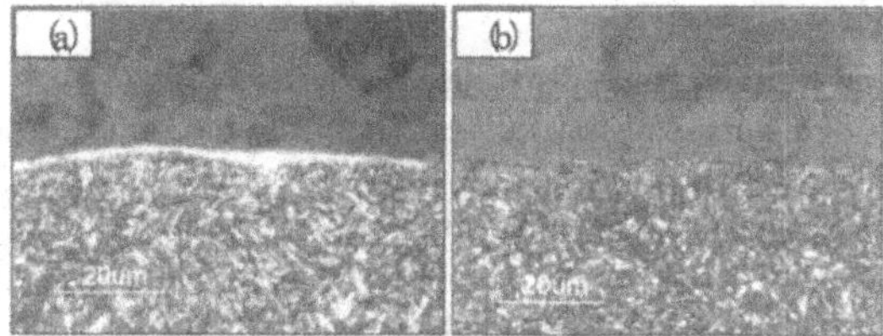

Figure 3 Cross-sectional optical microscope image of the surface section on manufactured springs; (a) spring A, (b) spring B.

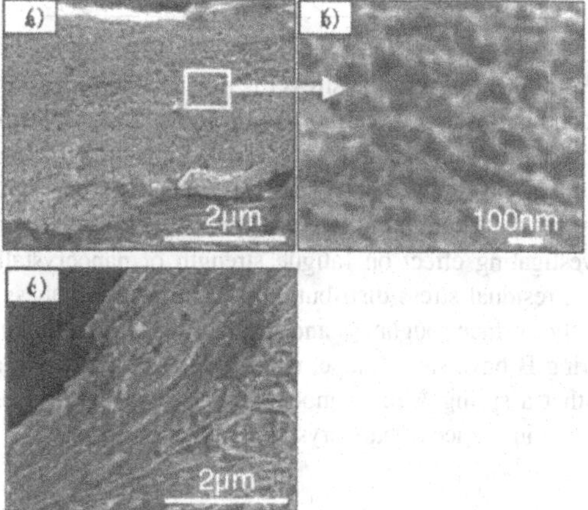

Figure 4 Cross-sectional Field-Emission Scanning-Electron-Microscope image of the surface

section on manufactured springs; (a)(b) spring A, (c)(d) spring B (Backscattered electron image).

Matrix phase was tempered martensite phase. FE-SEM image of surface section on these springs were shown in Figure 4. Figure 4 (a) (b) shows white-layer section of spring A. Result of observation white-layer based on Figure 4 (a) and (b). The microstructure of 50-100 nm crystal particle diameter was observed by the white-layer directly under the surface. Thickness of this layer was 2-3 μm. It seems that this nanocrystalline surface layer was produced by dynamic recrystallization in large strain deformation of shot peening. Furthermore, it is observed that metal flow layer was in the next to the nanocrystalline later. Thickness of metal flow layer was about 5 μm. Matrix phase was in the next of metal flow layer.

On the other hand, Image of Figure 4 (c) confirmed that the microstructure of spring B consist of metal flow phase and matrix phase. But, the nanocrystalline layer is absence in observed image.

We measured Vickers hardness of test springs. Vickers Hardness of the nanocrystalline layer was 8–10 GPa. And, Vickers hardness of Matrix phase was 5–6 GPa. These results confirmed that hardness of nanocrystalline phase was about 50 % higher than matrix phase. This increment was beyond hardening of work-harden. Therefore, it is likely that refining of grains size increased hardness. Hardness of surface section on spring B was about 6 GPa. The increase in metal flow phase was produced by work-harden.

Surface roughness and residual stress distribution

Table 4 shows surface roughness of test springs. The measurement of surface roughness, surface roughness of spring A and spring B were almost same value. Because, spring A of the 2nd step shot effect on surface roughness much stronger than 1st step peening.

Table 4 Surface roughness of test springs.

	Rmax (μm)	Rz (μm)
spring A	11.8	8.5
spring B	10.2	8.0

Figure 5 shows the measurement of residual stress distribution. The measurement of residual stress distribution of spring A and spring B were almost same value. Because, similar surface roughness, spring A of the 2nd step shot effect on residual stress distribution stronger than 1st step peening.

The shot conditions of the 2nd step shot of a spring A are the same as spring B. As such, surface roughness and residual stress distribution of these test springs were almost same level.

In the interest of investigating effect on fatigue strength of nanocrystalline layer, it is needed springs have same shape, residual stress distribution and surface roughness except nanocrystalline layer in surface. Especially, surface roughness and residual stress distribution very effect on fatigue stress. Spring A and spring B have same shape, residual stress distribution and surface roughness. Only one difference is that a spring A has nanocrystalline layer in surface. Testing spring A and spring B make it appear that influence of nanocrystalline layer on spring.

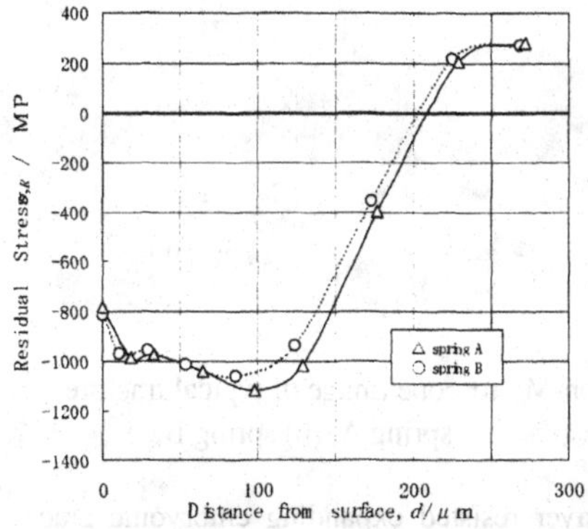

Figure 5 Residual stress distribution of test springs.

Fatigue test

Result of fatigue test shows Figure 6. In the case of testing stress was $\tau_m \pm \tau_a$=600±548 MPa. Many number of spring B, these springs don't have nanocrystalline layer in surface, fractured in about 2x10^6 cycles. Spring A, these springs have nanocrystalline layer, fractured in the longer life side than spring B. The fatigue limit of spring A was $\tau_m \pm \tau_a$=600±535 MPa at 10^7 cycles. And, spring B was $\tau_m \pm \tau_a$=600±490 MPa. From these results we confirmed that the nanocrystalline surface layer improved fatigue strength 9 %.

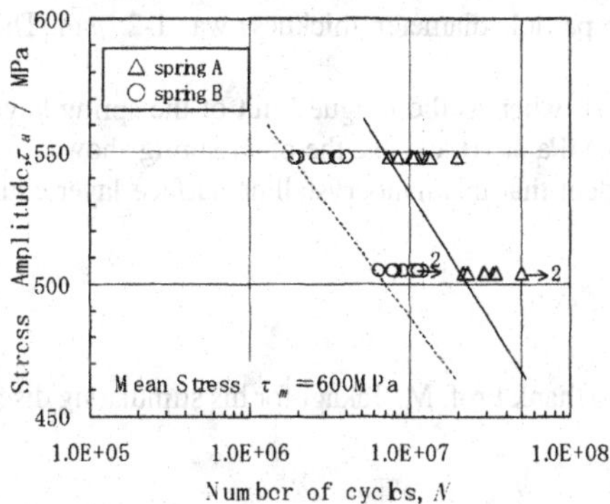

Fig.6 Stress-Number diagram of test springs.

Figure 7 shows Scanning-Electron-Microscope (SEM) image of typical fracture of broken springs. Spring A and spring B had same destruction cause. Surface facet of wire direction was observed in starting point of fracture of both springs. Because, Nanocrystalline surface layer of spring A was very thin, 1-2 μm of thickness.

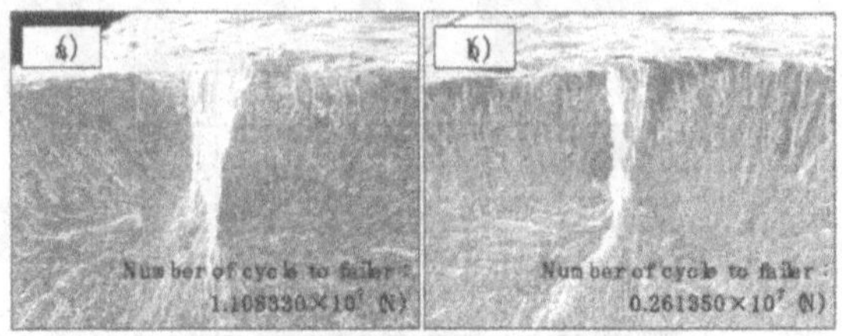

Figure 7 Scanning-Electron-Microscope image of typical fracture, $\tau_m \pm \tau_a$=600±548 MPa ; (a) spring A, (b) spring B.

Nanocrystalline surface layer resisted expanding embryonic crack as such, was probably the cause of high fatigue strength of spring A. It is possible that nanocrystalline surface layer functions surface hardening.

Conclusions

Shot peening process could produce a nanocrystalline layer on the surface of carbon steel. However, there has been little information on the effect this nanocrystalline surface layer has on fatigue properties. In this study, two types of springs were manufactured by different shot peening conditions: These springs had the same shape, surface hardness and residual stress distribution, with one having a nanocrystalline surface layer, and the other without. Nanocrystalline surface layer consist of 50-100 nm grain particle diameter, thickness was 1-2 μm. This layer had high hardness, 8-10 GPa.

The result of the fatigue test, whereas the fatigue limit of the spring having a nanocrystalline layer was $\tau_m \pm \tau_a = 600 \pm 535$ MPa at 10^7 cycles, the other spring showed $\tau_m \pm \tau_a = 600 \pm 490$ MPa. In other words, it was evident that this nanocrystalline surface layer could increase the fatigue life by 9%.

Acknowledgments

The authors would like to thank Prof. M. Takagi for his stimulating discussions.

References

1. G.Liu,S.C.Wang,X.F.Lou,J.Lu and K.Lu,*Script Mater.***44**,1791-1795(2001).
2. M.Umemoto,B.Haung,K.Tsuchiya and N.Suzuki,*Script Mater.***46**,383-388(2002).
3. N.R.Tao,Z.B.Wang,W.P.Tong,M.L.Sui,J.Liu and K.Lu,*Acta Mater.***50**,4603-4616(2002).
4. M.Umemoto,Z.G.Liu,Y.Xu and K.Tsuchiya,*Materials Science Forum,***386-388**,323-330(2002).
5. N.Tsuji,R.Ueji ,Y.Minamino and Y.Saito,*Scrip.Mater.***46**,305-310(2002).

Mechanical Properties of Surfaces, Interfaces and Thin Films

Mater. Res. Soc. Symp. Proc. Vol. 843 © 2005 Materials Research Society T4.10

UV Raman Scattering Analysis of Indented and Machined 6H-SiC and β-Si_3N_4 Surfaces

Jennifer J.H. Walter,[1] Mengning Liang,[1] Xiang-Bai Chen,[3] Jae-il Jang,[4] Leah Bergman,[3] John A. Patten,[2] George M. Pharr,[4,5] and Robert J. Nemanich[1]

[1]*Department of Physics, North Carolina State University, Raleigh, NC 27695-8202*
[2]*Department of Manufacturing Engineering, Western Michigan University, Kalamazoo, MI 49008*
[3]*Department of Physics, University of Idaho, Moscow, ID 83843*
[4]*Department of Materials Science and Engineering, University of Tennessee, Knoxville, TN 37996, and*
[5]*Metal and Ceramic Division, Oak Ridge National Lab, Oak Ridge, TN 37830*

ABSTRACT

UV Raman scattering is employed as a nondestructive structure sensitive probe to investigate the vibrational properties of the wide bandgap, machined and indented surfaces of 6H-SiC and β-Si_3N_4. In these materials, the short absorption depth of UV light allows for accurate probing of the surface, and the transparency to visible light allows for analysis of the bulk material. The study on 6H-SiC (0001) included measurements of indentations, and of machined circular (0001) wafer edges. The indentation analysis indicates the response of the material to localized pressures. Machined 6H-SiC wafer edges and machined β-Si_3N_4 surfaces indicate a ductile response and ductile material removal for machining at cutting depths on a nm and μm scale. Raman scattering measurements of the ductile surfaces and ductile material removed indicate residual structure changes. The residual surface structures could indicate that a high-pressure phase transformation is the origin of a ductile response on machined brittle materials.

INTRODUCTION

It has been suggested that the brittle materials, β-Si_3N_4 and 6H-SiC (0001), can be machined in a ductile regime [1,2]. In this study, the residual structures of the machined surfaces are investigated. These residual structures provide evidence of the origin of the ductile response. During the machining process, high pressures can be achieved at the contact interface between the machining tool and the surface of the material. The high pressures can cause transformation to occur through a series of phases. Upon releasing the pressure, the material can again be transformed to a new phase. Based on prior studies it has been suggested that β-Si_3N_4 may transform to an unidentified high-pressure phase [1]. Additionally, previous work of 6H-SiC indicates transformation in phase with high pressures [3,4]. A transition from covalent to metallic bonding that is caused from high pressures at the contact interface of the machining tool and the material may produce a ductile surface on machined brittle materials. In the case of point pressures, the non-hydrostatic stress component can influence the transition process. An understanding of the structural transformation process that can occur during machining can lead to improved manufacturing process by establishing parameters that will produce a ductile surface on brittle materials.

EXPERIMENTAL DETAILS

The 6H-SiC samples used for indentation were supplied by Cree Inc., and the wafers used for machining were purchased form SiCrystal AG. The wafers are circular with a 50 mm diameter and 250 μm thickness. Both the indented, and machined SiC wafers were transparent to visible light and had a light green color.

The β-Si_3N_4 sample surfaces were prepared first by grinding and then machining over the ground surface. The machining was employed for depths of cut of 5 μm. The presence of the Si inclusion in the bulk material was deduced from visible Raman spectra, which show a peak at 520 Rcm^{-1} (Rcm^{-1} represents the Stokes shift of the Raman scattering peak).

The indentation experiments were performed on a polished (0001) 6H-SiC surface using a Nanoindenter – XP (MTS, Oak Ridge, TN). This surface corresponds to the Si-face of an ideally terminated structure. A Berkovich diamond indenter tip shaped as a three-sided pyramid with a centerline-to-face angle, Ψ, of 65° was used to create triangular shaped indentations. The tip was brought into contact with the material at a constant loading rate of 5 mN/sec under a maximum load of 200 mN. The 6H-SiC (0001) machined wafer edges were machined with a sharp tip at rake angles of -45° and depths of cut of 50 nm. A machining rate of 20 rev./min was employed. Tool wear can change the sharpness of the tip with use, and therefore, alter the pressures that are achieved at the contact interface of the tool and the material. For the limited machining distances in these experiments, tool wear was not expected to significantly influence the machining results, and was therefore, not considered an issue [5].

The Raman scattering experiments were performed with a micro-focus UV spectrometer, and a micro-focus visible spectrometer. In the micro UV Raman system, a 180° backscattering geometry was employed and the laser was focused to a spot size of ~600 nm on the sample. The Raman experiments were also performed with the 244 nm line of a frequency doubled Ar ion laser (Lexel Inc.) and the 325 nm line of a HeCd laser (Kimmon Electric), and the scattered light was focused on the entrance slit of a triple spectrometer (Dilor). CCD detection was used to collect the spectra. In the visible macro system, a 45° backscattering geometry was employed, and the laser was focused to a spot size of ~100 μm x ~2 mm on the sample. The Raman experiments were performed with the 514.5 nm line of an argon ion laser (Coherent Inc.), and the scattered light was focused onto the entrance slit of a double spectrometer (U1000). Photomultiplier detection was used to collect the spectra. The probing depth into crystalline 6H-SiC with a 244 nm incident wavelength is ~35 nm and non-crystalline β-Si_3N_4 is ~1.4 μm (the values are calculated for the depth in which the intensity of the light is 37% of the incident intensity) [6].

RESULTS AND DISCUSSION

Indentation experiments were performed on a 6H-SiC (0001) wafer to investigate the local bonding properties from controlled point pressure conditions. A SEM image of an indented region and the corresponding micro UV Raman measurements are presented in Fig.1. The indentation was performed with a 200 mN maximum load, and the SEM of the indented region indicates an inelastic response with limited macro cracking and some chipping. Micro-focus Raman scattering measurements were obtained from the indent and the surrounding region. Within the indent, the TO (atomic vibrations perpendicular to the c-axis), and the LO (atomic vibrations parallel to the c-axis) modes are displaced to higher relative wavenumbers when compared to the reference spectrum, which was obtained far from the indent. The peak displacements indicate compressive stresses. A maximum displacement was observed in the

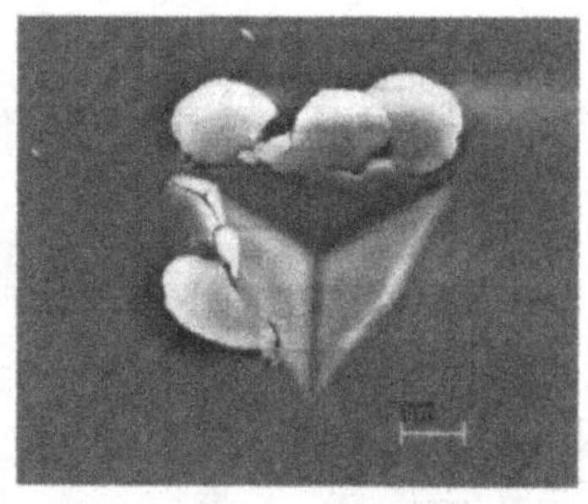

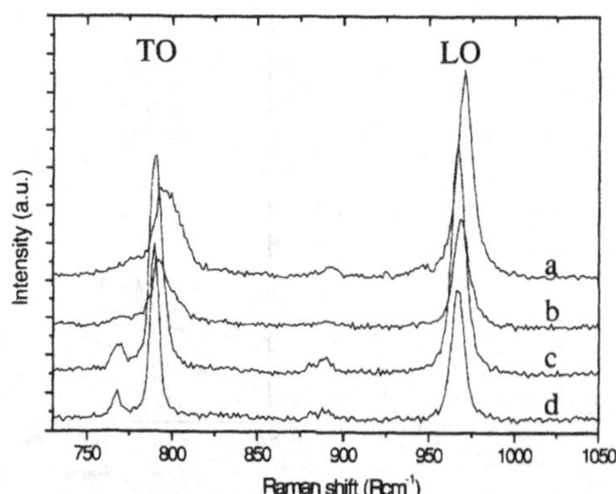

FIG 1: SEM image after indentation with a maximum load of 200 mN. The UV micro Raman scattering spectra display the TO, and LO modes for the (a) center of the indent (b) edge of the indent (c) surrounding area (d) reference area.

center of the indent. The peak displacements signify that the region at the contact point of the diamond tip, and the 6H-SiC is the zone of maximum stress. The stress decreases outward from the center. The center peak position of the TO mode is shifted by a larger amount than the center peak position of the LO mode, this may indicate that the material is under biaxial compression. The Raman scattering spectra do not show evidence of residual structural changes within the indentation. It is possible that a transformed region under the indenter may be very small or may have been extruded as debris. It is notable that the stresses indicate that the surrounding regions have not been completely relaxed with micro cracking.

Optical microscope inspection of the 6H-SiC machined wafer edges also indicates a ductile response from the machining process. The observation of a ductile response was found to be dependent on several parameters of the machining process including tool rake angle, cutting speed, surface orientation, and depth of cut. For a range of cutting depths between 50 and 500 nm, the 50 nm cutting depth produced almost completely ductile surfaces, while the 500 nm cutting depth produced almost completely brittle surfaces for all of the machined surfaces. The material removed during machining indicated elongated, ductile chips, in addition to the fractured and brittle chips. The percentage of ductile chips removed increased with shallower depths of cut, which is probably a direct consequence of the increased ductile surface area with the shallower depth of cut.

Raman scattering measurement of a machined 6H-SiC wafer edge acquired from the $[11\bar{2}0]$ wafer direction is shown in Fig. 2. Optical inspection of the machined surface indicated a ductile response. A spectrum of crystalline 6H-SiC obtained from a reference wafer edge, and the calculated phonon density of states for SiC is shown for comparison [7]. The spectrum of the machined edge indicates broadened peaks, and the TO mode center frequency is displaced to a higher relative wavenumber while the LO mode center frequency is displaced to a lower relative wavenumber from their crystalline peaks. The broadened displaced TO peak may be attributed to random oriented polycrystalline material since the TO mode frequency will shift with phonon propagation direction [8]. Because of the width of the features, the presence of residual stress

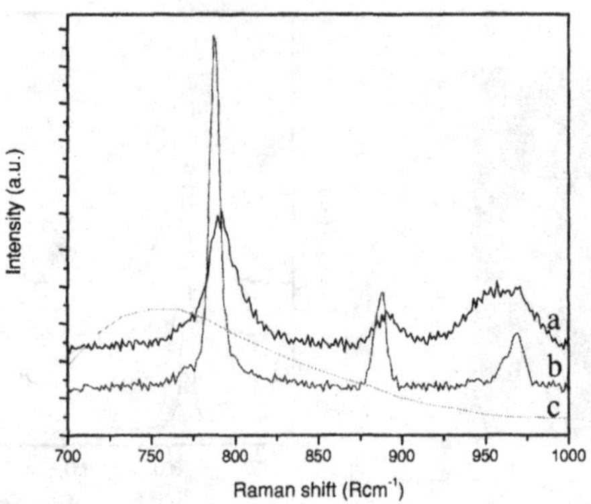

FIG 2: UV micro Raman scattering spectra of (a) machined ductile 6H-SiC edge, and a (b) non-machined 6H-SiC wafer edge. For comparison, the (c) calculated vibrational density of states (DOS) of SiC is presented [7].

cannot be excluded from these measurements. The TO and LO mode splitting is derived from both the anisotropy of the material, and an electrostatic component generated from the charge transfer between the Si and C atoms. A decrease in the TO-LO splitting could indicate small grain size polycrystalline material since the long range electrostatic interaction would be limited by the crystal domain size.

The β-Si_3N_4 surfaces were machined to further investigate the response of brittle materials to point pressures. It was found that for machining depths on a μm and nm scale, a smooth, ductile surface could be obtained. Fig. 3 presents the Raman scattering spectra of the machined β-Si_3N_4 surface for a 5 μm machining depth of cut and incident wavelengths of 514.5, 325, and 244 nm. The 514.5 nm visible Raman scattering spectrum indicates sharp crystalline peaks. The 325 nm incident wavelength indicates slightly broader peaks, but the broad background is enhanced at shorter wavelengths. The spectrum obtained with 244 nm excitation indicates completely broadened features. It has been reported that the broadened density of states of Si_3N_4 contains two broadened peaks centered at about ~400 and ~900 Rcm^{-1} [9], thus suggesting the presence of amorphous Si_3N_4 at the machined surface. The depth of the amorphous region can be estimated from comparison between the spectra obtained with different incident wavelengths. The spectrum obtained with 324 nm excitation shows significant amorphous and crystalline signatures suggesting that the amorphous layer thickness is similar to the absorption depth of the light.

Similar to the 6H-SiC, the β-Si_3N_4 material removed during the machining process also indicates a ductile response. A SEM image of a β-Si_3N_4 chip removed during the machining process and micro and macro Raman scattering measurements obtained from a large quantity of the chips pressed together are presented in Fig. 4. The SEM image indicates that the chips were bent and distorted without fracture. The broadened peaks indicated in the micro and macro Raman scattering spectra indicate that the chips are amorphous. There is some extent of crystalline

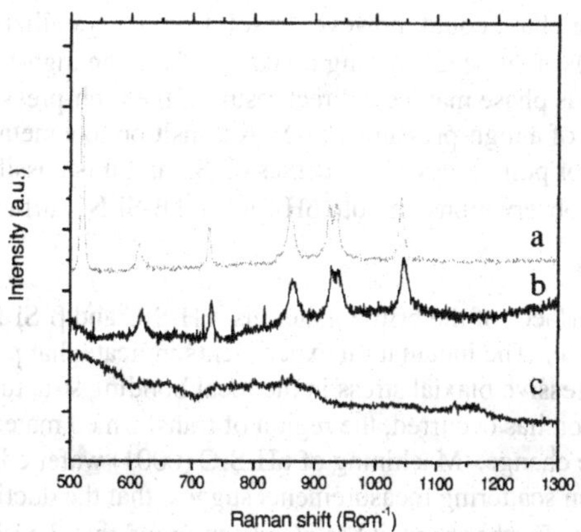

FIG 3: (a) Macro Raman scattering spectrum of machined β-Si_3N_4 obtained with 514.5 nm excitation, and micro Raman scattering spectra obtained with (b) 325 and (c) 244 nm excitation.

peaks observed in the micro Raman scattering spectrum indicating that the power density produced by the small spot size in micro Raman scattering spectrum was high enough to heat and crystallize the chips.

The residual surface layers on ductile-machined 6H-SiC and β-Si_3N_4 could be relaxed structures of high-pressure phases that were generated in the local regions due to the point pressures from the indenting and machining processes. Therefore, the residual structures may provide evidence of the origin of the ductile behavior. In our measurements, the transformed regions are observed on machined surfaces that exhibit a ductile response. Furthermore, it is not likely that the polycrystalline surface layer on the machined 6H-SiC wafer edges is a direct transformation from the parent crystalline structure since the transformation would involve randomly orienting

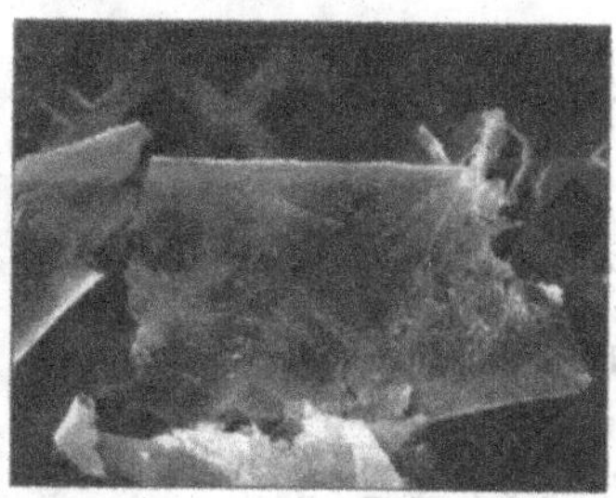

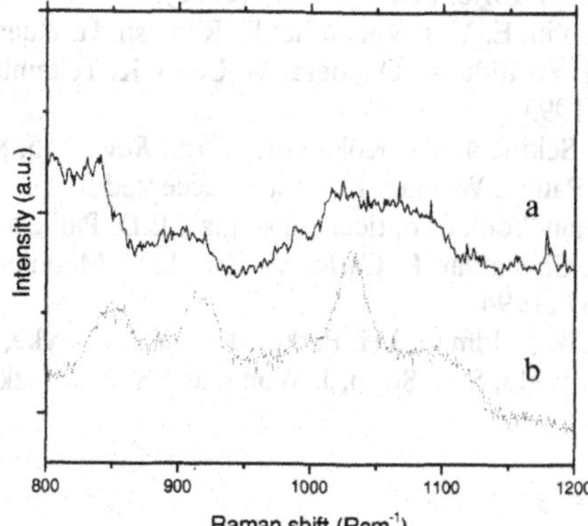

FIG 4: SEM image of a β-Si_3N_4 chip, and (a) macro and (b) micro Raman scattering spectra of the chips produced during machining at a 5 μm depth of cut.

the crystal. A polycrystalline phase could, however, result from recrystallization during a nucleation and growth process that occurs during relaxation from the high-pressure phase. In β-Si_3N_4, the residual amorphous phase may be a direct result of the high-pressure process or it could result from relaxation of a high-pressure phase. A transition to a metallic high-pressure phase has been considered for point pressure processes on Si, and it is possible that the same effect could account for the observations on both 6H-SiC and β-Si_3N_4 surfaces.

CONCLUSIONS

It has been found that the surfaces of the brittle materials, 6H-SiC and β-Si_3N_4, can be precision engineered in a ductile regime. The indentation experiments indicate that point pressures on 6H-SiC (0001) produce a compressive biaxial stress in the local bonding structure. In the indentations, if transformation has occurred, the region of transformed material may be too small to detect a residual structure change. Machining of 6H-SiC (0001) wafer edges also indicates ductile response. UV Raman scattering measurements suggest that the ductile surfaces have a polycrystalline surface layer. Furthermore, it has also been found that β-Si_3N_4 can be machined in a ductile regime. The ductile surfaces indicated an amorphous surface layer. The material removed during machining also indicates a ductile response, and an amorphous phase. We have suggested that the transformation to a polycrystalline phase on the 6H-SiC wafer edges occurs through the nucleation of growth of the crystalline phase. Therefore, these structural transformations are suggestive that a high-pressure phase transformation to a metallic phase during the machining process is the origin of the ductile response of the machined surfaces of these brittle materials.

ACKNOWLEDGEMENTS

The research was sponsored by The National Science Foundation under grant numbers FRG 0203552 and DMR 0403650.

REFERENCES

1. J. Patten, R. Fesperman, S. Kumar, S. McSpadden, J. Qu, M. Lance, J. Huening, R. Nemanich, *Appl. Phys. Lett.* 83, 4740 (2003).
2. L. Yin, E. Y. J. Vancoille, K. Ramesh, H. Huang, *Int. J. Manh. Tool Manu.* 44, 607 (2004).
3. M. Yoshida, A. Onodera, M. Ueno, K. Tekemura, and O. Shimomura, *Phys. Rev. B* 48, 10587 (1993).
4. T. Sekine and T. Kobayashi, *Phys. Rev. B* 55, 8034 (1997).
5. J. Patten, W. Gao, K. Yasuto, accepted for publication *J. Manuf. Sci. E.-T. ASME* (2004).
6. "Handbook of optical constants", E.D. Palik, ed. (Academic, Orlando, Fla. 1985).
7. A. Chehaidar, R. Carles, A. Zwick, C. Meunier, B. Cros, J. Durand, *J. Non-Crys. Solids* 169, 37 (1994).
8. D.W. Feldman, J.H. Parker, JR., W.J. Choyke, Lyle Patrick, *Phys. Rev.* 173, 787 (1968).
9. N. Wafa, S.A. Solin, J. Wong, and S. Prochazka, *J. Non-Crys. Solids* 43, 7 (1981).

Mater. Res. Soc. Symp. Proc. Vol. 843 © 2005 Materials Research Society T4.4

Structural Nanocomposite Bonding Reinforced by Graphite Nanofibers with Surface Treatments

L. Roy Xu[1], Charles M. Lukehart[2], Lang Li[2], Sreeparna Sengupta[1] and Ping Wang[1]
[1]Department of Civil and Environmental Engineering
[2]Department of Chemistry
Vanderbilt University, Nashville, TN 37235, USA

ABSTRACT

Graphitic carbon nanofibers were used to reinforce epoxy resin to form nanocomposite adhesive bonding. GCNFs having a herringbone atomic structure are surface-derivatized with bifunctional hexanediamine linker molecules capable of covalent binding to an epoxy matrix during thermal curing and are cut to smaller dimension using ultrasonication. Good dispersion and polymer wetting of the GCNF component is evident on the nanoscale. Tensile and shear joint strength measurements were conducted for metal-metal and polymer-polymer joints using pure epoxy and nanocomposite bonding. Very little bonding strength increase, or some bonding strength decrease, was measured.

INTRODUCTION

Since carbon nanotubes have extraordinary mechanical properties, they tend to be used as reinforcements in polymers and other matrices to form so-called nanocomposite materials [1-3]. Nanocomposites are a novel class of composite materials where one of the constituents has dimensions in the range between 1 and 100 nm [4]. Nanocomposite materials garner most of their material improvements from interactions at the molecular scale, influencing physical and material parameters at scales inaccessible by traditional filler materials. Wagner [5] reported that load transfer through a shear stress mechanism was seen at the molecular level. It has been reported that nanotubes increased the composite strength by as much as 25% [6]. However, multi wall nanotubes (MWNTs) are limited in their applications because of weak inter-shell interactions [7]. Single wall nanotubes (SWNTs) on the other hand are quite expensive and difficult to manufacture. Alternative reinforcement materials for nanocomposites include graphitic carbon nano-fibers (GCNFs) and graphite nanoplatelets etc [8]. GCNFs also have excellent properties and can be used as reinforcements in various kinds of matrices. They offer chemically facile sites that can be functionalized with additives thereby resulting in a strong interfacial bond with the matrix. Generally, the three main mechanisms of interfacial load transfer are micromechanical interlocking, chemical bonding and the weak van der Waals force between the matrix and the reinforcements [9]. In order to form a nanocomposite material with excellent mechanical properties, strong chemical bonding between the reinforcement and the matrix is a necessary condition, but might not be a sufficient condition. From the length-scale argument it is known that the effective toughening may not be energetically favorable at the nano length-scale [10]. This generally necessitates a filler size greater than 100 nm [11]. However, there might be significant difference in mechanical behaviors between a continuous fiber reinforced composite (e.g., carbon fiber reinforced composites, reinforced concrete) and a nanofiber reinforced composite, even if both have very strong interfacial bonding. It has been proved that a continuous fiber reinforced composite can effectively arrest the propagation of a major crack (which determines the material strength and toughness), while the short nanofiber/nanotubes might not have this kind of effect [12]. Here the length of nanofibers/nanotubes plays an important role in the toughening mechanism of nanocomposites as reported in some recent investigations [13-15].

Carbon fibers are widely used as reinforcements for advanced polymeric matrix composites in many high-technology applications because of their high specific stiffness, strength and excellent electrical and thermal properties. Graphitic carbon nanofibers (GCNFs) are attractive additives for fabricating carbon fiber/polymer composite materials of enhanced strength and electrical conductivity due to their unusual atomic structure. Many studies show that efficient wetting and high dispersion of carbon nanofibers within a polymer matrix continues to be problematic. Greatly enhanced performance of nanocomposite materials reinforced with nanotube or nanofiber additives has not been fully achieved because of difficulties in achieving efficient dispersion and wetting of the nanoscale component within the matrix material, even when using surface-functionalized additives. We now report the fabrication of GCNF/epoxy nanocomposites using CGNFs to reinforce an epoxy resin. As illustrated in Fig.1 (a), nanofiber-reinforced adhesives will be used to bond hybrid joints of metals and polymers. Processing conditions have been optimized to give a hybrid material containing derivatized carbon nanofibers highly dispersed within a thermoset matrix. In the present study, ultrasonic methods, including low-power sonication using an ultrasonic cleaner and high-power sonication using a commercial sonifier, are investigated as a means to disperse nanofibers in epoxy resin and to initiate nanofiber/epoxy resin covalent binding. The dispersion quality of the resulting hybrid nanocomposite is clearly seen in a TEM picture as seen in Fig. 1(b) [15].

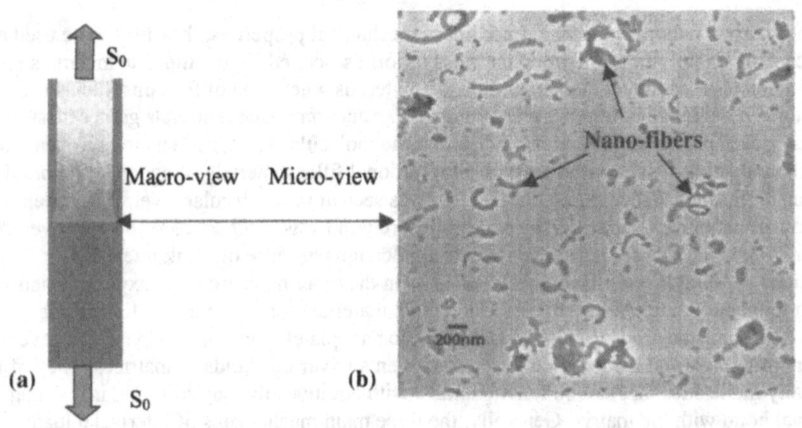

Figure 1 (a) Nano-fiber reinforced adhesive bonding applied to a hybrid metal-polymer joint (b) Transmission Electron Microscope (TEM) image of nanofibers as uniformly dispersed in epoxy matrix

However, the possible mechanical property change due to nanocomposites was not positive as seen in recent investigation. It is generally true that the stiffness of nanocomposites could be improved although the waviness and agglomeration of nanotubes/nanofibers will significantly affect the stiffness properties. The strengths or fracture toughness values of nanocomposite materials could be slightly lower than that of pure matrices [15]. Hence, we should pay great attention to the presence of stress singularity at the ends of nanofibers/nanotubes inside a matrix. It is well known that a high stress singularity/concentration occurs at the locations where discontinuity either in material or geometry exists [16]. It was found that interfacial shear stress concentration is quite high when the moduli ratio of the fiber and the matrix E_f / E_m is high [17]. These results based on previous traditional composites are still applicable to nanocomposite materials although the Young's moduli of nanofibers or nanotubes are much higher than the Young's moduli of any other traditional reinforcements. Therefore, the interfacial shear stress singularity/concentration of the nanocomposite should be much severe than that of traditional composites. In order to characterize the stress singularity/concentration at the interface between the matrix and reinforcements, we mainly use two Dundurs' parameters (α and β) to characterize the Young's modulus and bulk modulus mismatch of the fiber and the matrix. Since no one has measured the bulk modulus for nanotubes or nanofibers so far, we only employ one Dundurs' parameter to analyze our nanofiber composites under plane stress condition:

$$\alpha = \frac{E_f - E_m}{E_f + E_m} \tag{1}$$

We compare two composite systems with the same epoxy matrix (Young's modulus E=2.6 GPa [18]) reinforced by E-glass fibers (E =72 GPa) and nanofibers (E=600 GPa). Obviously, $\alpha_{nanofiber/epoxy}=0.99 > \alpha_{glass/epoxy}=0.93$ and β, which can be calculated if we know the bulk modulus or Poisson's ratio of nanofiber/nanotubes, should be quite large too for nanofiber/nanotubes composites. Since these two material constants are related to property mismatch and interfacial stress, we can conclude that the interfacial stress level is quite high in nanocomposite materials. Higher interfacial stress will be directly related to the final failure strain of nanocomposites so it is probably the major reason contributing to the low failure strains in nanocomposites over their matrices as explained in a simple stress-strain curve analysis in Fig. 2.

Although the composite stiffness can be increased by some degree, the final strength of the composites is determined by the failure strain of the nanocomposites. Therefore, it is not surprising that the nanocomposite strength is even less that that of the pure matrix if the failure strain is determined by interfacial failure between the matrix and the nanoscale reinforcement.

EXPERIMENTAL DETAILS

Materials and processing

Herringbone-type carbon nanofibers were grown by the interaction of a carbon source gas with mixed-metal powder growth catalyst. Iron-copper powder with atomic ratio of 7:3 was prepared by the co-precipitation of the respective metal nitrate solutions with ammonium bicarbonate.

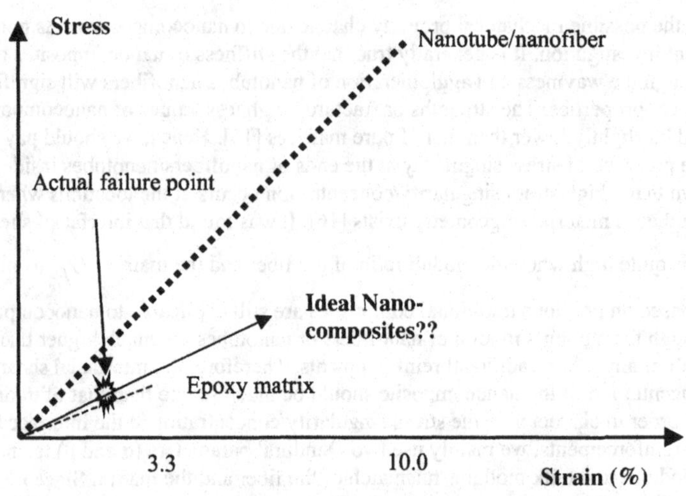

Figure 2. Nano-composite design based on stress-strain curves

The precipitate was dried in an oven at 110 °C and ground into fine powder. This powder was put into a quartz boat in a horizontal tubular furnace and was converted into metal-oxides mixture by calcining in air at 400 °C for 4 h. The growth of carbon nanofibers was completed at 600°C after 90 min. The product was cooled to room temperature under helium. The structure of the carbon nanofibers was confirmed by transmission electronic microscopy. The collected oxidized GCNFs were washed with deionized water until the filtrate reached a pH value of about 7 and then were dried in vacuum at room temperature. The dry oxidized GCNFs were activated by reaction with thionyl chloride at 70 °C for 24 h with a small amount of dimethylformamide (DMF). The reaction mixture was cooled and then filtered. The collected fibers were washed with tetrahydrofuran (THF) under nitrogen until the supernatant was clear. The as-prepared GCNFs were used as reinforcement to nanocomposite bonding. A commercial epoxy resin, DER 736, was used as the epoxy matrix material. Blends of epoxy, curing agent and nanofibers were mixed at room temperature, sonicated at controlled power levels and duration, filtered to remove any residual large agglomerated particles. After vacuum degassing for one hour at 40°C, the nanofiber-reinforced adhesive was applied onto the bonding surfaces of the joint specimens fixed in a specially designed fixture. Then, the fixture was put into an oven and the specimens were cured at 60°C overnight.

Mechanical property characterizations

Due to their simplicity, the Iosipescu shear and butt-joint tension tests were conducted for the characterization of shear and tensile bonding strengths. An MTS 810 testing machine was used and the loading rate was 1 mm/min. In order to understand the possible stress field change between a standard Iosipescu shear specimen (one material) and a bonded Iosipescu shear specimen (dissimilar materials), we made a direct comparison of experimental and numerical photoelastic patterns as shown in Fig. 3. The numerical photoelasticity fringe patterns in the Homalite specimens were obtained by use of the plotting software Tecplot. After the principal stresses at each node were obtained from finite element analysis, the numerical fringe order, *N*, was computed.

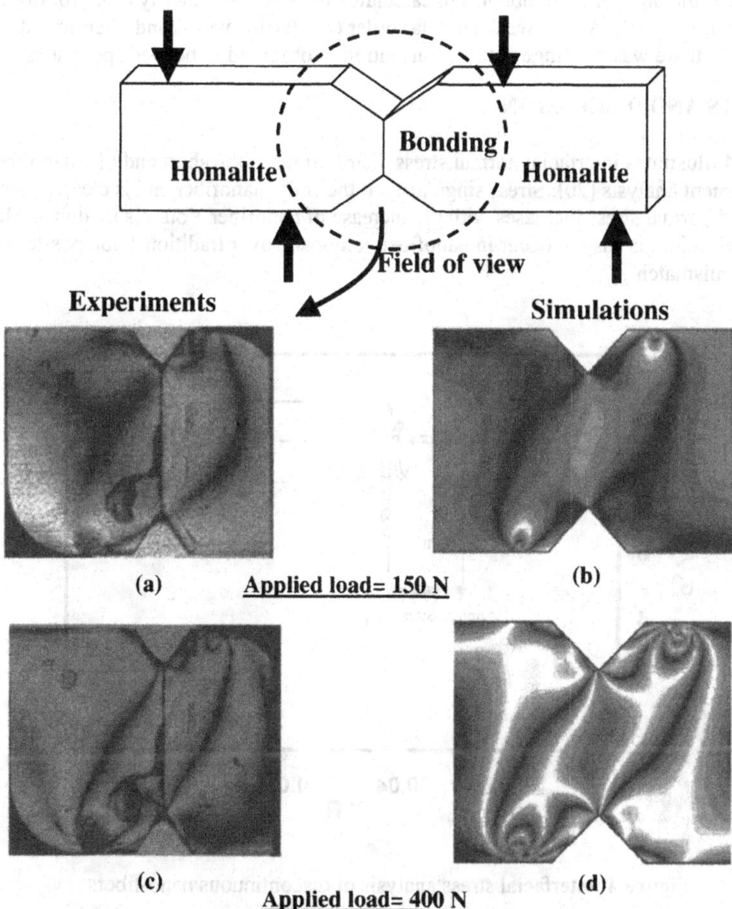

Figure 3. Direct comparison of photoelastic pictures and finite element simulations for a bonded Homalite Iosipescu shear specimen.

These fringe orders were then converted to gray-scale values by assignment of a gray-scale value of 255 to full fringe orders (e.g. 0, 1, 2 etc.) and a gray-scale value of 0 to half fringe orders (e.g. 0.5, 1.5, 2.5 etc.). The gray-scale values were then plotted with Tecplot to get the numerical fringe patterns. It may be noticed that our finite element analysis captured major experimental features as seen in Figs. 3(a) and (c). At the upper and right part of the specimen, there is a clear fringe pattern concentration, which is caused by direct compression of the upper load block. As the applied load was increased from 300N to 500N, the fringe patterns became more severe, as expected. Both experimental pictures and finite element simulation did not show any fringe pattern concentration at the round V-notch. No significant fringe pattern concentration was observed at the bonded sharp notch. We calculated the stress singularity order for our bonded shear specimens [19]. A very weak singular order ($\lambda = 0.014$) was found; therefore it seems logical that there was no fringe pattern concentration observed in bonded specimens.

RESULTS AND DISCUSSION

Fig. 4 illustrates interfacial normal stress distributions at the short end of a nanofiber, using finite element analysis [20]. Stress singularity at the finite nanofiber end is clearly seen and the interfacial normal stress increases with the increase of nanofiber Young's modulus. Hence, interfacial failure is easy to occur in nanofiber composite over traditional composites with less property mismatch.

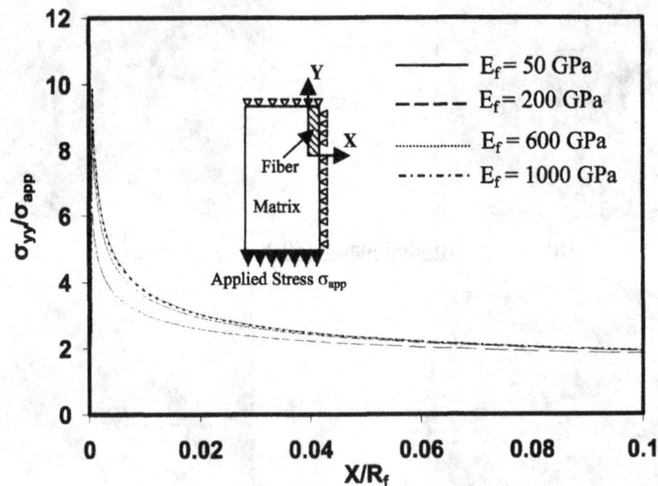

Figure 4. Interfacial stress analysis of discontinuous nano-fibers

Fig. 5 shows tensile strength comparison of two joints using nanocomposite bonding. It is not surprising to see a decrease in strength of aluminum/aluminum joints since initial defects were easily introduced in the processing of nanocomposite materials. However, some joint strength increase was achieved in PMMA/PMMA joints. Shear strength tests also revealed a similar trend. More bonding strength measurements will be conducted for further investigation. The low strength increase of nanofiber composites is related to interfacial stress transfer and interfacial failure. As discussed beforehand, there is a severe property mismatch between nanofibers and matrices. This mismatch may lead to higher interfacial stresses.

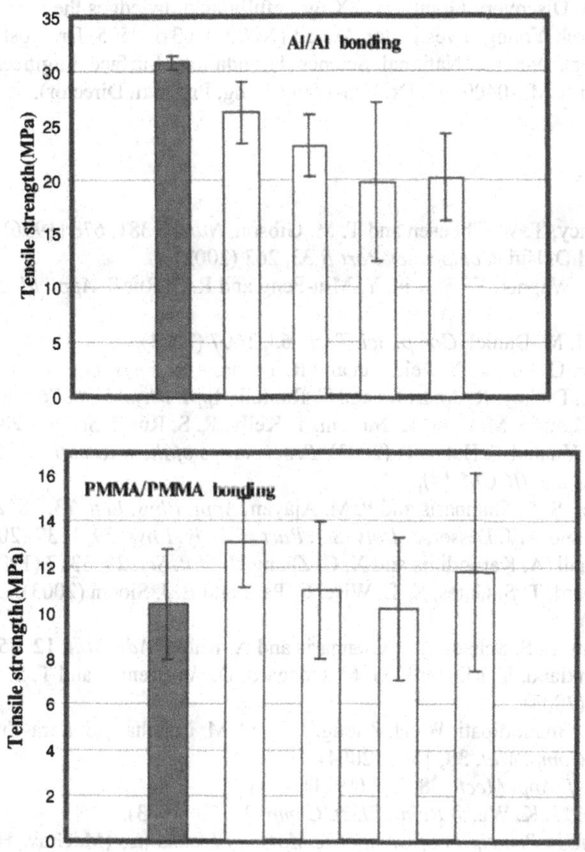

Figure 5. Tensile strength comparison of Aluminum/Aluminum and PMMA/PMMA bonds featuring nanofiber reinforced composites with different fiber weight percents and processing conditions (left bar—pure epoxy bonding)

ACKNOWLEDGEMENTS

LRX and CML gratefully acknowledge the support of Vanderbilt University through an Interdisciplinary Discovery Grant. L.R.X. gratefully acknowledges the support from the Office of Naval Research Young Investigator Award (N00014-03-1-0505, Dr. Roshdy G. S. Barsoum, Program Officer), and the National Science Foundation, Surface Engineering and Materials Design Program (CMS-0409665, Dr. Yip-Wah Chung, Program Director).

REFERENCES

1. M. M. J. Treacy, T. W. Ebbesen and T. M. Gibson, *Nature* **381**, 678 (1996).
2. K. T. Lau and D. Hui, *Composites Part B* **33**, 263 (2002).
3. D. Qian, G. J. Wagner, W. K. Liu, Y. Min-Feng and R. S. Ruoff, *Appl. Mech. Rev.* **55**, 495 (2002).
4. J. J. Luo and I. M. Daniel, *Comp. Sci. Tech.* **63**, 1607 (2003).
5. H. D. Wagner, O. Lourie, Y. Feldman and R. Tenne, *App. Phys. Lett.* **72**, 188 (1998).
6. D. Qian, E. C. Dickey, R. Andrews and T. Rantell, *Appl. Phys. Lett.*, **76**, 2868 (2000).
7. M. F. Yu, O. Lourie, M. Dyer, K. Moloni, T. Kelly, R. S. Ruoff, *Science* **287**, 637 (2000).
8. J. Y. Kim, L. Yu and T. Hahn, T. (2003). *Proceedings of the International Conference on Composite Materials (ICCM-14).*
9. L. S. Schadler, S. C. Giannaris and P. M. Ajayan, *Appl. Phys. Lett.* **73**, 3842 (1998).
10. A. S. Zerda and A. J. Lesser, *J. Poly. Sci. Part B: Poly. Phys.* **39**, 1137 (2001).
11. C. B. Bucknall, A. Karpodinis and X. C. Zhang, *J. Mat. Sci.* **29**, 3377 (1994).
12. G. M. Odegard, T. S. Gates, K. E. Wise, C. Park and E. J. Siochi (2003), *Comp. Sci. Tech.* **63**, 1671 (2003).
13. P. M. Ajayan, L. S. Schadler, C. Giannaris and A. Rubio, *Adv. Mat.* **12**, 750 (2000).
14. S. J. V. Frankland, V. M. Harik, G. M. Odegard, D. W. Brenner and T. S. Gates, *Comp. Sci. Tech.* **63**, 1655 (2003).
15. L. R. Xu, V. Bhamidipati, W.-H. Zhong, J. Li, C. M. Lukehart, E. Lara-Curzio, K. C. Liu and M. J. Lance, *J. Comp. Mat.* **38**, 1563 (2004).
16. D. B. Bogy, *J. App. Mech.* **38**, 377 (1971).
17. C. T. Sun and J. K. Wu, *J. Reinf. Plast. Comp.* **3**, 130 (1983).
18. R. F. Gibson, in *Principles of Composite Material Mechanics* (McGraw-Hill, Inc., New York, 1994).
19. L. R. Xu, S. Sengupta and H. Kuai, *Int. J. Adh. Adh.* **24**, 455 (2004).
20. L. R. Xu and S. Sengupta, *J. Nanosci. Nanotech.* (2004) (in press).

Mater. Res. Soc. Symp. Proc. Vol. 843 © 2005 Materials Research Society T4.6

Tribological Applications of Shape Memory and Superelastic Effects

Wangyang Ni[1,2], Yang-Tse Cheng[1], and David S. Grummon[2]
[1]Materials and Processes Laboratory, General Motors Research and Development Center, Warren, Michigan
[2]Department of Chemical Engineering and Materials Science, Michigan State University, East Lansing, Michigan.

ABSTRACT

We provide an overview of our recent studies on novel tribological applications of shape memory and superelastic effects: (i) the use of shape memory NiTi alloys as self-healing surfaces, and (ii) the use of the superelastic NiTi as an interlayer between a hard coating and a soft substrate to improve interfacial adhesion, decrease friction coefficient, and improve wear resistance.

1. INTRODUCTION

NiTi alloys are well-known for the shape memory and superelastic effect [1]. Since the first discovery of shape memory effect in NiTi in 1960's, there have been extensive research efforts on both the fundamental understanding and industrial applications of NiTi alloys. Recent years have seen an increasing interest in its wear applications. For example, shape memory alloys exhibit desirable wear properties under cavitation erosion [2-4] and dry sliding wear conditions [5-12]. Our recent studies, summarized in this paper, show that the shape memory alloys may be used in the following two new ways for tribological applications: (i) self-healing surfaces based on the shape memory effect, and (ii) superelastic interlayers between hard coating and soft substrate to reduce friction coefficient and wear loss.

2. SELF-HEALING SURFACES

In many applications, contact induced surface damages are unavoidable. It is thus highly desirable that a tribological system can detect and heal such damages automatically. The shape memory property offers a possible means to achieve self-healing in tribological systems. However, it was unknown whether shape memory effect exists at microscopic length scales and under complex loading conditions that are frequently encountered in tribological applications, although the shape memory effect was well established at macroscopic length scales and under simple loading conditions such as tensile, compression, or shear. To establish the basis for using shape memory materials as self-healing tribological surfaces, we have conducted a systematic investigation of the magnitude of shape recovery on the surfaces of a martensitic NiTi alloy using indentation and scratch tests. The indentation and scratch experiment can simulate an asperity contact under normal and tangential loading, respectively. Fig. 1 (a) shows the 3-D profiles of a spherical indent on the martensitic NiTi before and after recovery [13]. The recovery of the indent was induced by heating the specimen to a temperature above the austenite finish temperature. The specimen almost completely recovers to its original shape after being heated. The shape memory effect under uniaxial loading has been intensively studied. Here, we demonstrated that the shape memory effect also exists under indentation loading conditions.

Compared with the conventional uniaxial tension experiments, the advantages of using indentation techniques are that: (i) it is a non-destructive method, (ii) it can be conducted at micrometer or nanometer scale, and (iii) it is especially useful for the characterization of films on substrate, where obtaining free-standing films is very often challenging.

The self-healing capability of the indents can be quantitatively characterized by defining a thermal-induced recovery ratio

$$\delta = (h_f - h_f^{'})/h_f, \qquad (1)$$

where h_f is the residual indentation depth recorded immediately after unloading and $h_f^{'}$ is the final indentation depth after the completion of thermal-induced recovery [13,14]. Our experimental study shows that for spherical indentation δ is dependent on both the indenter radius and indentation depth. The results can be rationalized using the concept of the representative strain defined as

$$\varepsilon_r = 0.2a/R, \qquad (2)$$

where a and R are the contact radius and indenter radius, respectively. Fig. 1(b) shows the relationship between the thermo-induced recovery ratio and the representative strain together with the stress-strain relationship [14]. It shows that self-healing remains constant and almost complete until a critical strain is reached. This critical strain, which coincides with the end of the stress plateau in stress-strain curve, is the maximum recoverable strain in the uniaxial tensile test. Therefore, the maximum recovery strain can be measured using the spherical indentation techniques.

The self-healing effect exists in sliding contact as well. Fig. 2 shows the profile of a scratch before and after recovery. The scratch was generated by a spherical indenter with tip radius 213.4μm using a progressive load from 0 to 10N. It clearly demonstrates that the scratch scar can be self-healed when the temperature is raised above its austenite finish temperature. The recovery is almost complete when the scratch scar is shallow. The above observations demonstrate that the NiTi shape memory alloy can be used as a self-healing surface under contact or sliding loading conditions.

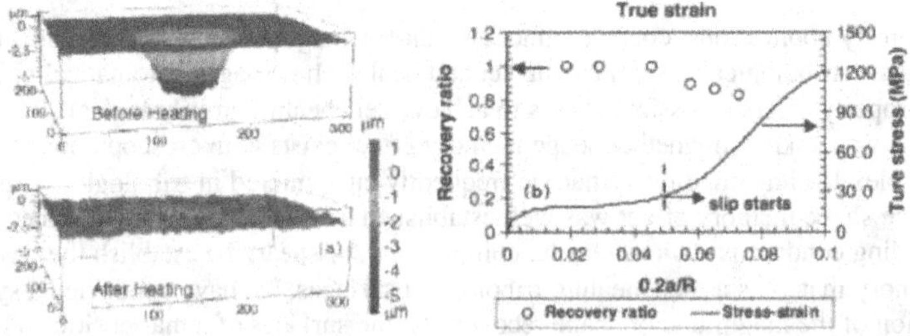

Figure 1. Shape memory NiTi: (a) recovery of spherical indents, and (b) relationship between thermally activated recovery ratio and representative strain (0.2a/R), and the relationship between the true stress and true strain.

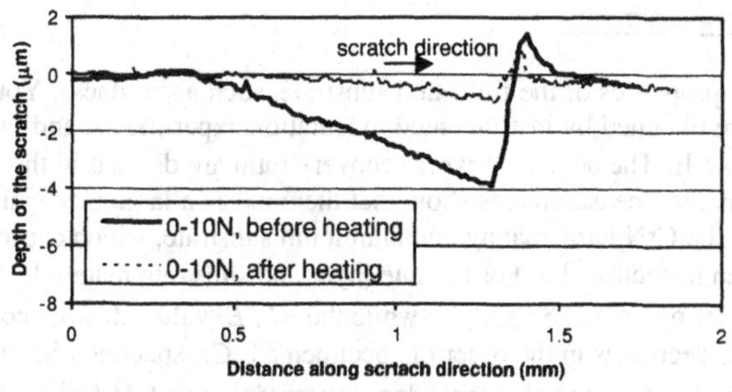

Figure 2. Self-healing of a scratch scar on a shape memory alloy.

3. TRIBOLOGICAL APPLICATIONS OF SUPERELASTICITY

Similar to the shape memory effect, the superelastic effect also exists under indentation loading conditions. The superelastic effect is manifested by a large elastic recovery during unloading [14,15]. It has been suggested that the superelasticity is beneficial to its wear resistance [8]. However, due to its low hardness NiTi does not excel in the wear resistance as compared to other hard materials. For example, our study shows the superelastic NiTi (H=4.4GPa) is less wear resistant than the amorphous NiTi (H=7.9GPa) although the former shows a much larger elastic recovery [16]. This is not surprising since hardness also plays an important role in determining wear resistance. We believe that, ideally, a good wear resistant material should have a combination of high hardness and large elastic recovery. For coating on substrate, good adhesion is also required, which is often achieved by using an interlayer material such as Cr and Ti. Base on the above considerations, we made a layered composite structure where the superelastic NiTi was used as an interlayer between the hard coating and soft substrate. The effects of the superelastic interlayer on adhesion, friction, and wear were studied [17].

A detailed description of the samples is presented in Table I: the austenitic NiTi (designated as S1), martensitic NiTi (designated as S2), and pure Cr films were used as the interlayer materials between a CrN hard coating and an aluminum substrate; for the samples without interlayer, two thickness were used. The NiTi, Cr, and CrN layers were deposited by magnetron sputtering. The detailed deposition conditions are given in Ref. 17.

Table I. Description of specimens.

Specimen	**CrN1-Al**	**CrN5-Al**	**CrN-S1-Al**	**CrN-S2-Al**	**CrN-Cr-Al**
CrN thickness (μm)	1	5	1	1	1
Interlayer material	No	No	$Ni_{51.8}Ti_{48.2}$	$Ni_{49.5}Ti_{50.5}$	Cr
Interlayer thickness (μm)	/	/	4	4	4
Average roughness, R_a (nm)	148	57	138	201	53

3.1 Indentation measurements

Indentation properties of the films and substrate, such as hardness, Young's modulus and recovery ratio, were obtained by instrumented indentation experiments, and the results are summarized in Table II. The depth and work recovery ratio are defined in the same way as in Ref. 15. The indentation measurements show that there exists a large mechanical property mismatch between the CrN hard coating and aluminum substrate, which causes poor adhesion at the interface as seen in section 3.2. For the interlayer and substrate materials, the hardness decreases in the order of Cr, S1, S2 and Al, while the H/E^* value, depth recovery ratio, and work recovery ratio decreases in the order of specimen S1, Cr, specimen S2, and aluminum.

For CrN layered coatings, the dependence of hardness and H/E^* on indentation depth is presented in Fig. 3(a) and (b), respectively. It was found that hardness and H/E^* are strongly dependent on the interlayer material, coating thickness, and indentation depth. Note that the specimen with an austenitic NiTi interlayer (specimen CrN-S1-Al) has a larger H/E^* than the specimen with a martensitic NiTi interlayer (specimens CrN-S2-Al) and a Cr interlayer (specimen CrN-Cr-Al), although specimen CrN-Cr-Al has higher hardness. The high H/E^* in specimen CrN-S1-Al arises from the superelasticity of the NiTi interlayer [14,15,17].

Table II. Nanoindentation properties of aluminum substrate, interlayer materials (S1, S2 and Cr), and CrN hard coating.

Materials	Al	S1 ($Ni_{51.8}Ti_{48.2}$)	S2 ($Ni_{49.5}Ti_{50.5}$)	Cr	CrN
H (GPa)	0.94	4.7	2.4	6.7	23.0
E (GPa)	77.6	95.0	89.3	318.4	257.0
Poisson's ratio	0.34	0.3	0.3	0.21	0.3
E^*	81.5	95.7	90.4	258.1	226.7
H/E^*	0.012	0.049	0.027	0.026	0.102
Depth recovery ratio	0.067	0.33	0.23	0.16	0.64
Work recovery ratio	0.070	0.35	0.23	0.17	0.70

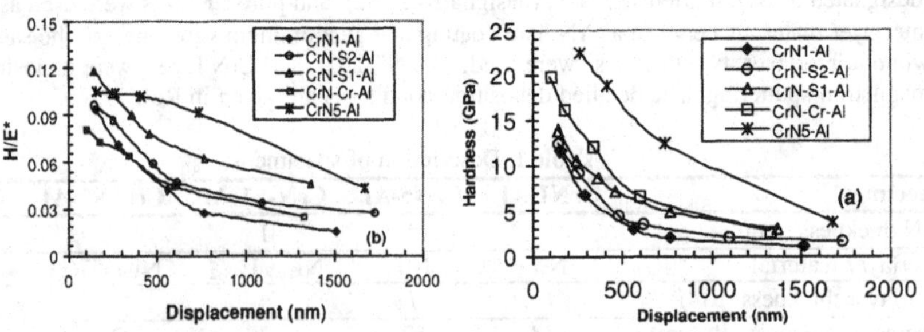

Figure 3. Mechanical properties of the coatings at various depths: hardness (a), and ratio of hardness to reduced modulus (b).

3.2 Interfacial adhesion

The interfacial adhesion strength of the CrN coated specimens was compared qualitatively by scratch adhesion test. The SEM images of the end of the scratches are shown in Fig. 4. Flakes of delaminated CrN coating at the end of the scratches are observed on specimen CrN1-Al, CrN5-Al, and CrN-S2-Al, as shown in Figs. 4(a), (b), and (d). Fig. 4 (a) and (b) show that increasing the thickness of CrN hard coating has little beneficial effect on improving interfacial adhesion. However, as shown in Fig. 4(c) and (e), using a Cr interlayer or superelastic NiTi interlayer can effectively improve adhesion since no coating delamination is observed in these two samples.

Interestingly, the specimen with an austenitic interlayer (CrN-S1-Al) shows much better adhesion than the specimen with a martensitic NiTi interlayer (CrN-S2-Al), although there is little chemical composition difference between these two interlayer materials. The difference in

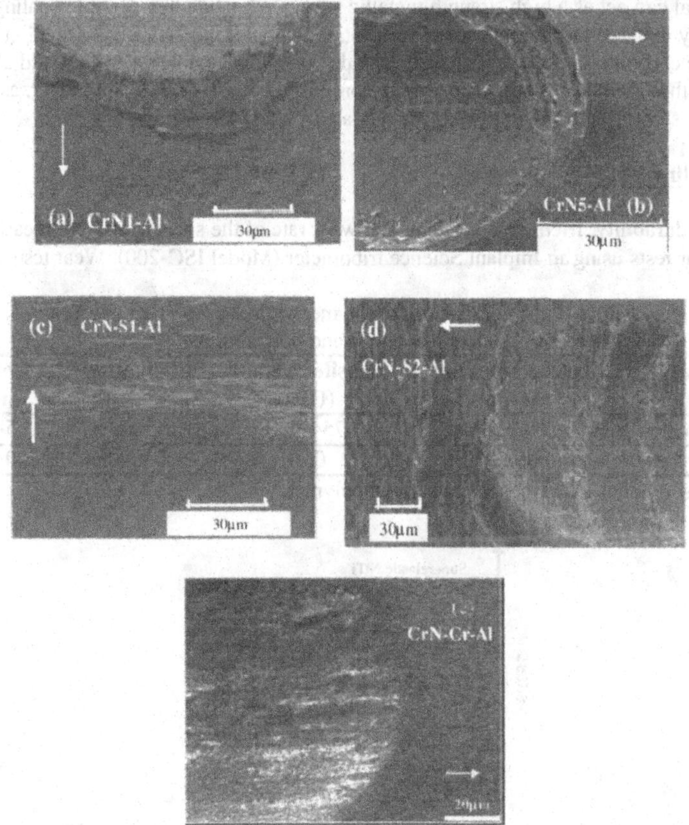

Figure 4. SEM images of the end of scratches on CrN1-Al (a), CrN5-Al (b), CrN-S1-Al (c), CrN-S2-Al (d), and CrN-Cr-Al (e). The arrow indicates the scratch direction.

interfacial adhesion strength must be due to their difference in mechanical properties. For the martensitic NiTi alloy, strain is not recoverable upon unloading when it exceeds the elastic limit of less than one percent. However, the austenitic NiTi alloy is superelastic and it has a large reversible strain. In fact, the stress-strain curve of the superelastic NiTi alloy bears some similarities to that of elastomeric polymer adhesives, as illustrated in Fig. 5 [1,18]. The comparison of the mechanical properties of the superelastic NiTi with a typical elastomeric polymer (e.g., polyisoprene) is listed in Table III. Fig. 5 shows that both of them have large recoverable strain [1,18]. However, the elastic modulus and strength of the superelastic NiTi alloy are several orders of magnitude greater than that of the polymeric adhesives. In addition, a hysterisis loop, which arises from the internal friction, is associated with the stress-strain curve of the superelastic NiTi. A part of the energy is thus consumed due to the impedance to the movement of the phase boundaries between austenite and martensite phases [1], which decreases the driving force for coating delamination and cracking. We thus suggest that the superelastic NiTi material can act as a high strength metallic adhesive for bonding ceramic coatings to ductile substrates by its large recoverable strain, large stain tolerance, and energy dissipation.

The chromium interlayer can also provide good adhesion between CrN and aluminum substrate, although the wear resistance of the composite coating with Cr interlayer, as seen in section 3.3, is not as good as that with a superelastic NiTi interlayer.

3.3 Dry sliding wear

The durability, friction coefficient, and wear rate of the specimens were measured by pin-on-disk wear tests using an Implant Science tribometer (Model ISC-200). Wear tests were

Table III. A comparison of the mechanical properties between superelastic NiTi and polyisoprene.

Materials	Young's modulus (GPa)	Tensile strength (GPa)	Elongation	Recoverable strain
Superelastic NiTi	75	754-960	15.5%	5~8%
Polyisoprene*	0.002~0.1	0.01	100~800%	100~800%

Note: Polyisoprene is a thermoset elastomeric polymer.

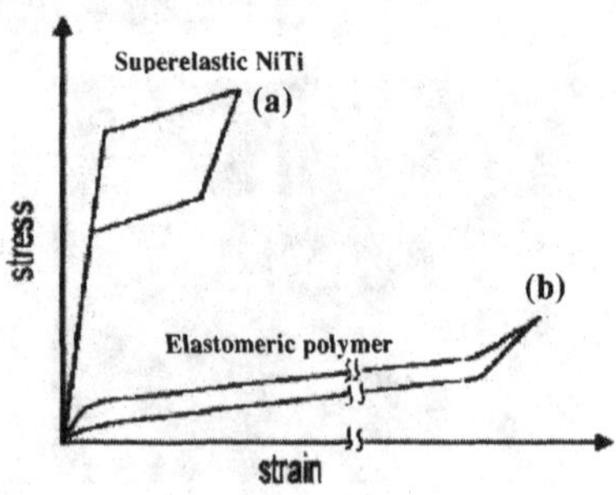

Figure 5. Schematic illustration of the stress-strain curves of superelastic NiTi (a), and elastomeric polymer (b).

carried out at room temperature of 25°C in air using a 3.175mm diameter cemented carbide ball in an unlubricated condition. All the samples were tested at the same relative humidity of 48%. The applied load was 1N and the sliding speed was 0.08m/s. Figs. 6(a) and (b) show the friction coefficient during pin-on-disk experiments. Specimen CrN-S2-Al and CrN1-Al failed after 1100 and 3000 cycles, respectively, accompanied by a sudden increase in the friction coefficient. Surface profile measurements and microscopy observations revealed that the CrN coating was worn through, and the aluminum substrate and S2 interlayer were exposed in sample CrN-Al and CrN-S2-Al, respectively. Poor adhesion between the hard coating and soft substrate is believed to be the failure mechanism.

The coating durability can be improved by using a Cr or superelastic NiTi interlayer, as shown in Fig. 6(b). In particular, the layered composite coating with a superelastic NiTi interlayer (specimen CrN-S1-Al) shows a lower friction coefficient than that with the Cr interlayer (specimen CrN-Cr-Al), although they have the same surface chemistry and specimen CrN-Cr-Al has a higher hardness. We believe that elastic recovery, characterized by H/E^*, plays an important role in determining the friction coefficient.

The effect of H/E^* on friction can be qualitatively understood as follows. We assume that multi-asperity contacts can be modeled locally by either conical or spherical indentations. It is well known that the deformed surface around the indenters can exhibit "piling-up" or "sinking-in." Furthermore, it has been recently shown that the degree of piling-up and sinking-in is determined by the mechanical properties of materials, as well as the indenter geometry [19-21]. For materials with a large ratio of yield strength to elastic modulus (such as the present case), sinking-in is expected and the tendency of sinking-in increases with H/E^*. The friction force consists of two parts: adhesion force and plowing force [22]. Since the plowing term is the force needed to deform materials in the sliding direction, the plowing contribution to friction is expected to be smaller for sinking-in than that for piling-up. Thus, the plowing contribution to friction is expected to decrease with increasing H/E^*, as observed in the present set of experiments.

The wear losses, defined as the volume loss per sliding distance per applied normal load, are shown in Fig. 7. The specimen with a superelastic interlayer (CrN-S1-Al) shows the lowest wear loss, followed by that with a Cr interlayer (CrN-Cr-Al), although the later has a higher hardness and smaller roughness. Specimens CrN-Al and CrN-S2-Al have much higher apparent wear loss primarily due to poor adhesion between coatings and aluminum substrates. In general, wear of tribological coating can be categorized into two types: (i) wear caused by gradual removal of coating materials or (ii) by cracking and delamination [23]. Wear of specimen CrN-Al and CrN-S2-Al belongs to the second category, while wear of the CrN-S1-Al and CrN-Cr-Al belongs to the first category. The gradual removal of coatings is caused by adhesive and abrasive wear. In this set of experiments, the contribution of adhesive wear to total wear should be the same since there is no difference in contact surface. For abrasive wear, Archard's equation predicts that specimens with high hardness are more wear resistant [22]. However, a comparison of the wear behavior between specimen CrN-S1-Al and CrN-Cr-Al shows that materials with larger H/E^* have better wear resistance. This observation is consistent with the observations by Leyland and Matthews [24-26], although their conclusion was based on wear couples of different

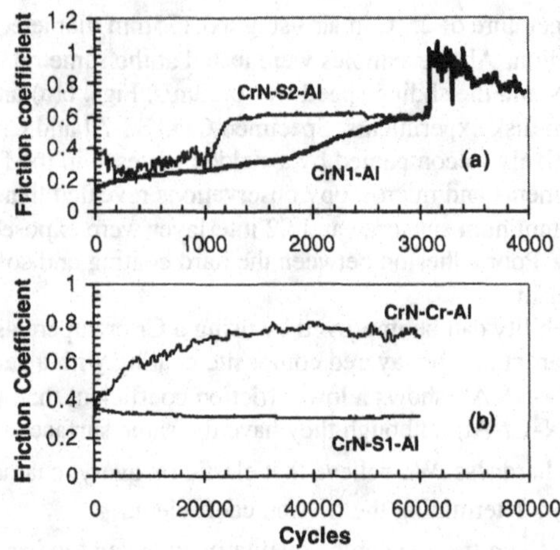

Figure 6. Friction coefficient of the coatings measured by pin-on-disk wear test: specimen CrN1-Al and CrN-S2-Al (a), and specimen CrN-S1-Al and CrN-Cr-Al (b).

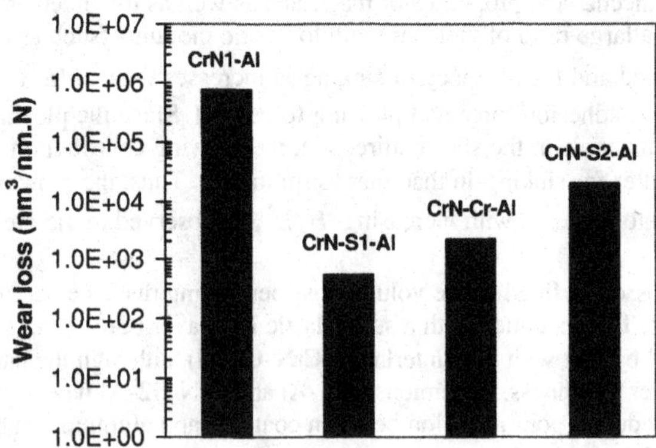

Figure 7. Wear loss of specimens. Note that the Y axis is in logarithmic scale.

materials. By keeping the chemical composition of sliding interface the same, the present set of experiments unambiguously demonstrates the H/E^* effect on wear.

The effect of H/E^* on wear can be rationalized by assuming that gradual material removal or gradual wear is caused by plastic deformation and not by elastic deformation or fracture. It follows that, first, for materials with high H/E^*, the "plasticity index" [27], which equals E^*/H multiplied by a factor determined by surface morphology, is small. Consequently, the deformation under contact is more likely to be elastic for high H/E^* materials [28]. Second, the ratio of reversible work, W_e, to total work, W_{tot}, under conical and pyramidal indentation has recently been shown to be proportional to H/E^*, i.e. [19,20,29],

$$W_e / W_{tot} \propto H / E^* . \tag{3}$$

Consequently, a larger fraction of the work is consumed in plastic deformation when H/E^* is smaller for the same degree of deformation. Third, the unrecoverable strain, measured by the ratio of final indentation depth, h_f, to maximum indentation depth, h_{max}, is found to be related to H/E^* through a relationship that [19,29]

$$h_f / h_{max} \propto (W_{tot} - W_e) / W_{tot} . \tag{4}$$

Thus, larger plastic strain is expected when contacting a material with smaller H/E^*. Assuming similar relationships hold for multi-asperity contact during sliding, materials with higher H/E^* are expected to have smaller accumulative strain, smaller accumulative strain energy, and thus better wear resistance.

4. CONCLUSIONS

Microscopic shape memory and superelastic effects exist under indentation and sliding contact conditions. The shape memory effect can be exploited for making self-healing tribological surfaces. By using the superelstic NiTi alloy as an interlayer between a hard coating and a soft substrate, it is possible to improve interfacial adhesion, decrease friction coefficient, and improve wear resistance. The decreased friction coefficient and improved wear resistance are rationalized using the concept of H/E^*.

ACKNOWLEDGEMENT

The authors would like to thank Anita M. Weiner, Michael J. Lukitsch, Lenoid C. Lev, Curtis Wong, Robert K. Cubic, Michael P. Balogh, Yue Qi, Louis G. Hector, Thomas A. Perry, and Mark W. Verbrugge at GM Research and Development Center for their technical assistance and valuable discussions.

REFERENCES

1 K. Otsuka and C. M. Wayman, *Shape Memory Alloys* (Cambridge University Press, Cambridge, 1998).

2 R. H. Richman and W. P. McNaughton, Journal of Materials Engineering and Performance **6,** 633-641 (1997).

3 T. Zhang and D. Y. Li, Materials Science and Engineering **A293,** 208-214 (2000).

4 H. Hiraga, T. Inoue, H. Shimura, and A. Matsunawa, Wear **231,** 272-278 (1999).

5 J. Jin and H. Wang, Acta Metall. Sinica **24,** A66-A70 (1988).

6 P. Clayton, Wear **162-164,** 202-210 (1993).
7 Y. N. Liang, S. Z. Li, Y. B. Jin, W. Jin, and S. Li, Wear **198,** 236-241 (1996).
8 R. Liu and D. Y. Li, Materials Science and Technology **16,** 328-332 (2000).
9 J. Singh and A. T. Alpas, Wear **181-183,** 302-311 (1995).
10 H. C. Lin, H. M. Liao, J. L. He, K. C. Chen, and K. M. Lin, Metallurgical and Materials Transaction **28A,** 1871-1877 (1997).
11 R. Liu and D. Li, Wear **250-251,** 956-964 (2001).
12 F. T. Cheng, P. Shi, and H. C. Man, Scripta Materialia **45,** 1089 (2001).
13 W. Ni, Y.-T. Cheng, and D. S. Grummon, Applied Physics Letters **80,** 3310-3312 (2002).
14 W. Ni, Y. T. Cheng, and D. S. Grummon, Surface and Coatings Technology, **177-178,** 512-517 (2004).
15 W. Ni, Y.-T. Cheng, and D. S. Grummon, Appl. Phys. Lett. **82,** 2881 (2003).
16 W. Ni and D. S. Grummon, in *Nanoindentation and wear behavior of nickel titanium alloys*, Mat. Res. Soc. Symp. Proc. **697**, p. P2.9.1-P2.9.6 (2002)
17 W. Ni, Y.-T. Cheng, M. J. Lukitsch, A. M. Weiner, L. C. Lev, and D. S. Grummon, Applied Physics Letters **85,** 4028-4030 (2004).
18 D. Brandon and W. D. Daplan, *Joining Process: An introduction* (John Wiley & Sons Ltd, 1997).
19 Y.-T. Cheng, Z. Y. Li, and C.-M. Cheng, Philosophical Magazine **82,** 1821-1829 (2002).
20 W. Ni, Y.-T. Cheng, C.-M. Cheng, and D. S. Grummon, J. Mater. Res. **19,** 149-157 (2004).
21 Y.-T. Cheng and C.-M. Cheng, Philosophical Magazine Letters **78,** 115-120 (1998).
22 I. M. Hutchings, *Tribology: Friction and Wear of Engineering Materials* (CRC Press, Boca Raton, USA, 1992).
23 S. Hogmark, s. Jacobson, and M. Larsson, Wear **246,** 20-33 (2000).
24 A. Leyland and A. Matthews, Wear **246,** 1-11 (2000).
25 A. Matthews and A. Leyland, Key Engineering Materials **206-213,** 459-466 (2002).
26 A. Leyland and A. Matthews, Surface and Coatings Technology **177-178,** 317-324 (2004).
27 J. A. Greenwood and J. B. P. Williamson, Pro. Roy. Soc. Lond. **A295,** 300-319 (1966).
28 T. L. Oberle, J. Metals **3,** 438 (1951).
29 Y.-T. Cheng and C.-M. Cheng, Appl. Phys. Lett. **73,** 614-616 (1998).

Mater. Res. Soc. Symp. Proc. Vol. 843 © 2005 Materials Research Society **T4.8**

Control of Stress in Surface Engineered Silicon

Y. Ma, R. Job, B. Zölgert, W. Düngen, Y. L. Huang, W.R. Fahrner
University of Hagen (LGBE), P.O. Box 940, D-58084 Hagen, Germany

ABSTRACT

Hydrogen plasma (H-plasma) treatments applied on (100)-oriented standard Czochralski (Cz) silicon wafers cause a structuring of the surface regions in the sub-100 nm scale. The reconstruction of the 'nano-structured' surface layers by high temperature vacuum annealing induces strong tensile stress (~ GPa) at the surface and in the subsurface regions of the wafers, as can be verified by depth resolved μ-Raman spectroscopy (μRS). Although the H-plasma caused 'nano-structured' surface layer is very thin (100 – 200 nm), the stressed subsurface regions of the annealed samples are quite extended, i.e. up to ~ 10 μm depth. The impacts of several process parameters including 1) the doping type of the wafer substrate, 2) the frequency of the H-plasma, 3) the power and duration of the H-plasma, 4) the temperature and duration of the vacuum annealing on the stress are investigated. It is found that the stressed region in the surface engineered silicon wafer can be controlled by the process parameters. The presented results might be important for various applications in semiconductor technology. For example, the H-plasma exposure could be done in local areas near the electrically active regions on the forefront of the wafer. After appropriate annealing, stressed local regions for external getter purposes can be created. Such a method might especially be useful for the formation of getter regions on SOI substrates.

INTRODUCTION

It is well known that hydrogen plays an important role in modern semiconductor industry and has prompted many experimental and theoretical studies on this subject since the last three decades [1-3]. The presence of hydrogen can seriously alter the properties of the semiconductor material such as passivating shallow and deep level impurities, catalyzing thermal donor formation (see [3] and references quoted therein), and facilitating the smart-cut® process [4]. Our recent investigations concentrate on the development of environmentally friendly, low cost and low thermal budget processes taking advantage of some catalytic properties of hydrogen in silicon. For such investigations hydrogen is incorporated into silicon wafers by H-plasma treatments at moderate temperatures (~ 260 °C). Due to its high activity, a hydrogen exposure causes a strong impact on the treated Cz silicon substrates. The hydrogen related effects on the Cz silicon wafers can be separated into either surface, subsurface or bulk effects depending on the local regions, where the hydrogen caused actions occur, i.e., a three-layer structure of the H-plasma treated wafers can be defined [5].

Our previous investigations have revealed that H-plasma treatments applied on the Cz silicon wafers cause a structuring of the surface regions in the sub-100 nm scale. The reconstruction of the 'nano-structured' surface layers by high temperature annealing induces strong tensile stress (~ GPa) at the surface and in the subsurface regions of the wafers, as can be verified by depth resolved μRS [6-8]. Although the H-plasma caused 'nano-structured' surface layer is rather thin (100 – 200 nm), the stressed subsurface regions of the annealed samples are quite extended, i.e. up to ~ 10 μm depth. In this article the impacts of various process parameters on the properties of the stressed subsurface regions of Cz silicon wafers are investigated.

EXPERIMENTAL

(100)-oriented p-type (1 – 30 Ωcm) and n-type (1 – 10 Ωcm) Cz silicon wafers (Ø = 3 inch) were investigated. The H-plasma treatments were carried out in a PECVD-setup operating at a frequency of either 13.56 or 110 MHz, and a pressure of 2000 and 400 mTorr, respectively. The hydrogen flux was 200 sccm. The applied plasma powers were varied between 5 and 50 W. The substrate temperature during the plasma hydrogenation was about 260 °C. After the H-plasma treatment the samples were annealed in a vacuum furnace at temperatures between 600 °C and 1200 °C, applying the following temperature profile: During 1 hour the samples were heated up to the desired maximum temperature; then this temperature was kept constant for 1, 2, 4, or 8 hours; finally the furnace was switched off. At temperatures below 200 °C (i.e. after a cooling period of ~ 1 – 1.5 hours) the samples were removed from the furnace.

Some samples were then beveled with a mechanical polishing machine under a bevel angle of about 0.2°. The accurate value of the bevel angle was obtained based on the reflection of the He-Ne laser both on the original and the beveled surfaces. The Raman measurements were carried out with a μRS setup, where a microscope (objective: ×100) is coupled to a spectrograph with 300 mm focal length. The excitation was supplied by an Ar^+ ion laser (488 nm, 44 mW). The spectra were collected at room temperature by a Peltier cooled CCD detector (grating: 1800 mm^{-1}, collection time: 5 s, averaging factor: 30). For details of the depth resolved μRS see reference [5]. The SEM analysis was done with a digital field emitter device. During these investigations the applied voltage was 1 kV; and the working distance between the cathode and the surface of the sample was about 3 mm.

RESULTS AND DISCUSSION

The observed tensile stress in H-plasma treated and vacuum annealed Cz silicon wafers has been explained with scanning electron microscopy (SEM). SEM analysis gives a strong hint that the reconstructed wafer surface is (111)-oriented, in contrast to the (100)-orientation of the underlying wafer substrate. Due to this mismatch the appearance of stress is evident [6-8].

The stress can be quantitatively verified by μRS analysis, since the Raman shift depends on the elastic material properties. The relation between the components of the strain tensor and the Raman frequency has been calculated [9]. The results are simple for uniaxial or biaxial stress:

$$\sigma = k \times \Delta\omega \qquad (1)$$

or

$$\sigma_{xx} + \sigma_{yy} = k \times \Delta\omega \qquad (2)$$

where

$$k = -434\ (\mathrm{MPa/cm^{-1}})$$

$$\Delta\omega = \omega_{stress} - \omega_0 \qquad (3)$$

and ω_0 is the silicon stretch mode without stress (around 520 cm^{-1}) and ω_{stress} the one shifted by stress. Tensile stress causes a decrease of the Raman mode ($\Delta\omega < 0$) and compressive stress an increase ($\Delta\omega > 0$).

1. Effect of substrate doping type

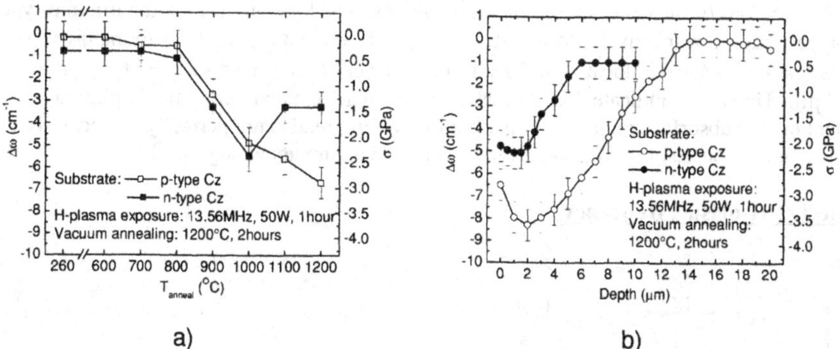

Figure 1. (a) The relative Raman shift (*Δω*) and tensile stress (*σ*) measured on the surface of the sample depend on the vacuum annealing temperature for the p-type and n-type Cz silicon substrate. (b) Depth profile of the relative Raman shift and tensile stress for the p-type and n-type Cz silicon substrate.

Both p-type and n-type Cz silicon wafers are treated with a 13.56 MHz H-plasma at ~ 260 °C for 1 hour. After vacuum annealing at different temperatures the relative Raman shifts measured on the surfaces of the samples are presented in Fig. 1(a) (the sample "annealed" at 260 °C refers to the as-plasma treated sample). The calculated tensile stress is also shown. The stress (relative Raman shift) on both samples does not occur up to 800 °C; above this temperature both samples shows tensile stress. This result is related to the SEM analysis on these samples, which shows the reconstruction of the surface starts above 800 °C (see Fig. 2). The tensile stress monotonously increases with the annealing temperature for p-type samples. However, for n-type samples the curve shows some scattering around at 1000 °C.

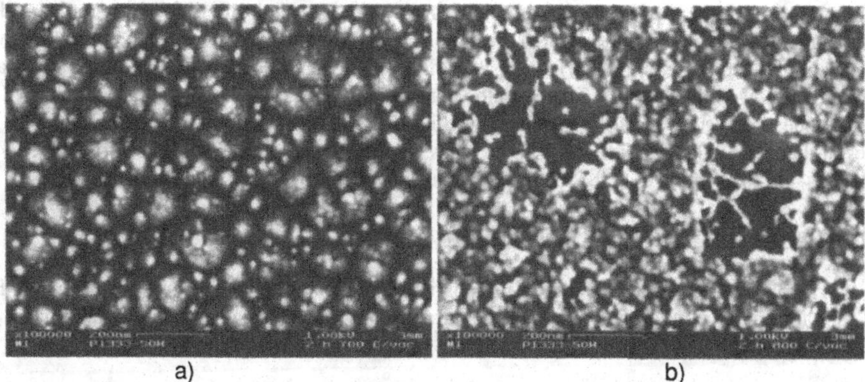

Figure 2. SEM pictures of plasma hydrogenated p-type silicion after vacuum annealing at (a) 700 °C and (b) 800 °C for 2 hours. The SEM pictures of the samples annealed below 700 °C are similar as (a).

The depth distribution of the relative Raman shift has also been investigated with depth resolved µRS, as shown in Fig. 1(b). The annealing temperature is 1200 °C. Although the H-plasma caused 'nano-structured' surface layer is very thin (100 – 200 nm), the stressed subsurface regions of the annealed samples are quite extended, i.e. up to 14 and 5 µm in p-type and n-type samples, respectively. Moreover, Fig. 1(b) shows that the region with the highest tensile stress (~ 3.7 GPa) is located in the subsurface layer of the p-type sample at a depth of about 2 µm. This might indicate that the platelets, which are created during the H-plasma treatment in the subsurface regions [~ 1 µm] of the wafer, are also not perfectly healed and cause a supplementary source of tensile stress after high temperature annealing [5, 8].

2. Effect of H-plasma frequency

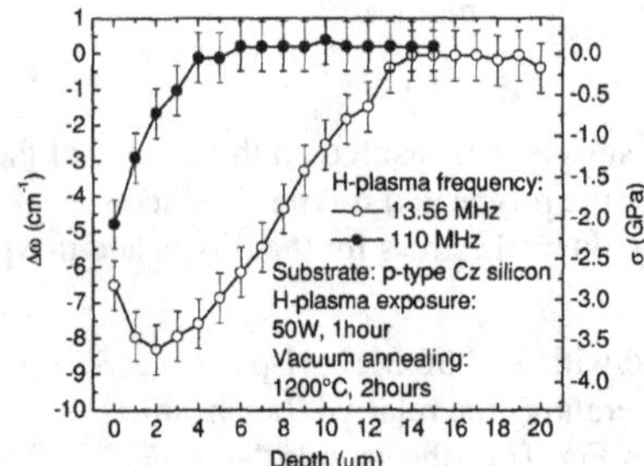

Figure 3. Depth profile of relative Raman shift ($\Delta\omega$) and tensile stress (σ) depend on H-plasma frequency.

The H-plasma frequency (either 13.56 or 110 MHz) is another factor to determine the stress distribution. As show in Fig. 3, the stress in the sample treated with 13.56 MHz H-plasma is stronger and much more extended than in the samples treated with 110 MHz H-plasma. Generally speaking, the H-plasma with higher frequency is more "softer" for the sample, so that less damages and platelets are created during plasma process. As a consequence, after vacuum annealing the stress with 110 MHz is weaker than with 13.56 MHz. Also, a maximum stress at ~ 2 µm depth, which is possibly related to the platelets (as discussed above and in reference [8]), cannot be observed in the sample treated with 110 MHz. This is consistent with the previous result that the H-plasma with 110 MHz induces less platelets – quasi-two-dimensional open void defects – in the sample than the 13.56 MHz plasma under otherwise same condition [10].

3. Effect of H-plasma power and duration

The plasma power dependence of the relative Raman shift and tensile stress is shown in Fig. 4. It is found that neither the power nor the duration of the H-plasma has a significant impact on the stress.

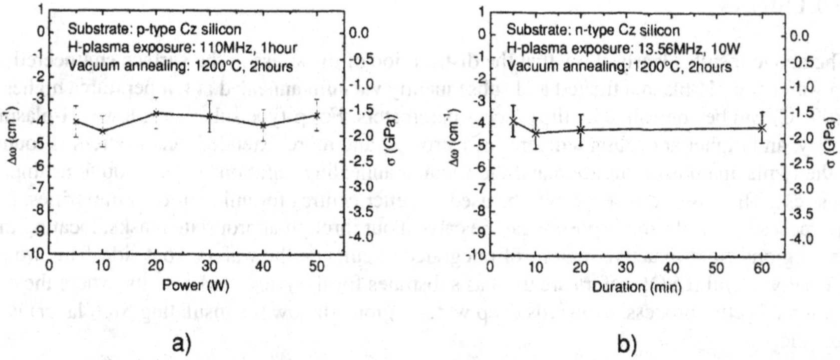

Figure 4. Relative Raman shift ($\Delta\omega$) and tensile stress (σ) measured on the surface of the sample dependence on the H-plasma power (a) and the duration of the plasma exposure (b).

4. Effect annealing temperature and duration

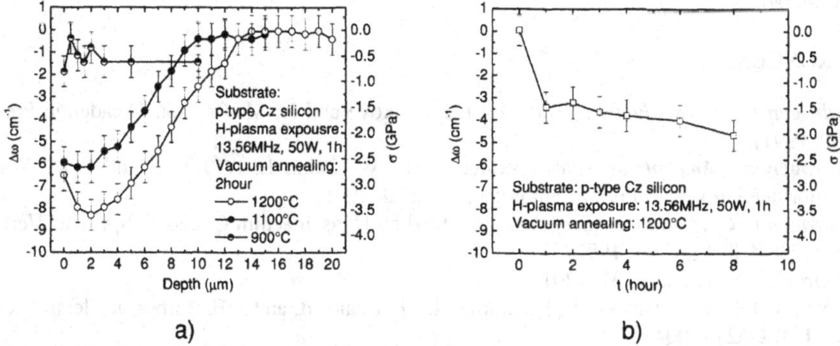

Figure 5. Depth profile of the relative Raman shift ($\Delta\omega$) and tensile stress (σ) in dependence on vacuum annealing temperature; (b) relative Raman shift and tensile stress measured on the surface of the sample in dependence on the annealing duration; the first point in (b) (annealing duration of 0 hours) represents the as-plasma treated sample.

Fig. 5(a) shows the depth profile of the relative Raman shift and tensile stress of the samples with different annealing temperature. The stress of the sample annealed at 900 °C is weak and shows some scattering with the depth. When the annealing temperature increases, the stress also increases and extends into deeper wafer regions (from ~ 10 to ~ 14 μm when the temperature increases from 1100 to 1200 °C). The stress slightly increases with the annealing durations, as shown in Fig. 5(b).

CONCLUSIONS

The above results clearly show that the distribution of the stress in the surface engineered silicon wafer (i.e., H-plasma treated and subsequently vacuum annealed at temperatures higher then 800 °C) can be controlled by the process parameters. For p-type substrates, lower H-plasma frequency, and higher annealing temperatures stronger and more extended tensile stress is found; while the H-plasma power, duration and the vacuum annealing durations have no obvious impact on the stress. The stressed layers might be used as getter centres for unintentional impurities. For such purposes the H-plasma exposure can be carried out through appropriate masks, locating the stressed regions close to active regions of integrated circuits on the wafers front side. This would be especially useful if SOI wafers are used as substrates for the integrated circuits, where the application of getter processes towards deep wafer regions (below the insulating SiO_2 layer) is problematic.

ACKNOWLEDGEMENT

Katrina Meusinger and Boguslaw Wdowiak are acknowledged for the technical support with the PECVD setup. The authors Y. Ma and Y. L. Huang would like to thank the German Academic Exchange Service (DAAD) for the financial support. Finally, the partial financial support of the "Deutsche Forschungsgemeinschaft" (DFG Project No. Jo297/7-1) is gratefully acknowledged.

REFERENCES

1. *Hydrogen in semiconductors*, edited by J. I. Pankove and N. M. Johnson (Academic Press, USA 1991).
2. *Hydrogenated Amorphous Silicon*, edited by R. W. Cahn, E. A. Davis, and I. M. Ward (Cambridge University Press, Cambridge, England, 1991).
3. *Hydrogen in Crystalline semiconductors*, edited by Hans-Joachim Queisser (Springer-Verlag Berlin Heidelberg, USA, 1992).
4. M. Bruel, Electron. Lett. **31,** 1201 (1995).
5. Y. Ma, R. Job, Y. L. Huang, W. R. Fahrner, M. F. Beaufort, and J. F. Barbot, J. Electrochem. Soc. **151,** G627 (2004).
6. R. Job, Y. Ma, and A. G. Ulyashin, MRS Symposium Proceedings Series, Vol. **788**, 571 (2004).
7. R. Job, Y. Ma, Y. L. Huang, and A. G. Ulyashin, XII International Workshop on the Physics of Semiconductor Devices (IWPSD 2003) Delhi, India, p. 100, December 16-20, 2003.
8. R. Job, Y. Ma, Y. L. Huang, and W. Düngen, Proc. *High Purity Silicon VIII*, Eds. C.L. Claeys, M. Watanabe, R. Falster and P. Stallhofer, Electrochem. Soc. Ser. PV 2004-05, 407 (2004).
9. V. Senez, A. Armigliato, I. De Wolf, G. Carnevale, R. Balboni, S. Frabboni, and A. Benedetti, J. Appl. Phys. 94, 5574 (2003).
10. Y. Ma, Y. L. Huang, R. Job, and W. R. Fahrner, Phys. Rev. B (accepted for publication).

Mater. Res. Soc. Symp. Proc. Vol. 843 © 2005 Materials Research Society T4.9

Buckling failure of compressive loaded hard layers in flexible devices

Piet C.P. Bouten and Marcel A.J. van Gils[1]
Philips Research Laboratories, Eindhoven, The Netherlands.
[1]Philips Centre for Industrial Technology, Eindhoven, The Netherlands.

ABSTRACT

Substrate materials for flexible devices are multi-layer composite structures. On top of a base polymer functional inorganic layers such as permeation barriers and conductive layers are applied. Due to the thermal mismatch between polymer and inorganic layer, the layer is compressive loaded at ambient conditions. A characteristic failure mode, occurring with compressive loaded thin layers, is buckling failure. The interface between adjacent layers fails locally, and the thin top layer bends outwards. The buckles have a characteristic width and height. These sizes are used to analyse the compressive strain, present in the layer before failure, and the adhesion quality of the failed interface. A buckle map is introduced to guide this analysis.

INTRODUCTION

Substrate materials for flexible devices are multi-layer composite structures. On top of a base polymer functional layers such as permeation barriers, hard coats and conductive layers are applied. In flexible microelectronic devices and flexible displays, high performance functional layers are needed. Flexible displays require a transparent well-conducting electrode material, such as Indium Tin Oxide (ITO). In OLED (Organic Light Emitting Devices) displays, transparent (inorganic) hermetic barrier coatings are necessary. Thin functional inorganic layers (30-500 nm) are applied on top of a polymer substrate. Due to the thermal mismatch between polymer and layer, the functional inorganic layer is compressive loaded at ambient conditions. Bending of the (flexible) substrate during manufacturing or application can result in an additional compressive load.
A characteristic failure mode, occurring with compressive loaded thin layers, is buckling failure. The interface between adjacent layers fails locally, and the thin top layer bends outwards. It is schematically represented in figure 1. The buckles have a characteristic width L and height w. These sizes might be used in the analysis of the compressive strain ε, present in the layer before failure, and the adhesion quality of the failed interface.
The present paper shows buckling patterns observed in ITO layers on a polymer substrate. Different experimental techniques are used to characterise the patterns.

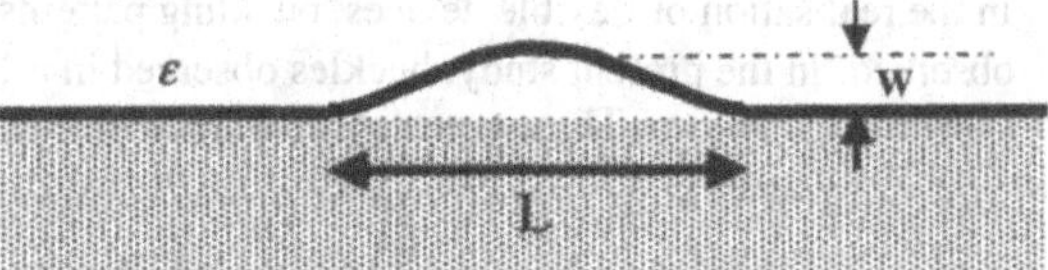

Figure 1. A schematic representation of a buckling layer on a substrate.

THEORY

Basic theory for buckling of thin compressive loaded layers is presented in [1]. Buckling can occur above a critical strain ε_c. For an edge clamped sheet of thickness h and length L it is

$$\varepsilon_c = (\pi^2/3)\,(h / L)^2 \qquad (1).$$

Buckling of an edge clamped sheet leads to an energy release G_{ss} in the buckled part:

$$G_{ss} = (1 - \varepsilon_c / \varepsilon)^2\, G_o \qquad (2),$$

with

$$G_o = \{E/(1-\nu^2)\}\,(h/2)\,\varepsilon^2 \qquad (3)$$

G_o is the elastic energy content of a layer of thickness h, loaded at compressive strain ε. From an expression for the relative buckle height (eq. 6.11 in [1]) a relation between compressive strain ε, prior to buckling, and the maximum buckle height w is obtained:

$$\varepsilon = (\pi^2/12)\,\{(3w^2 + 4\,h^2)/L^2\} \qquad (4).$$

The shape of the edge clamped buckle [2] is described by

$$y = (w/2)\{1+\cos(2\,\pi x\,/L)\} \qquad \{-L/2 \le x \le L/2\} \qquad (5).$$

Upon buckling the strain in the compressed layer decreases. In the theory presented in [1] the position of the clamped edges does not change upon buckling. Then it can be shown that the strain decrease upon buckling, $\Delta\varepsilon$, is related to the maximum buckling height w and width L:

$$\Delta\varepsilon = (\pi^2/4)\,(w/L)^2 \qquad (6).$$

The basic theory, summarised in equations 1 to 4 above, is derived for an edge-clamped buckle; equation (2) predicts $G_{ss}/G_o < 1$. In flexible devices high modulus inorganic layers are frequently applied on a significantly softer substrates. In these situations the basic theory is no longer valid [2, 3]: the ratio $G_{ss}/G_o > 1$ is obtained, since elastic energy from adjacent regions contributes to the delamination process. An important parameter influencing this ratio is the elastic mismatch between the two materials as described in [1] with Dunders' parameter α; $E_{1,\,eff}$ and $E_{2,\,eff}$ are the effective elastic moduli of layer and substrate respectively:

$$\alpha = (E_{1,\,eff} - E_{2,\,eff}) / (E_{1,\,eff} + E_{2,\,eff}) \qquad (7).$$

EXPERIMENTAL

In the realisation of flexible devices, buckling patterns of brittle layers on polymer substrates are observed. In the present study, buckles observed in a 100 nm ITO layer on a 100 μm polymer substrate are shown. Three types of measurements are performed on these samples:

- From the curvature radius of the coated substrate a compressive strain in the ITO layer is estimated using Stoney's formula.
- A sample is heated in a hot stage microscope (rate 10 K/min, T_{max} = 210 °C), and the evolution of the buckling pattern is observed.
- AFM images are made from the buckling patterns. From the cross-section of these buckles, experimental values for height w and width L are obtained.

Finite element model calculations (Marc Mentat) are used to study buckling of a 100 nm ITO layer on a compliant substrate. In a two-dimensional plane-strain model a crack of a given length L is present at the coating-substrate interface. The ITO layer behaves elastically (Young's modulus E = 112 GPa, Poisson's ratio ν = 0.20). For the polymer substrate both elastic and elasto-plastic material behaviour are used. For the material used in the experiments, E = 3.2 GPa and ν = 0.40 are used in the elastic calculations, corresponding to Dunders' elastic mismatch parameter (eq. 7) α = 0.937. The experimental determined stress-strain curve is used in the elasto-plastic model calculations. The effect of the compliance of the substrate is studied in elastic calculations, using Dunders' parameter α = 0.90 and 0.99, corresponding to a substrate modulus E = 5.2 GPa and E = 0.49 GPa (ν = 0.40). Compressive strains ε up to 1.5 % are applied in the model calculations. Crack sizes L $\leq$ 8 μm are used at the coating-substrate interface. A J-integral approach is used to analyse the energy release in the system upon buckling.

RESULTS

From radius of curvature measurements on ITO-coated polymer substrates a compressive strain of about 0.6 % in the ITO layer is estimated. The experimental data show a significant scatter. Figure 2 shows a buckling pattern of an ITO layer on a polymer substrate. The grey edges at the image are polymer walls, applied on top of the ITO layer to define the separate pixels (300 μm) in a flexible display. The buckling pattern is observed in a hot stage microscope. Upon heating the compressive strain in the ITO layer decreases. The buckling pattern disappears gradually at a temperature increase of 120-160 K. The mismatch in thermal expansion coefficients between the presently used polymer substrate and the ITO layer is 60 10^{-6} K^{-1}. It is assumed that the compressive strain in the layer, which drives buckling, disappears completely at a characteristic temperature increase of 140 K. This gives (compressive) strain ε = -0.84 % at room temperature. In the subsequent cooling step the pattern reappears partly. A closer observation of the images has shown a significant hysteresis; buckle formation upon cooling requires a new nucleation step.

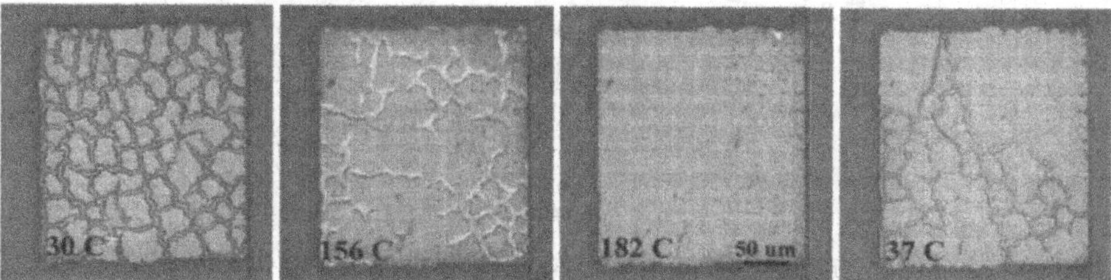

Figure 2. The behaviour of the buckling ITO layer upon heating and cooling.

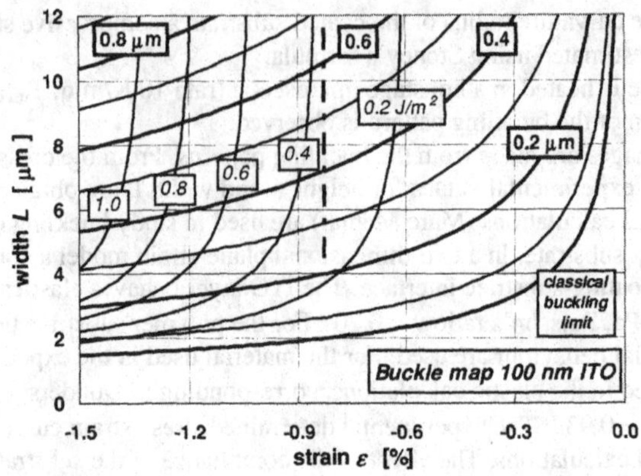

Figure 3. Buckling map for a 100 nm ITO layer on a rigid substrate.

From AFM images, cross sectional profiles of the buckles are obtained. These profiles are well described with the shape of the buckle, given in equation 5. Characteristic values for width L and height w are 7.5 μm and 0.65 μm respectively. From these values, a length difference Δε of 1.85 % is determined for trajectories along the buckle and along the base line (eq. 6). Different experimental techniques give quite different estimates for the compressive strain, responsible for the buckling process. In order to account for these differences, finite element method calculations are preformed on an ITO layer on compliant substrates. Before presenting these results, model predictions for a 100 nm ITO layer on a rigid substrate are given in figure 3. This figure represents a "buckling map": contour lines for constant buckling energy ($0.2 \leq G_{ss} \leq 1.0$ J/m^2) and constant buckling height ($0.1 \leq w \leq 0.9$ μm) are shown as a function of the compressive strain ε and the buckling width L. These contour lines are obtained from equations (2) and (4) respectively.

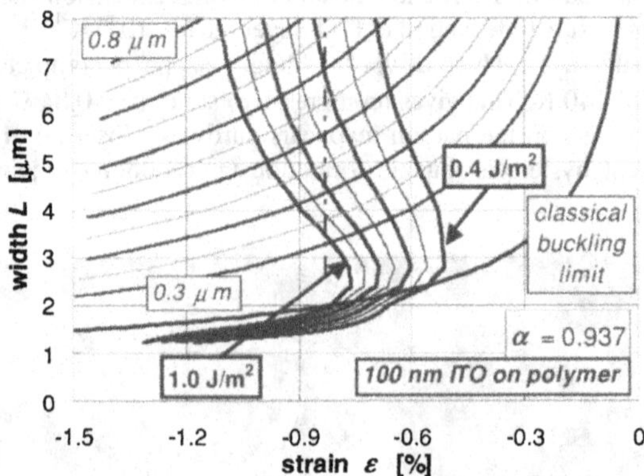

Figure 4. Buckling map for a 100 nm ITO layer on a polymer substrate (α = 0.937).

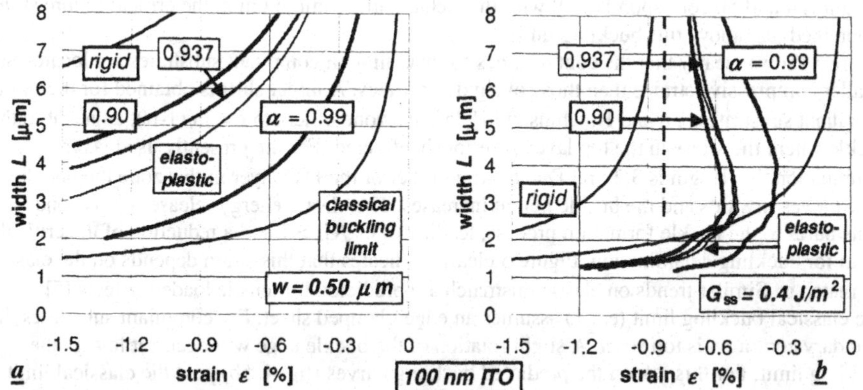

Figure 5. The effect of elastic mismatch on the buckling map for 100 nm ITO. (a) At constant buckle height (w = 0.50 μm). (b) At constant adhesion energy (G_{ss} = 0.4 J/m^2).

The lowest curve included in the graph represents the classical buckling limit ε_c (eq. 1). This latter curve bound both the contour lines for buckling energy G_{ss} and buckle height w. The vertical dashed line at a compressive strain 0.83 % represents G_o = 0.40 J/m^2 for the 100 nm ITO layer. It is the asymptotical limit for G_{ss} = 0.40 J/m^2 curve at large buckling width L.
Figure 4 represents the buckling map obtained from a set of finite element method calculations on a 100 nm ITO layer on an elastic polymer substrate. In this case Dunders' elastic mismatch parameter (eq. 7) α = 0.937. Contour lines for constant buckling height ($0.3 \leq w \leq 0.8$ μm) and constant buckling energy ($0.4 \leq G_{ss} \leq 1.0$ J/m^2) are presented as a function of the compressive strain ε and the buckling width L. The curve for the classical buckling limit ε_c and the dashed line for G_o = 0.40 J/m^2 at a compressive strain of 0.83 % are also included in this figure.
In figure 5 the effect of the elastic mismatch on the contour lines for constant buckling height (w = 0.5 μm) and constant adhesion energy (G_{ss} = 0.4 J/m^2) is shown. Curves are included for the rigid substrate model (figure 3), elastic finite element calculations at α = 0.90, 0.937 and 0.99, and a curve for the elasto-plastic substrate model. These latter curves should be compared with the α = 0.937 curves.

DISCUSSION

From hot stage microscopy and the measured buckle height quite different estimates are obtained for the residual strain in the thin ITO layer on the polymer substrate (0.84 % and 1.85 %). The latter value is also obtained from the analytical model for rigid substrates, represented in the buckling map of figure 3: a residual compressive strain before buckling, ε = 1.91 % and an adhesion energy G_{ss} = 2.0 J/m^2 are derived from the buckling height w (0.65 μm) and length L (7.5 μm). The difference between Δε = 1.85 % and ε = 1.91 % is strain ε_c = 0.06 % (eq. 1).
In recent publications [3, 4] attention is paid on the shape of the G_{ss} curves for a buckling layer on compliant substrates. Information on the buckling height is not included in these papers. The ratio G_{ss}/G_o shows a distinct maximum just above the buckling limit. The shape of the G_{ss} curves

in figures 4 and 5b corresponds well with this behaviour. A minimum in the critical failure strain is obtained just above this buckling limit.
Figure 5 shows clearly that the contour lines for buckling on compliant substrates are located at smaller compressive strains than those of rigid substrates. A larger shift is obtained for the more compliant substrates. Yu and Hutchinson [4] define a zone of length $l = 2h/(1-\alpha)$, adjacent to the buckle where the stress in the top layer is seriously affected. For the presently used layer combination this length is 3.2 μm. Due to strain relief of the ITO layer in this zone the buckle edge moves inwards and the buckle height increases. The elastic energy release in this zone contributes to the buckle formation process, leading to $G_{ss}/G_o > 1$ and a reduction of the critical strain for buckling delamination. Figure 5 clearly indicates that this strain depends on the elastic mismatch α. Similar trends on elastic mismatch are predicted for tensile loaded defects [5].
The classical buckling limit (eq. 1) assumes an edge-clamped sheet. For compliant substrates this boundary condition is too strict. A slight rotation of the buckle edge will occur; it lowers the buckling limit. For this reason the predicted buckling curves (fig. 4, 5b) pass the classical limit.
From AFM profile measurements on he ITO layer, $L = 7.5 \pm 1.0$ μm and $w = 0.65 \pm 0.08$ μm are reported. Using the buckle map for the elastic substrate (fig. 4), $\varepsilon = -0.85$ % and $G_{ss} = 0.6$ J/m^2 are derived. This compressive strain, obtained from the AFM profile measurements, compares well with the value obtained from hot stage microscopy (0.84 %). Due to the poor adhesion, a low value for G_{ss} is obtained in the present case.

CONCLUSION

Buckling is an important failure mode for compressive loaded layers; the interface fails locally. The height and the width of buckles can be used to determine the residual strain and the adhesion quality of the layer. In a buckle map the relation between these quantities is given. Buckle maps for layers on rigid and compliant substrates are compared.

ACKNOWLEDGEMENT

The authors acknowledge Hans Buijk, Ferrania, Unaxis Displays and Vitex for providing samples, Harry Nulens, Kees Mutsaers, Giovanni Nisato, Jaap den Toonder and Peter Slikkerveer for experimental support and discussions during model development, and the European Commission for financial support (FLEXled project, grant IST-2001-34215).

REFERENCES

1 J.W. Hutchinson and Z. Suo, *Advances in Applied Mechanics* **29** 63 (1992).
2 W. Fang and J. A. Wickert, *J. Micromech. Microeng.* **4** 116 (1994).
3 B. Cotterell and Z. Chen, *Int. J. Fracture* **104** 169 (2000).
4 H.H. Yu and J.W. Hutchinson, *Int. J. Fracture* **113** 39 (2002).
5 J.M. Ambrico and M.R. Begley, *Thin Solid Films* **419** 144-153 (2002).

Mater. Res. Soc. Symp. Proc. Vol. 843 © 2005 Materials Research Society T3.10

First-Principles Study of Mechanical Properties of Alumina-Copper Nano-Coating Interfaces

Shingo Tanaka[1], Rui Yang[2] and Masanori Kohyama[1]
[1] Research Institute for Ubiquitous Energy Devices, National Institute of Advanced Industrial Science and Technology (AIST), Ikeda, Osaka 563-8577, Japan.
[2] Department of Computer Science, Faculty of Engineering and Information Technology Australian National University, Canberra, ACT 0200, Australia.

ABSTRACT

Mechanical properties of alumina-copper nano-coating interfaces have been studied by the first-principles calculations. We have applied the *rigid-type* first-principles tensile tests for O-rich (O-terminated) and stoichiometric (Al-terminated) interfaces and for back Cu^{1st}-Cu^{2nd} interlayers. The O-terminated interface is twice as strong as the Cu^{1st}-Cu^{2nd} interlayer whereas the Al-terminated interface is twice as weak as the Cu^{1st}-Cu^{2nd} interlayer. We have performed the fitting of interlayer potentials by universal binding-energy relation (UBER). The interlayer potential of the Cu-Al interface is well reproduced by UBER in whole region, although those of Cu-O interface and Cu^{1st}-Cu^{2nd} interlayer are partially reproduced.

INTRODUCTION

Mechanical properties of nano-coating interfaces as ceramics/metal interfaces [1-9] are of great importance. In order to develop the high performance materials, it is desirable to understand the mechanical properties of such interfaces at the atomic and electronic level. Alumina-copper nano-coating interfaces are typical ceramics-metal systems frequently used in mechanical and electronic applications. Recently, we have performed the first-principles calculations of the O-rich (O-terminated) and stoichiometric (Al-terminated) interfaces using a pseudopotential method and the first-principles molecular dynamics [10,11]. The stable atomic configuration of the O-terminated interface is in good agreement with the recent HRTEM observation [10,12]. The O-terminated interface has a strong covalent and ionic bonding, whereas the Al-terminated interface has a weak metallic bonding with partially covalent character. The O-terminated interface is about five times as large as the Cu adhesive energy of the Al-terminated one [6,9-11]. These interfaces have quite different characters to each other.

The first-principles tensile test (FPTT) is a powerful method for an investigation of the interface mechanical and electronic properties in atomic scale. In a *rigid-type* FPTT, we can select cleavage planes for the relaxed configuration, and the changes in the energy and electronic structure are examined for the rigid cleavage between selected planes. The interlayer potential can be directly obtained from *rigid* FPTT. In this paper, we estimate an ideal tensile strength near the interface and an interface Young's modulus of alumina-copper system from interlayer potential curves. Universal binding-energy relation (UBER) analyses [13,14] have succeeded in explaining the atomic potential curves for simple covalent or metallic bonding nature. We perform the UBER fitting of interlayer potentials obtained from FPTT and examine the applicability of the present potentials. Also we develop the interatomic potentials.

THEORETICAL METHODS

Present calculations were performed by the first-principles pseudopotential method and the first-principles molecular dynamics. Soft-type norm-conserved pseudopotentials developed by Troullier-Martins [15] with Kleinman-Bylander separable form [16] were used. The residual minimization and direct inversion in the iterative subspace (RMM-DIIS) [17,18] method and the conjugate-gradient (CG) method [19] were used for the fast solution technique for the eigenstates in the ground states. Self-consistent charge densities were obtained from the efficient charge-mixing method [20,21]. A plane-wave cutoff energy of 70 Ry was selected based on the tests of total energy convergence for bulk Al_2O_3 and for bulk Cu. Four and 16 sampling k-points in the irreducible Brillouine zone of the supercell for self-consistent calculations and for band-structure calculations were used. Stable atomic configurations were obtained through the relaxation steps according to Hellmann-Feynman forces. The force convergence was approximately 0.01 eV/nm for each atom. We performed efficient parallel computations with respect to each band through our efficient program code [22].

The O-terminated and Al-terminated interfaces were treated as our previous works [10,11]. Figure 1 shows constructed supercells. The O-terminated interface was constructed by removing top Al layers of both sides of a stoichiometric $Al_2O_3(0001)$ slab, so that less two atoms than the Al-terminated one. We dealt with three high-symmetry models of Cu^{1st} adhesive site on $Al_2O_3(0001)$ surface: (i) atop of Al atoms (Al-site), (ii) atop of outermost O atoms (O-site), and (c) atop of second-layer O atoms (H-site). The atomic relaxation was performed under preserving the unit-cell symmetry, namely the three-fold rotation along the principal rotation axis and the inversion (C_{3i}). Because of the symmetry condition, Cu^{1st} atoms at the O-site or H-site have degree of freedom for in-plane displacement away from the strict adhesive site whereas the displacement of Cu atoms at the Al-site is fixed to be normal to the interface. In the preliminary calculations [10], the O-site and H-site of the O-terminated interface and the O-site of the Al-terminated one have been energetically stable, hence we only discuss these models hereafter.

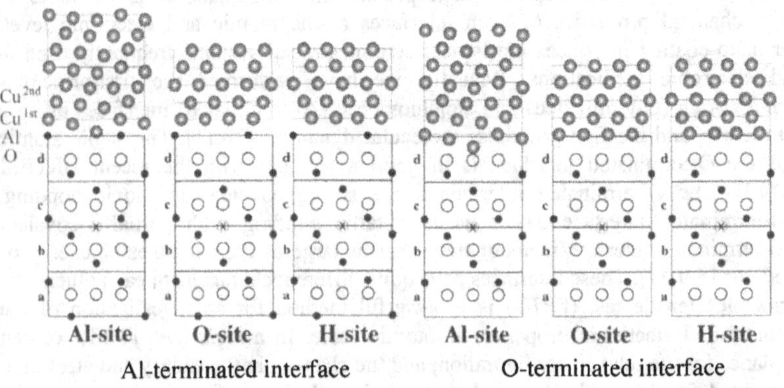

Figure 1. Calculated supercells of Al-site, O-site and H-site models for Al-terminated and O-terminated interfaces. The mark at the center of the Al_2O_3 slab indicates the inversion center of C_{3i} symmetry.

RESULTS AND DISCUSSIONS

In order to examine the mechanical properties, we have performed the *rigid-type* FPTT. The *rigid* FPTT is different from the usual *relaxed-type* one with full atomic relaxation. The reason why we selected the *rigid* FPTT is mainly two factors: i) be able to clarify the local strength of each selected interlayer or interface, ii) be able to provide rich information for the development of interlayer and interatomic potentials as discussed later. On the other hand, the *relaxed* FPTT can only deal with the weakest area with respect to failure. We have also performed the *relaxed* FPTT for the Al-terminated interface in our recent paper [23].

We have applied the *rigid* FPTT to the Cu-O interface, the Cu-Al interface and the Cu^{1st}-Cu^{2nd} layers of the both interfaces. Figure 2 shows snap shots of charge-density distributions in FPTTs. In figure 2 (a) and (b), each Cu-O interface has a large hybridization of the charge density at R = 0. This hybridization gradually decreases by increment of R, and finally disappears. This means that the Cu-O bonding break down electronically at this stage. On the other hand, the Cu-Al interface in figure 2 (c) does not have a large hybridization not only between Cu and Al but also between Cu and outermost O. However, there exists some charge-rich region above the interface Al atom and surrounded by the Cu atoms and some charge depletion region near the O and Cu atoms. With the increment of R, the depletion region extends to the region between Cu and O atoms. This region grows up into a void and finally the interface separates into two surfaces.

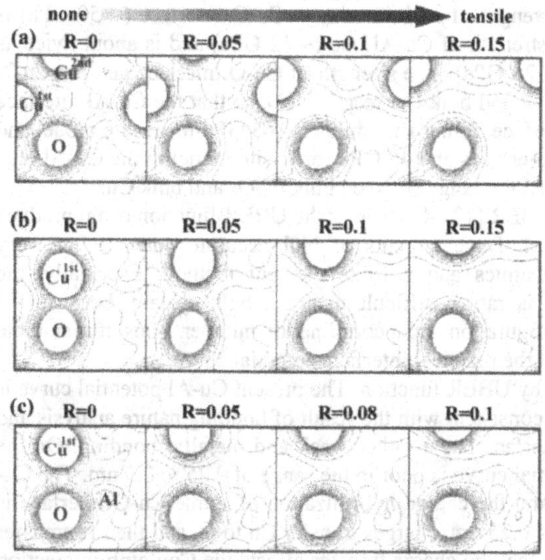

Figure 2. Charge-density distributions projected to the (11-20) plane normal to the interface for (a) H-site model and (b) O-site model of the Cu-O interfaces, and (c) O-site model of the Cu-Al interface through the *rigid-type* first-principles tensile tests. R denotes an increment of the interface distance from the stable atomic structure.

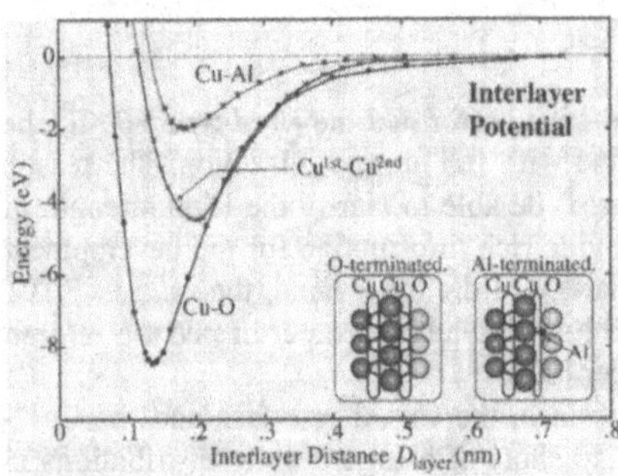

Figure 3. Calculated interlayer potentials for Cu-O, Cu-Al interfaces and Cu^{1st}-Cu^{2nd} layer.

Figure 3 shows interlayer potentials for Cu-O, Cu-Al interfaces and Cu^{1st}-Cu^{2nd} layer. The interlayer potential represents the total-energy change for stretching and compression of each selected interlayer normalized by the sum of the total energies of two surface slabs without relaxation, Al_2O_3 and Cu surface slabs in Cu-O and Cu-Al tensile tests and Al_2O_3 with Cu monolayer and Cu surface slabs in Cu^{1st}-Cu^{2nd} tensile test. Thus, the depth of each curve corresponds to the ideal fracture energy or the work of separation. The curve for the Cu^{1st}-Cu^{2nd} interlayer of the O-terminated interface is similar to that of the Al-terminated interface, which means that the interface stoichiometry effects are not strong. The maximum gradient of each curve corresponds to the ideal tensile strength of each interlayer. Cu-O interface (~ 50 GPa) is about four times as large as the tensile strength of Cu-Al one (~ 12 GPa) and is about twice as large as that of the Cu^{1st}-Cu^{2nd} layers (~ 25 GPa). The strength of Cu-O interface and the Cu^{1st}-Cu^{2nd} layers is similar to that of bulk Al_2O_3 and bulk Cu, respectively, so that the Cu-Al interface is extremely weak among them. The interface Young's modulus is 145 GPa for H-site model and 225 GPa for O-site model in the Cu-O interfaces and 45 GPa for O-site model in the Cu-Al one. These values of Cu-O interface are located the range between bulk Al_2O_3 and bulk Cu.

We have performed the analyses of UBER [13,14] whether the UBER function can reproduce well the present potential curves or not. Developments of high-accurate potential are very important to a classical molecular dynamics and a finite element method. Especially, the potential development of the interface is rather difficult than the bulk system, because the interface has variations as an atomic configuration and a coordination number. Thus, it is suitable to that the analytical function can well fit the practical interface potential curve.

Figure 4 (a) shows the fitting curves by UBER function. The present Cu-Al potential curve is well reproduced in whole range. This is consistent with the result of bonding nature analysis and ideal tensile strength that the Cu-Al interface has a rather weak and metallic bonding. On the other hand, the fitting of the Cu-O potential curve is poor in the range of 0.25 to 0.5 nm. The Cu-O bonding has an ionic character [10] and the charge hybridization of stable Cu-O interface is large as shown in figure 2. In the $D_{layer} < 0.25$ nm, charges transfer Cu to O, and then re-transfer O to Cu by increment of D_{layer}. It seems that the charge transfer affects the Coulomb interaction between Cu and O. Thus, there exists a possibility that the UBER cannot be enough to take into account such effects. The Cu^{1st}-Cu^{2nd} curve can be fitting in the stretched area, although poor fitting in the compressive area.

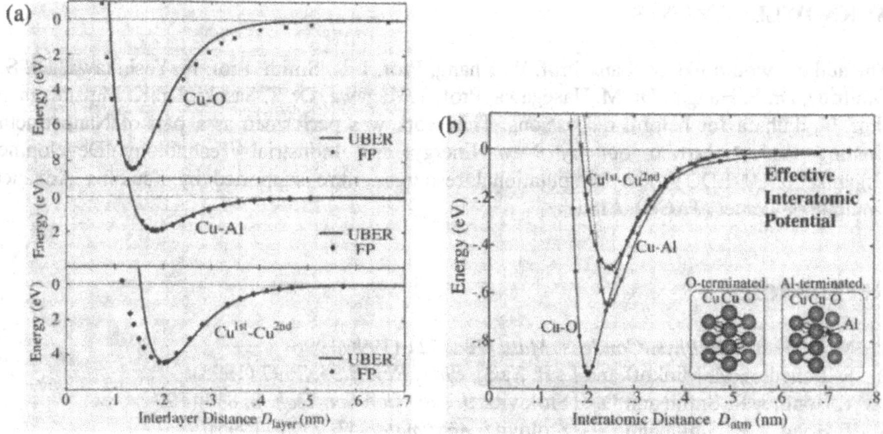

Figure 4. (a) Universal binding-energy relation (UBER) fitting curves with the first-principles interlayer potentials. FP indicates the result of the first-principles calculations. (b) effective interatomic potentials for Cu-O and Cu-Al interfaces and Cu^{1st}-Cu^{2nd} interlayer.

Dmitriev and co-workers have also developed the atomic potentials using the well-defined classical function [24,25]. The authors carried out the fitting procedure of the Cu-O interface and Cu^{1st}-Cu^{2nd} interlayer of the Al_2O_3/Cu interface. The fitting of the Cu-O potential [25] is the same trend as the present UBER fitting. One would expect that the fitting of Cu-O potential by analytical function is difficult because of complicated interactions of the Cu-O interface. On the other hand, the Cu^{1st}-Cu^{2nd} interlayer potential was not reproduced in whole range [25]. The reason why the fitting of Cu^{1st}-Cu^{2nd} potential was poor is unknown.

We construct effective interatomic potentials from interlayer potentials. Although the interatomic distances exist two and more kinds in the present models, we deal with an average interatomic distance for simplicity. Thus, the present interatomic potential corresponds to two-body pair potential distributed by the coordination number of referenced atoms. Figure 4 (b) shows the effective interatomic potentials. We now examine the transferability of such simple potentials [25] and can extend the three and more body potential. This process is important to develop the high-accuracy atomic potential using the multi-scale calculations.

CONCLUSIONS

We performed the first-principles calculations of the mechanical properties of alumina-copper nano-coating interfaces. The O-terminated interface has a strong bonding comparable to the bulk Al_2O_3 whereas the Al-terminated one has a weak bonding. The failure process will start at Cu-Al interface if both terminated interfaces exist in the system. UBER fitting can reproduce well the Cu-Al interlayer potential and shows some difficulty for Cu-O and Cu^{1st}-Cu^{2nd} interlayer potentials, which means clear need to the practical calculation as the first-principles one. We show the simple effective interatomic potentials.

ACKNOWLEDGMENTS

The authors would like to thank Prof. W. Zhang, Prof. J. R. Smith, Prof. N. Yoshikawa, Dr. S.V. Dmitriev, Dr. Y. Hangai, Dr. M. Hasegawa, Prof. Y. Kagawa, Dr. T. Sasaki, Dr. K. Matsunaga and Prof. Y. Ikuhara for helpful discussions. This work was performed as a part of Nanostructure Coating Project carried out by New Energy and Industrial Technology Development Organization (NEDO). The computational resources were supported by Tsukuba Advanced Computing Center (TACC), AIST.

REFERENCES

1. M. W. Finnis, *J. Phys. Condens. Matter.* **8**, 2721 (1996).
2. R. Benedek, M. Minkoff and L. H. Yang, *Phys. Rev. B* **54**, 7697 (1995).
3. T. Hong, J. R. Smith and D. J. Srolovitz, *J. Adhesion Sci. Tech.* **8**, 837 (1994).
4. T. Hong, J. R. Smith and D. J. Srolovitz, *Acta mater.* **43**, 2721 (1995).
5. W. Zhang and J. R. Smith, *Phys. Rev. Lett.* **85**, 3225 (2000).
6. W. Zhang, J. R. Smith and A. G. Evans, *Acta Mater.* **50**, 3803 (2002).
7. G. L. Zhao, J. R. Smith, J. Raynolds and D. J. Srolovitz, *Interface Sci.* **3**, 289 (1996).
8. I. G. Batyrev, A. Alavi, M. W. Finnis and T. Deutsch, *Phys. Rev. Lett.* **82**, 1510 (1999).
9. I. G. Batyrev and L. Kleinmann, *Phys. Rev. B* **64**, 033410 (2001).
10. S. Tanaka, R. Yang, M. Kohyama, T. Sasaki, K. Matsunaga and Y. Ikuhara, Mater. Trans. **45**, 1973 (2004).
11. R. Yang, S. Tanaka and M. Kohyama, *Phil. Mag. Lett.* **84**, 425 (2004).
12. T. Sasaki, K. Matsunaga, H. Ohta, H. Hosono, T. Yamamoto and Y. Ikuhara, *Sci. Tech. Adv. Mater.* **4**, 575 (2003).
13. J. H. Rose, J. R. Smith and J. Ferrante, *Phys. Rev. B* **28**, 1835 (1983).
14. A. Banerjea and J. R. Smith, *Phys. Rev. B* **37**, 6632 (1988).
15. N. Troullier and J. L. Martins, *Phys. Rev. B* **43**, 1993 (1991).
16. L. Kleinman and D. M. Bylander, *Phys. Rev. Lett.* **48**, 1425 (1982).
17. G. Kresse and J. Furthmüller, *Phys. Rev. B,* **43**, 1993 (1991).
18. P. Pulay, *Chem. Phys. Lett.,* **73**, 393 (1980).
19. D. M. Bylander, L. Kleinman and S. Lee, *Phys. Rev. B* **42**, 4021 (1990).
20. G. P. Kerker, *Phys. Rev. B,* **23**, 3082 (1981).
21. P. Pulay, J. Comp. Chem. **3**, 556 (1982).
22. T. Tamura, G. –H. Lu, R. Yamamoto, M. Kohyama, S. Tanaka and Y. Tateizumi, *Modell. Simul. Mater. Sci. Eng.* **12**, 945 (2004).
23. R. Yang, S. Tanaka and M. Kohyama, *Phys. Rev. B*, submitted.
24. S. V. Dmitriev, N. Yoshikawa, Y. Kagawa and M. Kohyama, *Surf. Sci.* **542**, 45 (2003).
25. S. V. Dmitriev, N. Yoshikawa, M. Kohyama, S. Tanaka, R. Yang and K. Kagawa, *Acta Mater.* **52**, 1959 (2004).

Mater. Res. Soc. Symp. Proc. Vol. 843 © 2005 Materials Research Society T3.11

Double Interface Coatings on Silicon Carbide Fibers

Jun C. Nable[1], Shaneela Nosheen[1], Steven L. Suib[1,2*], Francis S. Galasso[1], Michael A. Kmetz[3]
[1] University of Connecticut, Department of Chemistry, Storrs, CT 06269
[2] University of Connecticut, Institute of Materials Science, CT 06269
[3] Pratt & Whitney, East Hartford, CT

* Corresponding author

Abstract

Interface coatings on fibers are important in ceramic matrix composites. In addition to providing toughness, the interface coating must also protect the reinforcing ceramic fibers from corrosive degradation. A double interface coating has been applied onto silicon carbide fibers. The double interface coating is comprised of a combination of nitride and oxide coatings. Among the nitrides, boron nitride and titanium nitride were utilized. These nitrides were deposited by CVD. The metal oxides of choice were aluminum oxide and zirconium oxide which were applied onto the nitride coatings by MOCVD. The phases on the coated fibers were determined by XRD. The surface coating microstructures were observed by SEM. The effect of the coatings on the tensile strengths was determined by Instron tensile strength measurements.

Introduction

Ceramic matrix composites (CMCs) are widely utilized as high temperature application components. Ceramics are chosen due to their resistance to corrosion and to chemical deterioration at elevated temperatures. Common applications of CMCs include automobile and aircraft parts and components [1].

Ceramic fiber reinforcements such as SiC are typically embedded in a ceramic matrix by high temperature processes. These composites may be fabricated by chemical infiltration, hot pressing, and liquid phase sintering. However, a major issue encountered during this high temperature fabrication step, as well as during high temperature operation is the exposure of the reinforcing fiber material to corrosive conditions and material interdiffusion resulting in a decrease in the overall mechanical strength of the composite eventually leading to failure of the entire component. Such degradation can be minimized by providing an interfacial barrier coating between the fiber reinforcement and matrix material to prevent interdiffusion of materials and exposure to oxidative damage.

Interface coatings such as BN and C are ideal materials to provide toughness in ceramic matrix composites however they are prone to oxidation in the presence of air at high temperatures [2,3]. Alternative materials include metal oxides, aluminosilicates, monazites and xenotimes [4-6]. Common coating techniques to deposit an interface layer onto ceramic fibers include sol-gel and physical vapor deposition. The sol-gel technique is a wet chemical approach which subsequently requires a high temperature post treatment to densify the coating. Chemical vapor deposition (CVD) is another coating technique that can provide a dense, compact coating in a single deposition step. This method also offers the advantage of not being a line-of-sight deposition method.

Our present work focuses on the utilization of a double interface coating by combining nitride and oxide layers deposited onto SiC fibers. Boron nitride and titanium nitride were deposited by CVD of their respective metal chloride precursor. These nitride coatings have been used as diffusion barrier materials for ceramics. Aluminum oxide and zirconium oxide were deposited by metal-organic chemical vapor deposition (MOCVD) utilizing the metal acetylacetonate precursors at 500 and 450 °C respectively. These oxides should provide oxidation barrier protection for the nitride layer.

Experimental

The samples for the CVD double interface coatings are summarized in Table 1. The nitride coatings were deposited in single zone furnaces. Boron trichloride was reacted with NH_3 gas at 900 °C under reduced pressure conditions (0.050 Torr) conditions for 2 h. Titanium nitride was deposited at atmospheric conditions by bubbling N_2 gas into a container of $TiCl_4$ and this vapor mixture was combined with H_2 at 700 °C for 1 h.

The metal oxide coatings were deposited by MOCVD under atmospheric conditions using a two zone furnace system set-up. Aluminum oxide was coated at 500 °C and ZrO_2 at 450 °C. Their respective solid metal acetylacetonate precursors were vaporized and carried by Ar gas into the reactor furnace together with O_2 gas. The metal oxide deposition was complete after 1 h.

A Scintag X-ray diffractometer equipped with a CuK α radiation source was utilized to determine the phase and structure of the deposited coating material. The interface coating microstructure and morphology were investigated with an Amray model 1810 scanning electron microscope equipped with an EDAX detector. The tensile strengths of the coated SiC fibers were tested with an Instron tensile tester model 1011 using a 1000 lb load cell.

Results

The SiC fibers (ceramic grade Nicalon™) were coated with either BN or TiN using their respective metal chloride precursor. The BN coating obtained after 2 h deposition was multi colored. The X-ray diffraction of the BN coating revealed a non-crystalline phase for this coating but elemental analysis shows the presence of the BN coating. The microstructure of this coating, shown in Fig. 1a, appears to be smooth and uniform with a thickness in the range of half a micrometer or less. In addition there is no fiber bridging between the coated fibers as indicated in Fig. 1b. The SiC fibers coated with TiN are completely covered with a golden brown material. The X-ray diffraction pattern in Fig. 2a shows that the TiN coating is crystalline osbornite. No other phases were found except for the SiC fiber substrate. A micrograph of the TiN coating is given in Fig. 2b. The TiN coating is not as smooth as the BN layer. The coating appeared to have a bumpy and rough texture. The thickness is about 1 μm.

After the deposition of the nitride, the coated fibers were additionally coated with either Al_2O_3 or ZrO_2 by MOCVD using metal acetylacetonate precursors. Both metal oxides gave multi-color appearance coatings. X-ray diffraction of the metal oxide coated substrates did not show any crystalline phase formations. Micrographs of the Al_2O_3 coating on SiC-B and SiC-T are shown in Fig. 3. Both metal oxide coatings are relatively smooth and uniform with thicknesses of half a micrometer.

Table 1 lists the relative tensile strength of the coated fibers. After deposition of the BN interface coating, the fiber strength decreases by almost 20%. The subsequent Al_2O_3 coating did

Table 1. Summary of double interface coatings and their tensile strength.

Sample	Nitride coating	Oxide coating	Strength in GPa (ksi)*#
SiC	None	None	*2.98±0.198 (432±28.7)*
SiC-B	BN	None	*2.46±0.195 (357±28.3)*
SiC-B-A	BN	Al_2O_3	*2.38±0.147 (345±21.4)*
SiC-B-Z	BN	ZrO_2	*2.63±0.135 (381±19.7)*
SiC-T	TiN	None	*1.65±0.197 (239±28.6)*
SiC-T-A	TiN	Al_2O_3	*1.48±0.120 (214±17.4)*
SiC-T-Z	TiN	ZrO_2	*1.49±0.254 (216±36.8)*

* The cross sectional diameter of the fibers was assumed to be 14 μm in all cases and each tow contains 500 individual filaments.
# Tensile testing done at least 6 trials

not drastically decrease the fiber strength further as sample SiC-B-A had a similar decrease of 20% in tensile strength. Sample SiC-B-Z which had a tensile strength of 2.63 GPa was slightly better. However, a large drop in fiber strength is observed for TiN coated SiC fibers. Sample SiC-T decreases the fiber strength by 45% from 2.98 to 1.65 GPa. The metal oxides coated on the TiN layer seemed to have little effect on the final tensile strength as the values decreased slightly but remained close to those of sample SiC-T.

Discussion

Silicon carbide ceramic fibers were coated with a duplex interface layer of a nitride layer and a metal oxide layer. Boron nitride had been used extensively as an interface between reinforcing fibers and the matrix material in CMCs [7,8]. Despite the stability of BN at elevated temperatures, this material is prone to pesting in the presence of moisture and oxygen. So a high temperature metal oxide such as Al_2O_3 or ZrO_2 was coated on top of this nitride coating. The BN layer obtained did not give any diffraction pattern signifying the presence of an amorphous phase. The tensile strength of SiC-B decreased by 20% from the reference value of 2.98 to 2.46 GPa. The slight decrease in strength could be due to degradation during the deposition process. Temperature has been known to impact the structure of the bulk fiber material and result in the loss of mechanical properties [9]. Despite this decrease, the coated fibers still retained significant strength to be used as reinforcements in composite materials. Aluminum oxide or zirconium oxide coating was subsequently deposited on the BN layer at deposition temperatures of 500 or 450 °C respectively depending on the oxide of choice. Both metal oxides were also amorphous as no diffraction patterns were observed. The non-crystalline phase was expected because of the low process temperature for MOCVD of these metal oxides. Typical crystalline forms of Al_2O_3 and ZrO_2 are obtained at deposition temperatures of around 1000 °C. The surface microstructures of the metal oxide coatings are also smooth and uniform similar to that of the BN coating.

Titanium nitride is another fiber coating of interest as an interface for CMCs. This nitride coating is substituted for the BN layer. The TiN interface obtained at 700 °C was the crystalline osbornite phase. The microstructure of TiN at this temperature has a grain morphology. This type of surface morphology can be converted into facetted crystals by increasing the process temperature [10]. Significant decrease of 45% in tensile strength was obtained for sample SiC-T.

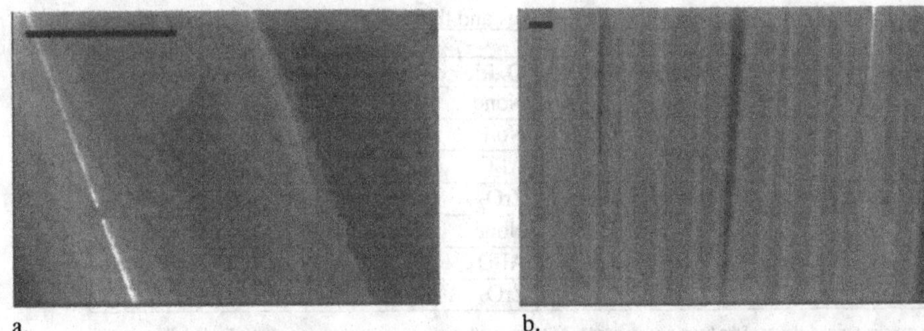

a. b.

Figure 1. Micrographs of BN coated on SiC (bar denotes 10 μm)

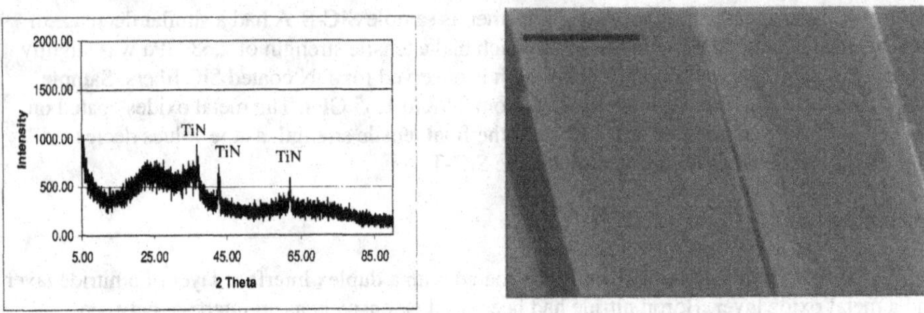

a. b.

Figure 2. (a) X-ray diffraction pattern of TiN and (b) micrograph of TiN coated on SiC (bar denotes 10 μm)

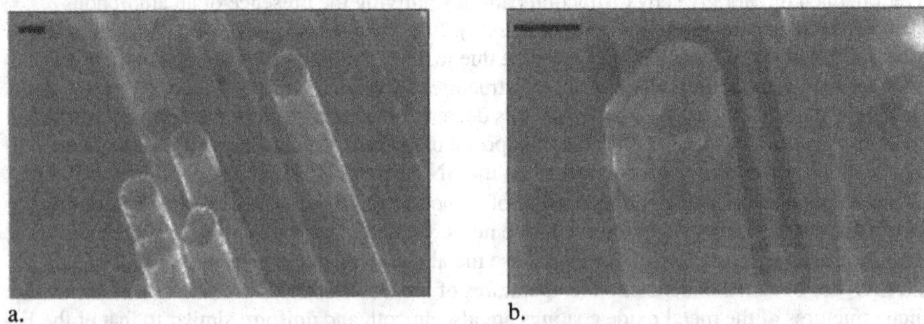

a. b.

Figure 3. Micrographs of (a) SiC-B-A and (b) SiC-T-A (bar denotes 10 μm)

A crucial degradation most likely had occurred during the deposition process. Titanium nitride was coated onto the SiC fibers at atmospheric conditions which could have substantial amount of oxygen or moisture present in the reactor chamber. The presence of oxygen or water vapor during the coating step can cause damage to the bulk fiber material, changing the microstructure leading to changes and loss of mechanical strength and properties.

The metal oxide layer on top of the nitride layer had minimal impact on the overall coated fiber tensile strength as the values remain similar to those of nitride coated SiC fibers. It may be that the presence of the nitride layer already deposited provided a barrier coating against the deposition process environment used for the metal oxide layer. The microstructures for the metal oxides were smooth and uniform without the observed roughness of the TiN layer.

Conclusion

Both BN and TiN coatings on SiC reinforcing fibers were successfully deposited by CVD. The BN coating obtained was amorphous and TiN was crystalline. The nitride coated fibers decreased the tensile strength of the SiC fibers by 20% for the BN coated and 45% for TiN coated fibers. The Al_2O_3 and ZrO_2 coated onto the nitride layer were both amorphous and did not significantly further affect the tensile strength of the nitride coated ceramic fibers.

References

1. *Ceramic Matrix Composites*, edited by R. Warren, (Chapman & Hall, 1992) pp. 11-21
2. L. Shen, B.J. Tan, S.L. Suib, F.S. Galasso, *Mater. Res. Soc. Symp. Proc.* 227-232 (1992)
3. M.A.Kmetz, J.M. Laliberte, W.S. Willis, S.L. Suib, F.S. Galasso, *Ceram. Eng. Sci. Proc.* **12**, 2161-2174 (1991)
4. R.J. Kerans, R.S. Hay, T.A. Parthasarathy, *Curr. Opin. Solid State Mater. Sci.* **4**, 445-451 (1999)
5. I.A.H. Al-Dawery, E.G. Butler, *Composites Part A* **32**, 1007-1012 (2001)
6. S. Varadarajan, A.K. Patanaik, V.K. Sarin, *Surf. Coat. Technol.* **139**, 153-160 (2001)
7. L.U.J.T. Ogbuji, *J. Eur. Ceram. Soc.* **23**, 613-617 (2003)
8. R.D. James, R.A. Lowden, K.L. More, *Ceram. Trans.* **19**, 925-935 (1991)
9. R. Colmet, I. Lhermitte-Sabire, R. Naslain, *Adv. Ceram. Mater.* **1**, 185-191 (1986)
10. H.E. Cheng, Y.W. Wen, *Surf. Coat. Technol.* **179**, 103-109 (2004)

Mater. Res. Soc. Symp. Proc. Vol. 843 © 2005 Materials Research Society T3.22

Effects of High-Speed Deformation on the Phase Stability and Interdiffusion in Ultrasonically Joined Aluminum and Zinc Foils

Ibrahim Emre Gunduz[1], Emily D. Shattuck[2], Teiichi Ando[1], Peter Y. Wong[2], Charalabos C. Doumanidis[3]

[1]Mechanical and Industrial Engineering Department, Northeastern University, Boston MA 02115

[2]Department of Mechanical Engineering, Tufts University, Medford, MA 02155

[3]Department of Mechanical and Manufacturing Engineering, University of Cyprus, Cyprus

ABSTRACT

The effects of high strain-rate deformation on the phase stability and interdiffusion were investigated for Al-Zn welds produced by ultrasonic welding at 513 K. The welds exhibited three distinct regions: a featureless region indicative of local melting on the zinc side, a solidified mushy layer and a layer of fcc grains enriched with zinc. Al-Zn phase diagrams calculated from vacancy-modified Gibbs free energy curves indicate that local melting at the weld interface may result even at 513 K if the vacancy concentrations in the fcc and hcp solutions approach 0.07 as a result of high strain-rate deformation. EDS analysis of the weld interface yielded an interdiffusivity of 1.9 $\mu m^2/s$, which is five orders of magnitude larger than the normal diffusivity of zinc in aluminum at 513 K. Application of the mono-vacancy diffusion mechanism to the diffusion data also yields a vacancy concentration of 0.07, indicating that such a high vacancy concentration may indeed resulted during the ultrasonic welding at 513 K.

INTRODUCTION

Ultrasonic metal welding is a process for joining similar and dissimilar metals by introducing a high-frequency vibratory energy into the weld zone of the metals to be joined [1]. Two metals, usually in the form of thin foils or thin wires, are clamped together and displaced with respect to each other at ultrasonic frequencies to apply shear deformation at the interface to disperse surface contaminants and oxides, providing pure metal contact and subsequently forming a bond. Further application of the ultrasonic vibration results in repeated deformation of the metals. The metals experience high-rate cyclic plastic deformation with strain rates up to ~10^3. The process is ideal for investigating the effects of high-speed plastic deformation on metals due to the ability to repeatedly deform the metals with minimal heat generation.

Studies on high speed deformation of fcc metals [2,3] provide direct evidence for generation of excess vacancies in the metals, the concentration of which may reach as high as 10^{-1}. Such a huge amount of vacancies created should have a substantial effect on the thermodynamics and microstructural evolution of the metals being deformed. In the present study, the effects of high strain-rate deformation on the phase stability and interdiffusion in the Al-Zn system is investigated. The effects of excess vacancies on the stability of the phases involved in the Al-Zn system were assessed by thermodynamic calculations using a solution model that accounts for the presence of the excess vacancies.

EXPERIMENTAL

The ultrasonic welding machine used is a STAPLA Condor® unit working at a frequency of 20 kHz. The amplitude of vibration can be adjusted in the range 7.5-15 μm using a controller unit which employs a variable power input to keep the amplitude constant under any clamping load. The high-speed steel welding tip (sonotrode tip) has the overall dimensions 3200 x 3200 μm with a matrix of tapered teeth each with a base size of 120 x 120 μm and tip size of 80 x 80 μm. The metals used are 50 μm thick 1100 H19 aluminum foil and 250 μm thick 99.98% pure zinc sheet. The temperature of the samples is monitored by a 100 μm wire K-type thermocouple press-fitted at the interface between the metals to be welded.

The welding was carried out with an amplitude of 10 μm and a duration of 1 s at 513 K. Samples were preheated in air to 573 K without contact and brought together to produce the weld, and subsequently quenched in water. Upon contact of the sonotrode tip, the interface temperature dropped to 513 K and remained constant during welding. Cross sections of the as-welded samples were etched with 2% HF in water and examined by scanning electron microscopy (SEM). Energy-dispersive X-ray spectroscopy (EDS) was performed at the center of the welds to determine the concentration profile of zinc, which was utilized to analytically determine the average diffusivity of zinc in aluminum during ultrasonic welding.

RESULTS AND DISCUSSION

The Al-Zn foils welded at 513 K exhibited three distinct regions at the interface as shown in Figure 1. Figure 2 shows an EDS profile of the Zn concentration across the weld interface determined along the path indicated by the broken arrow in Figure 1. On the zinc side of the interface is a featureless zone that has penetrated into the fcc grain boundaries. This region has a constant composition of 80% Zn, Figure 2. The morphology and the composition of this region strongly indicate that it was molten during welding. Next to this region is a layer that indicates partial melting of the fcc grains during welding. The third region next to the mushy layer exhibits fcc grains enriched with zinc enrichment as evident in Figure 2.

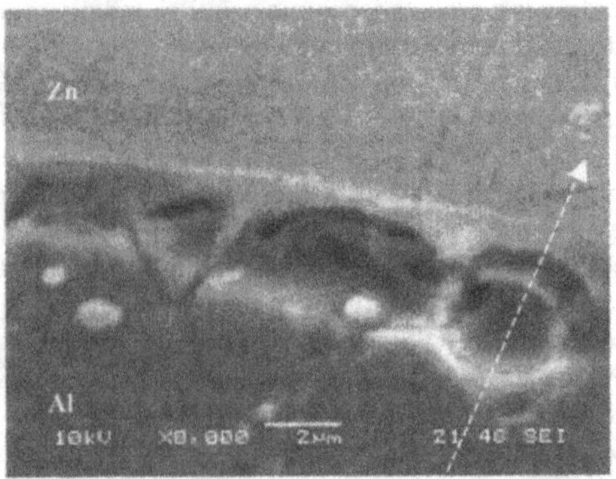

Figure 1: SEM micrograph of etched Al-Zn interface produced at 513 K. The EDS profile presented in Figure 2 was obtained along the white arrow.

The local welding temperature monitored using the 100 μm gage K-type thermocouple placed within close proximity indicated that there was no noticeable temperature rise above the intended welding temperature of 513 K. Thus, the observed local melting in the weld cannot be explained by an increase in flash temperature above the equilibrium melting temperatures of the hcp and fcc solutions.

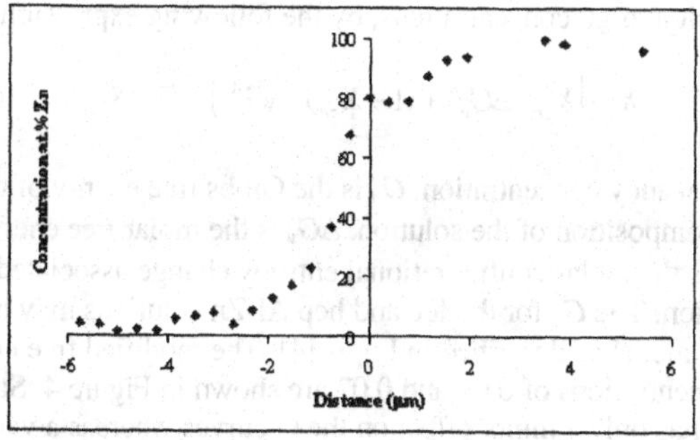

Figure 2: EDS profile obtained along the arrow indicated on Figure 1.

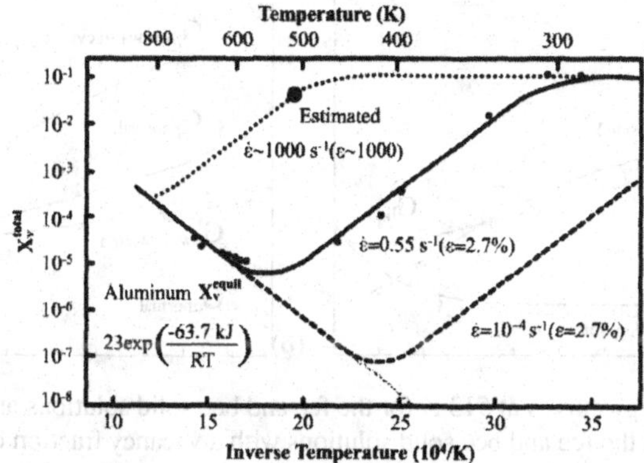

Figure 3: Variation of total (excess and equilibrium) vacancy concentration in aluminum with temperature and strain rate reproduced from Ref. [2]. The small solid circles show the NMR data from [2]. The large solid circle shows the vacancy concentration calculated from the diffusion profile at 513 K. The dotted line represents the total vacancy concentration schematically drawn for the strain rate in the present study, which is estimated as $\sim 10^3\ s^{-1}$.

The EDS profile in Figure 2 suggests that there has been substantial zinc penetration into the fcc solution. Such enhanced diffusion implies a high concentration of vacancies generated in the aluminum during the ultrasonic deformation. In fact NMR measurements [2], reproduced in Figure 3, have shown that vacancy concentration increases up to 0.1 in pure aluminum being deformed at sufficiently low temperature and high strain rates [2]. Therefore it was suspected that such a large increase in vacancy concentration might have a drastic effect on the stability of the fcc Al-Zn and hcp Zn-Al solid solutions.

The variation of Gibbs free energies of the solid solutions with vacancy concentration can be approximately expressed, assuming a rule of mixture and neglecting the vacancy interaction effects that might result at high concentrations, by the following expression:

$$G = (1 - X_v) \cdot G_p + X_v \cdot \left[X_{zn} \cdot \Delta G_v^{zn} + (1 - X_{zn}) \cdot \Delta G_v^{Al}\right] - T \cdot \Delta S_{mix} \qquad (1)$$

where X_v is the total vacancy concentration, G_p is the Gibbs free energy of the solution without vacancies, X_{zn} is the composition of the solution, ΔG_v is the molar free energy for vacancy formation, and ΔS_{mix} is the molar configurational entropy change associated with vacancy formation in the solution. The G_p for the fcc and hcp Al-Zn solutions may be approximated using a sub-regular solution model as described in Lupis [4]. The modified free energy curves for two different vacancy concentrations of 0.01 and 0.07 are shown in Figure 4. Strikingly, a vacancy concentration of 0.01 has only a minor effect on the G curves whereas a vacancy concentration of 0.07 has a major effect on relative stabilities of the solid solutions with respect to the liquid solution.

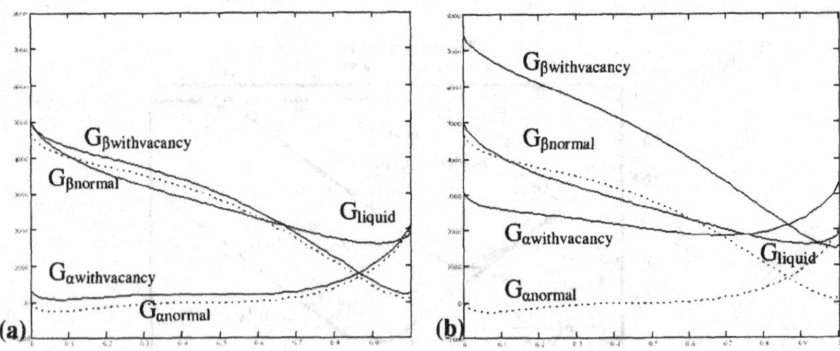

Figure 4: Free energy curves at 513 K for the fcc and bcc solid solutions and the liquid phase, along with those of the fcc and bcc solid solutions with a vacancy fraction of **(a)** 0.01 (1 at.%) and **(b)** 0.07 (7 at.%).

Figure 5(a) and (b) shows modified Al-Zn phase diagrams calculated with Eq. (1) for the two vacancy concentrations of 0.01 and 0.07. Clearly, melting may result at 513 K if the vacancy concentrations in the fcc and hcp solutions both increase to 0.07. This strongly suggests that the observed melting in the Al-Zn weld was caused by the shift in free energy curves caused by the excess vacancies. It also follows that the vacancy concentrations in the fcc and hcp solutions may have been as high as 0.07.

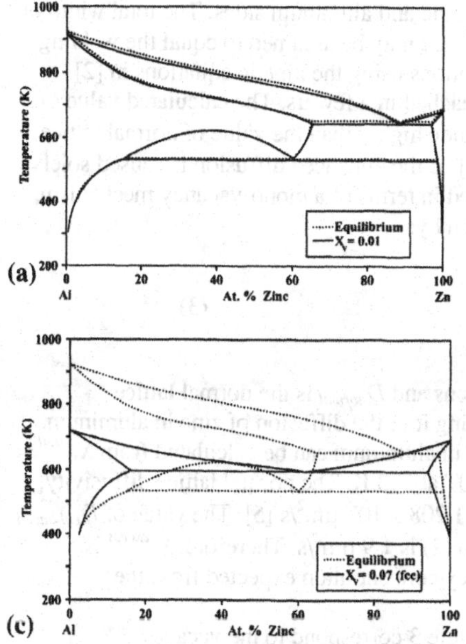

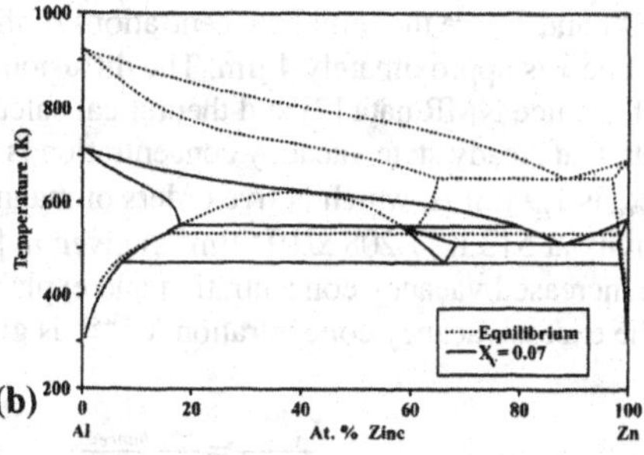

Figure 5: Al-Zn phase diagrams calculated for vacancy mole fractions of **(a)** 0.01 in both fcc and hcp solutions, **(b)** 0.07 in both fcc and hcp solutions and **(c)** 0.07 in fcc solution with no excess vacancy in hcp solution. The phase boundaries in the equilibrium phase diagram are shown by dotted curves.

However, no prior report exists as to whether the vacancy concentration in the hcp solution can indeed increase to the levels observed in aluminum during ultrasonic welding. The vacancy production in fcc metals under extensive plastic deformation is usually considered to be caused by the non-conservative motion of jogs on screw dislocations. Whereas in hcp metals it is not clear whether the same mechanism applies, although at elevated temperatures non-basal slip is activated in many hcp metals increasing the mobility of dislocations. To assess the effects of vacancy concentration in the hcp solution on the relative phase stability, another modified Al-Zn phase diagram was computed for a vacancy concentration of 0.07 in the fcc solution but with no excess vacancies in the hcp solution. The result, Figure 5(c), is a simple eutectic phase diagram where liquid exists only above 600 K, the eutectic temperature. A similar eutectic phase diagram results even if the vacancy concentration in the hcp solution is increased to 0.01. Therefore it is likely that vacancy concentration in the hcp solution actually exceeded 0.01 during the welding at 513 K.

Such high vacancy concentrations in the fcc and hcp solutions at the weld interface would enhance interdiffusion by many orders of magnitude. In fact the EDS profile in Figure 2 strongly indicates that such an enhancement of interdiffusion actually resulted. The effective enhanced interdiffusivity $D_{enhanced}$ may be determined from the EDS profile. To achieve this, it is assumed that the EDS profile approximates to an error-function solution of the form

$$X(x) = \frac{X_1 + X_2}{2} + \frac{X_1 - X_2}{2} \cdot erf\left(\frac{x}{2 \cdot \sqrt{D_{enhanced} \cdot t}}\right) \qquad (2)$$

where X_1 and X_2 are the initial concentrations of the zinc and aluminum sides. The total width of the profile x is approximately 4 μm. The diffusion time t may be assumed to equal the welding time, 1 s, since NMR data [2] and theoretical calculations using the kinetic equations in [2] indicate that steady state vacancy concentration is reached in a few μs. The calculated value of $D_{enhanced}$ is 1.9 $\mu m^2/s$, which is five orders of magnitude higher than the value of normal lattice diffusivity at 513 K (1.208 x 10^{-5} $\mu m^2/s$) given in [5]. If the enhanced diffusion is caused solely by the increased vacancy concentration and explained in terms of a mono-vacancy mechanism, then the excess vacancy concentration X_v^{excess} is given by:

$$\frac{X_v^{excess}}{X_v^{equil}} = \frac{D_{enhanced}}{D_{normal}} \quad (3)$$

where X_v^{equil} is the equilibrium vacancy concentrations and D_{normal} is the normal lattice diffusivity. This hypothesis may be tested by applying it to the diffusion of zinc in aluminum. The value of the equilibrium vacancy concentration in aluminum can be calculated from X_v^{equil} = 2.226•(exp(-59.82kJ/RT) [6], which gives 1.804 x 10^{-6} at 513 K. The normal lattice diffusivity, D_{normal}, of zinc in aluminum at 513 K is reported as 1.208 x 10^{-5} $\mu m^2/s$ [5]. The value of $D_{enhanced}$ at 513 K calculated from the EDS data and Equation (2) is 1.9 $\mu m^2/s$. Therefore, X_v^{excess} is calculated to be ~0.07, which agrees with the vacancy concentration expected from the observation of local melting at the weld interface.

The large solid circle and the dotted curve in Figure 3 correspond to the vacancy concentration for the strain rate during the ultrasonic welding experiment, which is estimated to be ~10^3 s^{-1}. The value of 0.07 at 513 K indicates that the strain rate at the weld interface during ultrasonic welding was not high enough to increase the vacancy concentration to the maximum realizable value of ~10^{-1} at the welding temperature of 513 K.

CONCLUSIONS

Aluminum-zinc foils ultrasonically welded at 513 K exhibited three distinct regions: a featureless region indicative of local melting on the zinc side, a solidified mushy layer and a layer of fcc grains enriched with zinc. Calculation with a thermodynamic model that accounts for the change in Gibbs free energy due to vacancy generation during the ultrasonic deformation indicates that substantial melting point depression may result in the fcc Al-Zn and hcp Zn-Al solutions if the vacancy concentrations in the fcc and hcp solutions both approach 0.07. Reported NMR data for aluminum undergoing plastic deformation support that such high vacancy concentrations are possible. An EDS zinc concentration profile determined across the weld interface yielded an interdiffusivity value of 1.9 $\mu m^2/s$, which is five orders of magnitude larger than the normal diffusivity of zinc in aluminum at 513 K. The determined value of interdiffusivity also yields a vacancy concentration of 0.07 if the mono-vacancy diffusion mechanism still applies. The consistency among the metallographic observations, the thermodynamic calculations and the diffusivity measurements suggests that ultrasonic deformation produced high concentrations of vacancies in the weld causing enhanced interdiffusion and large melting point depressions.

ACKNOWLEDGEMENTS

The authors acknowledge the financial support for the research by the National Science Foundation (DMI 0114309).

REFERENCES

1. "Welding Handbook", 4th edition sec. 3. Miscellaneous Metal Joining and Cutting Processes, A.W.S. pp. 52.1-52.20, 1959.
2. K. L. Murty, K. Detemple, O. Kanert, J. T. M. Dehosson, Metallurgical and Materials Transactions A **29A**, 153 (1998).
3. M. Kiritani, Y. Satoh, Y. Kizuka, K. Arakawa, Y. Ogasawara, S. Arai, Y. Shimomura, Philosophical Magazine Letters **79**, 797 (1999).
4. Lupis CHP. Chemical Thermodynamics of Materials. Amsterdam: North Holland Publishing Company; 1983. p. 458.
5. J. E. Hilliard, B. L. Averbach, M. Cohen, Acta Metallurgica **7**, 86 (1959).
6. A. Seeger, D. Schumacher, W. Schilling , J. Diehl, "Vacancies and Interstitials in Metals", Amsterdam: North Holland Publishing Company, p. 1, 1970.

Mater. Res. Soc. Symp. Proc. Vol. 843 © 2005 Materials Research Society T3.24

Improved Fatigue Properties of 316L Stainless Steel Using Glass-Forming Coatings

F. X. Liu*, C. L. Chiang*,**, L. Wu*, Y. Y. Hsieh**, W. Yuan*, J. P. Chu**, P. K. Liaw*, C. R. Brooks*, and R. A. Buchanan*

*Department of Materials Science and Engineering, The University of Tennessee, Knoxville, TN 37996, USA

** Institute of Materials Engineering, National Taiwan Ocean University, Keelung 202, Taiwan

Abstract

The effects of the glass-forming coatings on the fatigue behavior of 316L stainless steel were investigated. Films consisting of 47%Zr, 31%Cu, 13%Al and 9%Ni (atomic percent) were deposited onto the stainless steel by magnetron sputtering. The influences of the substrate condition, the surface roughness, the adhesion, and the compressive residual stresses on the fatigue behavior were studied. The applications of the glass-forming coating gave rise to significant improvements in both the fatigue life and the fatigue limit, in comparison with the uncoated steel. Depending on the maximum stress applied to the steel, the fatigue life can be increased by at least 30 times, and the fatigue limit can be elevated by 30%.

Introduction

Bulk-metallic glasses (BMGs) possess desirable properties, including extraordinarily high strengths, high hardness, low friction coefficients, and excellent wear and corrosion resistance, good bending ductility in the form of thin films, which makes them good candidates for coating materials to improve materials properties, such as fatigue and wear resistances [1,2]. The wide applications of BMGs are limited by the relatively small sample size and high cost [3]. Thus, the deposition of thin glass-forming coating materials onto structural materials to enhance the fatigue behavior becomes a challenging area for scientific studies.

The Zr-based metallic glasses with the composition of Zr-31Cu-13Al-9Ni (in atomic percent) have good glass-forming abilities and high strengths. The annealing treatment of Zr-31Cu-13Al-9Ni films results in crystallization and amorphization [4]. The Zr-based glass-forming coating with this composition was deposited on 316L stainless steel by magnetron sputtering. The effect of the glass-forming coating on the fatigue behavior of the 316L stainless steel was investigated by four-point-bend fatigue tests. Atomic-force microscopy (AFM) and scanning-electron microscopy (SEM) were employed to measure the surface roughness and study the fractography, respectively. Fatigue results such as fatigue life and fatigue-endurance limit were compared with those reported from similar steels coated with other hard coating materials [5, 6].

Experimental procedure

316L stainless steel with 20% cold work was used as the substrate material with the composition listed in Table 1. The samples, with the geometry of 3 x 3 x 25 mm^3 were ground with 800-grit sand paper, followed by electrical polishing to ensure a smooth surface. The bottom of the 316L specimen was coated with the Zr-based amorphous film to avoid the crack initiation and propagation in the tensile stresses, as shown in Figure 1. The composition of the glass-forming film was Zr-31Cu-13Al-9Ni, as shown in Table 2, with a thickness of ~ 200 nm. The specimen geometry for four-point-bend fatigue testing with glass-forming coatings is shown in Figure 1. A computer-controlled Material Test System (MTS) servohydraulic testing machine was used for fatigue testing. Samples were tested at various stress ranges with a R ratio ($R = \sigma_{min.} / \sigma_{max.}$; where $\sigma_{min.}$ and $\sigma_{max.}$ are the applied minimum and maximum stresses, respectively) of 0.1 under a load-control mode, using a sinusoidal waveform at a frequency of 10 Hz. Upon failures, samples were examined by scanning-electron microscopy (SEM). Atomic-force microscopy (AFM) was also employed to measure the surface roughness.

Table 1. The composition of the substrate material (in weight percent).

Material	Content*	C	Mn	Si	P	S	Cr	Mo	Ni	N	Fe
316L Stainless Steel	Min.	-	-	-	-	-	16.0	2.00	10.00	-	bal.
	Max.	0.03	2.0	0.75	0.045	0.03	18.0	3.00	14.00	0.10	

Table 2. The composition of Zr-based glass-forming films used in this study (in atomic percent).

Specimen	Composition			
Zr-based thin film	Zr	Al	Cu	Ni
	47	31	13	9

* atomic percent

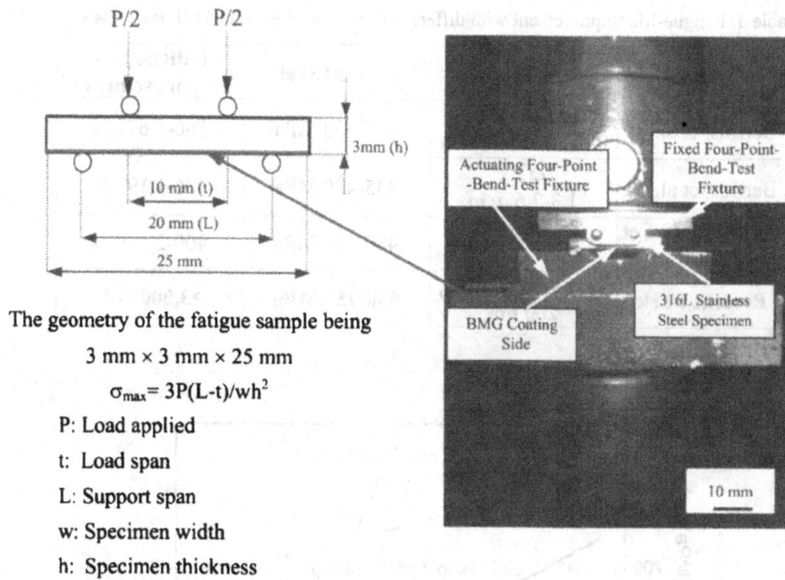

Figure 1. Specimen geometry for four-point-bend fatigue testing.

Results and discussion

Figure 2 presents fatigue test results of the coated material along with those of uncoated material. Depending on the maximum stress applied to the steel, the fatigue life can be increased by 30 times, and the fatigue limit can be elevated by 30%. At moderate stress levels of about 650 MPa to 700 MPa, the improvement is very significant, compared to higher stress levels. In coated materials, the fatigue life is usually controlled by the crack initiation at the surface of the coating and further propagation through the thickness of the sample. This trend indicates that the fatigue-fracture process is mainly controlled by the crack initiation at the surface of the coating. Berrios et al. [5] and Puchi-Cabrera et al. [6] have deposited some ZrN_x and TiN_x films on 316L stainless steel, with thickness ranging from 1.3μm to 3 μm. 400% to 2,100 % of fatigue life improvements are reported, as seen in Table 3. The improvement of the fatigue life in our present work with 200 nm-thick Zr-based glass-forming films seems to be greater than those reported by Berrios et al. and Puchi-Cabrera et al. During four-point-bend fatigue loading, the maximum stress within a component occurs at the surface of the tension side. And most cracks leading to fatigue failure originate at surface positions, especially at stress amplification sites. The high strength of the amorphous coating combined with its good bending ductility in the form of thin films, might retard the crack initiation process and thus lead to the improvement of the fatigue properties.

Table 3. Fatigue-life improvement with different films deposited onto 316L stainless steel.

Research Group	Films	Stress Level	Fatigue Life Improvement
Berrios, et al.[7]	TiNx (x= 0.55, 0.65, 0.75) ~3 μm	400-450 MPa	566-1,677 %
Berrios, et al.[5]	ZrNx 1.3-2.6 μm	435-480 MPa	406-1,192 %
Puchi-Cabrera, et al.[6]	TiN, 1.4 μm	480-510 MPa	400-2,119 %
Present work	**Zr-based coating, 200 nm**	**650-750 MPa**	**≥3,000% ↑**

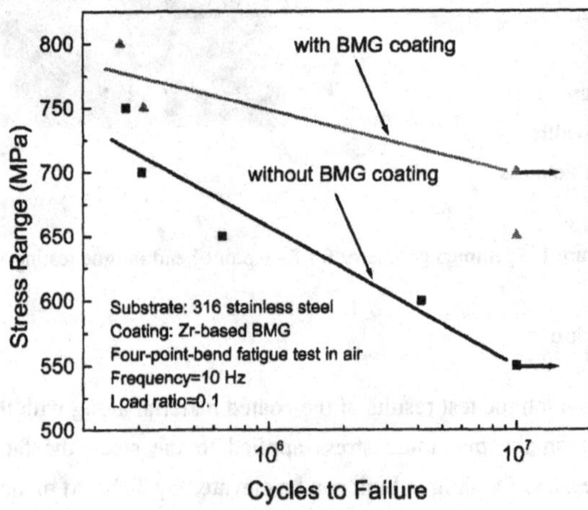

Figure 2. Stress versus fatigue life for materials with and without glass-forming coatings.

Figure 3(a) shows the fracture surface of the coated 316L stainless steel specimen tested at σ_{max} = 750 MPa. The whole fatigue fracture surface consists of three regions: the fatigue-crack-initiation, crack propagation and final-fast-fracture areas. Figure 3(b) presents the interface of the coating and the substrate. It is clear that the coating remained well adhered with the substrate even at this high stress level. The surface roughness was measured by atomic-force microscopy for materials with and without glass-forming coatings, as shown in Figure 4. The Ra value of the surface roughness drops from 4.81 nm to 2.55 nm due to the introduction of the coating, which is beneficial to the improvement of the fatigue behavior of the coating system.

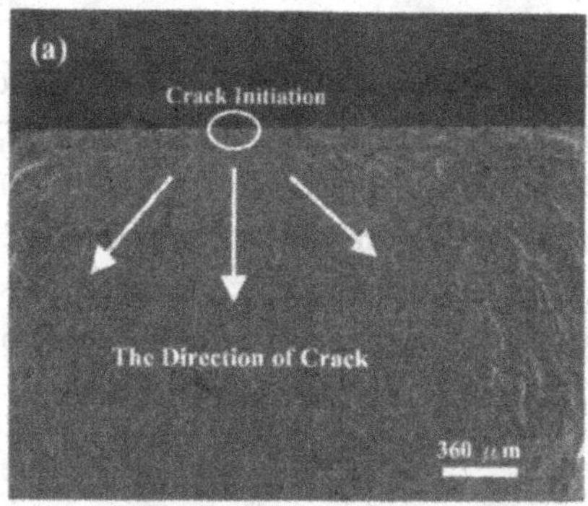

Figure 3. Fractography of the fatigued 316L stainless steel specimen with the glass-forming coating tested at $\sigma_{max.}$ = 750 MPa.

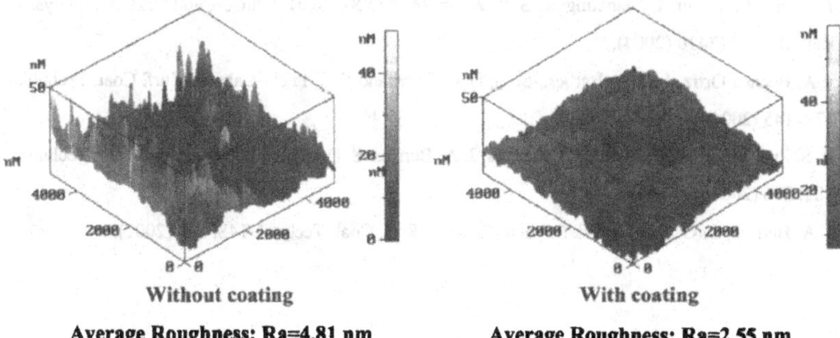

Figure 4. The surface roughness measured by atomic force microscopy for the 316L stainless steels without glass-forming coatings (left) and with coatings (right).

Conclusions

With the deposition of the Zr-based glass-forming coating, the fatigue life of 316L stainless steel could be increased by 30 times, and the fatigue limit could be elevated by 30%, depending on the maximum stress applied to the steel. The fatigue-resistance improvement was mainly due to the reduction of the surface roughness and the good adhesion of the glass-forming coatings with the substrate. The high strength and the good bending ductility might be other important factors for the improvement.

Acknowledgements

The present work is supported by (1) the Department of Materials Science and Engineering, The University of Tennessee, (2) the National Science Foundation (NSF) Integrative Graduate Education and Research Training (IGERT) Program (DGE-9987548), (3) the NSF Combined Research and Curriculum Development (CRCD) Program (DGE-0203415) with Dr. C. McHargue, Dr. W. Jennings, Dr. L. Goldberg, Dr. L. Clesceri, and Ms. M. Poats as the contact monitors, respectively, and (4) the National Science Council of Republic of China in Taiwan (NSC93-2216-E-019-007).

References

1. A. Inoue, H. M. Kimura, K. Sasamori, and T. Masumoto, Mater. Trans JIM. 35(2), 85 (1994).
2. Y. Kawamura, T. Shibata, A. Inoue, and T. Masumoto, Appl. Phys. Lett. 69(9), 1208 (1996).
3. A. Inoue, Acta. Mater. 48, 279 (2000).
4. J. P. Chu, C. T. Liu, T. Mahalingam, S. F. Wang, M. J. O'Keefe, B. Johnson, and C. H. Kuo, Phys. Rev. B. 69, 113410 (2004).
5. J. A. Berrios-Ortiz, J. G. La Barbera-Sosa, D. G. Teer, and E. S. Puchi-Cabrera, Surf. Coat. Technol. 179, 145 (2004).
6. E. S. Puchi-Cabrera, F. Matinez, I. Herrera, J. A. Berrios, S. Dixit, and D. Bhat, Surf. Coat. Technol. 182, 276 (2004).
7. J. A. Berrios, D. G. Teer, and E. S. Puchi-Cabrera, Surf. Coat. Technol. 148, 179 (2001).

Mater. Res. Soc. Symp. Proc. Vol. 843 © 2005 Materials Research Society T3.26

Structural, Mechanical and Tribological Properties of TiN and CrN Films Deposited by Reactive Pulsed Laser Deposition

A.R.Phani[1] and J.E.Krzanowski[2]
[1] Department of Physics, University of L'Aqula, INFM-CASTI Regional Laboratory, Via Vetoio-10, Copitto, L'Aquila-67010, ITALY
[2] Department of Mechanical Engineering, University of New Hampshire, Durham, NH 03824, U.S.A.

ABSTRACT

Nitride thin films have potential applications in different areas of silicon device technology, namely as diffusion barrier in metallization schemes, rectifying and ohmic contacts, and gate electrodes in field effect transistors. In the present investigation, TiN and CrN films have been deposited by reactive pulsed laser deposition technique using Ti and Cr targets at 10mTorr background pressure of N_2. Si (100) and AISI 440C steel substrates were used for the present study. Films were deposited at different temperatures in the range of 200°C to 600°C. The deposited films exhibited densely packed grain, with smooth and uniform structures. X-ray Photoelectron Spectroscopy (XPS) analysis of the films showed and 50% Ti and 40% of N in TiN films, 45% of Cr and 45% of N in CrN films deposited on Si (111), with the balance mostly oxygen, indicating near stoichiometric composition of the deposited nitride thin films. Hardness of the films changed from 22 GPa at 200°C to 30 GPa at 600°C for TiN, whereas for CrN we obtained 26 GPa at 200°C to 31 GPa at 600°C. The residual stress in the films showed a change from compressive stress at 200°C to tensile stress at 600°C in both the cases. Friction coefficient of the films were measured by pin-on-disk technique for all films, up to the tested limit of 10,000 cycles at 1 N load and found to be very high (≥ 1) in both cases.

INTRODUCTION

Hard coatings play an important role in industry for improving tool lifetime and performance. TiN is one of the most commonly used and studied coatings, and has many beneficial properties including high hardness, low friction and chemical intertness[1]. There is however, still an industrial interest in coatings with further improved wear resistance, hardness, fracture toughness. Other transition metal nitrides have generated practical applications in industry as protective coatings; in particular CrN films have also shown high hardness, enhanced wear and corrosion resistance, excellent diffusion barrier properties and can be utilized as anti-wear and anti-corrosion coatings or as selective solar absorbers[2]. Many investigators have deposited TiN and CrN films by various techniques such as sputtering[3] arc-evaporated[4] and pulsed laser deposition[5-6]. Layered coating structures such as superlattices or nanostructured multilayers have also been investigated, such as TiN/Ti[7], TiN/CrN[8], TiN/NbN[9], TiN/BN[10]. In the present study, we have deposited TiN and CrN films by reactive pulsed laser deposition and the deposited films are characterized for their structural, compositional, mechanical and tribological properties.

EXPERIMENTAL PROCEDURE

TiN and CrN films have been grown by reactive pulsed laser deposition in 10mTorr N_2 partial gas (99.99%). The pulsed laser deposition (PLD) chamber (base pressure 3×10^{-7} torr) and technique used to deposit the films has been described elsewhere[11]. We have used high purity titanium and chromium (both 99.99%) targets to deposit TiN and CrN films. For present study, we have used a KrF excimer laser with 248nm wavelength, and 650mJ laser power with 50Hz repetition rate. Si (100) and 400C steel we used as substrates. The substrate-to-target distance was 2.5 inches and both the target and the substrate were rotated during deposition to ensure good spatial uniformity. The substrate temperature was measured by thermocouple mounted on the substrate holder. After 2 hours of deposition, typical films thicknesses were on the order of 800 nm. The deposited films appeared smooth and continuous were golden yellow (TiN) and pinkish green (CrN) in color at all substrate temperatures. The adhesion of the films on to the substrates has been tested by the simple scotch tape test. No peeling or crack has been observed for the deposited films.

RESULTS

X-Ray Diffraction (XRD)

X-ray Diffraction patterns were collected using a Rigaku θ-θ diffractometer to determine the lattice parameters and peak width at half maximum, with the background and $K\alpha_2$ contributions subtracted numerically before the measurements. The instrumental peak broadening was measured using a silicon standard. Figure 1(a) and (b) represents the TiN and CrN films grown on to 440C steel at different substrate temperatures. JCPDS card No: TiN:38-1420, CrN: were used to verify the crystalline structures for both the TiN and CrN. TiN films were highly oriented in (200) direction at all temperatures. The CrN films were highly oriented in (111) orientation only when deposited at a substrate temperature of 600°C. The experimental diffraction peaks, grain size, calculated (Exp.) and literature (Lit.) "d" , 2θ, and lattice parameters (Lat. Param.) of TiN and CrN on 440C steel are tabulated in Table I and II, respectively. The crystallite sizes of the deposited films were in the range of 21 nm to 28 nm for TiN and 19 nm to 30 nm for CrN. There was no significant change in the lattice parameters of the deposited films at different substrate temperatures compared to TiN films obtained from TiN target[6] by PLD. Moreover, the crystallite size was 28 nm at 600°C in our case, when compared to TiN films (67 nm at 600°C) obtained directly from TiN target by PLD[6].

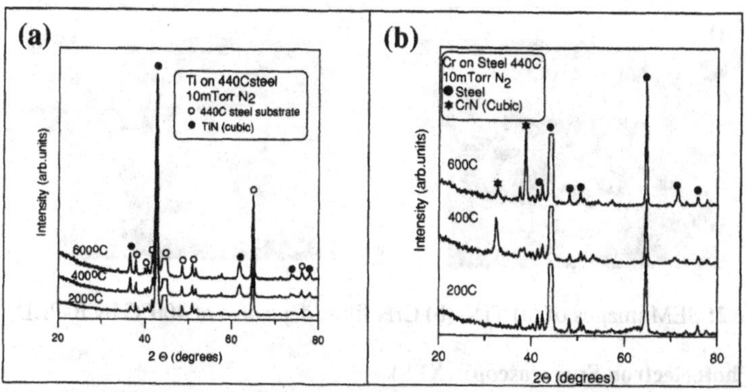

Figure 1: XRD patterns (a) TiN films (b) CrN films deposited by R-PLD.

Scanning Electron Microscopy (SEM)

The surface morphology of the deposited films was examined by scanning electron microscopy (SEM model: AMRAY 3300FE). The SEM images of TiN and CrN films deposited at 400°C are shown in Fig 2(a) and (b), respectively. Similar surface morphology has been observed for all the deposited films in case of TiN and CrN films. The deposited films were flat, showing densely packed grains. The particulates which usually form in the PLD process were not frequently observed in case of TiN and CrN films when compared to the TiC films, where we have observed many particulates (~25%) on the surface[12]. However, the exact role played by the background gas of N_2, in case of TiN films, to make the films smooth is still to be understood.

Table-I: Grain size, Lattice Parameters of TiN by R-PLD

TiN on 440C steel	I/Io (%)	(hkl)	Lit. 2θ (degrees)	Exp. 2θ (degrees)	Lit. "d" (nm)	Exp. "d" (nm)	Grain. Size (nm)	Lit. Lat. Param. (nm)	Exp.Lat Param. (nm) (a=b=c)
TiN (cubic) at 200°C	100	100	42.59	42.55	2.121	2.136	21.4	0.424	0.421
TiN (cubic) at 400°C	100	100	42.59	42.52	2.119	2.139	25.2	0.424	0.422
TiN (cubic) at 600°C	100	100	42.59	42.51	2.120	2.137	28.7	0.424	0.421

Table-II: Crystallite size, Lattice Parameters of CrN by R-PLD

CrN on 440C steel	I/Io (%)	(hkl)	Lit. 2θ (degrees)	Exp. 2θ (degrees)	Lit. "d" (nm)	Exp. "d" (nm)	Grain. Size (nm)	Lit. Lat. Par am. (nm)	Exp.Lat Param. (nm) (a=b=c)
CrN (cubic) at 200°C	80	111	37.54	37.5	2.393	2.394	8.5	0.4138	0.413
CrN (cubic) at 400°C	80	111	37.54	37.56	2.393	2.394	15.2	0.4138	0.413
CrN (cubic) at 600°C	80	111	37.54	37.9	2.393	2.381	17.2	0.4138	0.412

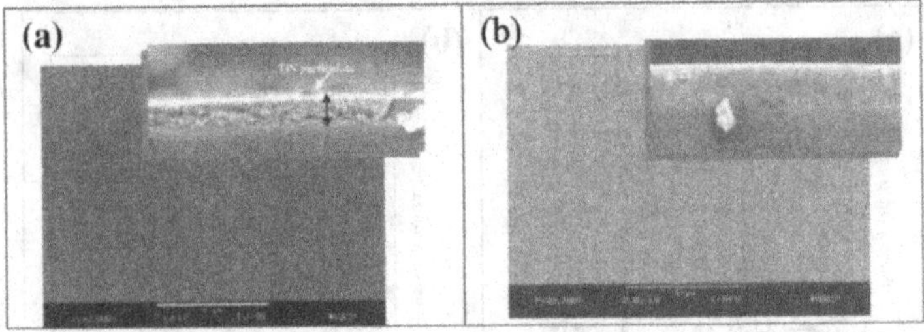

Figure 2: SEM images of (a) TiN (b) CrN films deposited at 400°C by R-PLD.

X-Ray Photoelectron Spectroscopy (XPS)

The compositions of the films were measured by X-ray Photoelectron Spectroscopy. We have obtained 50 at.% of Ti in TiN films and 45 at.% of Cr in CrN films and 40-45 at.% N and ~10 at.% O in the deposited films, indicating near stoichiometry of the deposited films. In brief, on laser ablation molecular nitrogen is either breaking into atomic nitrogen or reacting at the films of titanium or chromium to form TiN and CrN films, respectively. The formation of TiN or CrN is based on the background pressure of the N_2. In our case, we have used 10mTorr to obtain single cubic phase TiN and CrN structures. A more complete mechanism of nitride formation will require further research.

Atomic Force Microscopy (AFM)

Topographical atomic force microscopy images of TiN and CrN deposited at 400°C by R-PLD were shown in Figure 3a and 3b, respectively. More particluates covering the surface 10-15% with diameter 20-25 nm have been observed in case of TiN films compared to CrN films surface. The roughness of the TiN films ranged from 20-40 nm, and for CrN films from 15-20 nm.

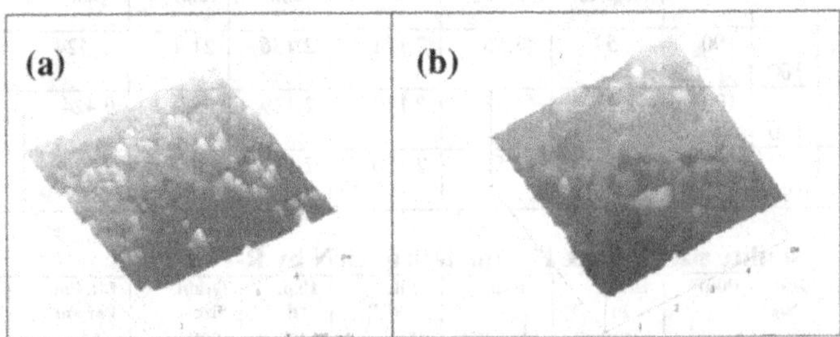

Figure 3: AFM images of (a) TiN (b) CrN films deposited at 400°C by R-PLD (scan area: 5 μm × 5 μm and Z axis: 100 nm)

Nanoindentation

Nanoindentation was used to measure film hardness and was conducted on a Hysitron nano-mechanical test instrument equipped with Berkovich indenter. Films were tested using a multiple loading sequence with loads ranging from 0.1 to 7 mN. For each film, approximately 3-5 indentations were made, with each of the multiple load sequence giving at least 10 hardness and elastic modulus values. Residual stress was measured using x-ray diffraction and the $\sin^2\psi$ method. Figure 4a and 4b shows the residual stress, hardness and elastic modulus of the TiN and CrN films deposited at different substrate temperatures. It was evident from the graph that the stress in the films changed from compressive to tensile as the substrate temperature changes from 200°C to 600°C. Films deposited at 600°C have shown highest hardness (30 GPa). The elastic modulus for these films was around 250 GPa. Similar kind of observation was found in the CrN films where the films stress changed from compressive to tensile as the substrate temperature changes

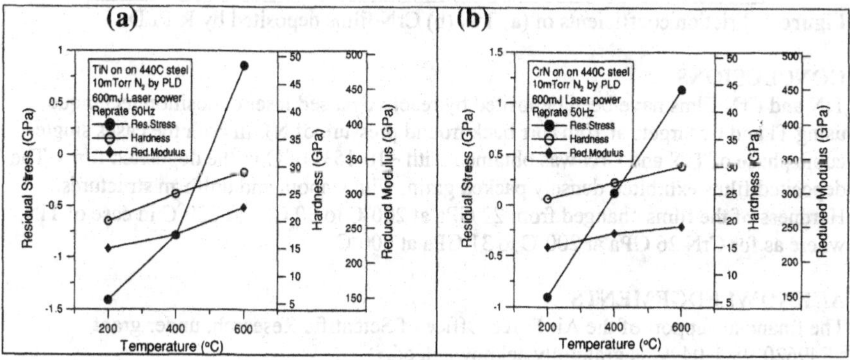

Figure 4: Hardness, reduced modulus and residual stress data of (a) TiN (b) CrN films deposited by R-PLD

from 200°C to 600°C. Films deposited at 600°C have shown highest hardness (31 GPa). The elastic modulus for these films is around 250 GPa. The films deposited at higher substrate temperatures are more dense than the films at 400°C, this has resulted an increase in the hardness for with increasing the substrate temperature.

Friction Test

Limited tribological evaluation of the samples was also carried out using a pin-on-disk apparatus. The test conditions were ambient atmosphere and temperature, with a 100 gm load on a ¼-in. (0.625 cm) 440C steel ball, a speed of 200 revolutions per minute, and a track diameter of 6mm. These tests were done with the disc horizontal, and any wear debris will likely remain in the track and influence the observed friction and wear behavior. All samples were run for a total of 10,000 revolutions. Pin-on-disk friction measurements were made on TiN and CrN films deposited on 440C steel are shown in Figure 5. The average friction values are 0.80 and 1.3. This indicates that all the films survived the test for 10,000 revolutions, since steel-on-steel friction values are usually

0.8-0.9 under the test conditions used. The best wear properties were observed for the film deposited at 400°C due to lower amount of oxygen in the films, which show friction coefficients in the range of 0.80 to 0.75.

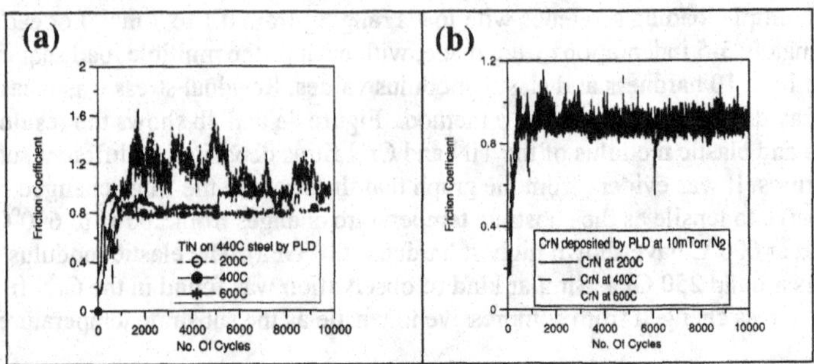

Figure 5: Friction coefficients of (a) TiN (b) CrN films deposited by R-PLD.

CONCLUSIONS

TiN and CrN films have been deposited by reactive pulsed laser deposition technique using Ti and Cr targets at 10mTorr background pressure of N_2. In both the cases single cubic phase of TiN and CrN was obtained with ~10-15% of O in the deposited films. The deposited films exhibited densely packed grain, with smooth and uniform structures. Hardness of the films changed from 22 GPa at 200°C to 30 GPa at 600°C in case of TiN, where as for CrN 26 GPa at 200°C to 31 GPa at 600°C.

ACKNOWLEDGEMENTS

The financial support of the Air Force Office of Scientific Research, under grant #F49620-98-1-0499, is gratefully acknowledged.

REFERENCES

1. B.Bhushan and B.K. Gupta, Hand book of Tribology, McGraw-Hill, New York, 1991, 4.57
2. P.Engel, G.Schwarz and G.K.Wolf, Surf.Coat.Technol. 98, 1002, 1998
3. X.M.He, N.Baker, B.A.Kehler, K.C.Walter, M.Nastasi and Y.Nakamura, J.Vac.Sci.Technol. A18, 30, 2000
4. J.Almer, M.Oden, L.Hultman and G.Hakansson, J.Vac.Sci.Technol. A18, 121, 2000
5. J.C.S.Kools, C.J.C.Nilensin, S.H.Brongersma, E.van de Reit and J.Dieleman, J.Vac.Sci.Technol. A10, 1809, 1992
6. R.Chowdary, R.D.Vispute, K.Jagannadham and J.Narayan, J.Mat.Res, 11, 1458, 1996
7. E.Kusano, M.Kitagawa, H.Nanto and A.Kinbara, J.Vac.Sci.Technol. A16, 1272, 1998
8. P.Yashar, S.A.Barnett, J.Rechner and W.D Sproul, J.Vac.Sci.Technol. A16, 2913, 1998
9. H.Ljungcrantz, C.Engstrom, L.Hultman, M.Olsson, X.Chu, M.S.Wong and W.D.Sproul, J.Vac.Sci.Technol. A16, 3104, 1998
10. J.Mmendez, B.D.Qu, M.Evstigneev and R.H.Prince, J.Appl.Phys.87, 1235, 2000
11. A.R.Phani and J.E.Krzanowski, Appl.Surf.Sci. 6917, 1, 2001
12. A.R.Phani and J.E.Krzanowski, J.J.Nainaparampil, J.Vac.Sci.Tech.A, 19,2252, 2001

Mater. Res. Soc. Symp. Proc. Vol. 843 © 2005 Materials Research Society T3.27

The Use of Closely Spaced Vickers Indentations to Model Erosion of Polycrystalline α-Al_2O_3

Adolfo Franco Júnior and Steve G. Roberts[1]
Núcleo de Pesquisa em Química, Universidade Católica de Goiás,
Caixa Postal 86, Setor Universitário, 74605-010 Goiânia-GO, Brazil
[1]Department of Materials, University of Oxford
Parks Road, Oxford OX1 3PH, UK

ABSTRACT

Arrays of closely spaced quasi-static indentation were made on specimens of polycrystalline α-Al_2O_3, mean grain size G=1.2, 3.8 and 14.1 μm. The critical indentation spacing to produce crack coalescence between indentations, and thus significant loss of material from the surface, was determined. These data are compared to results for low-impact-velocity wet erosive wear on the same materials; a good correspondence is found. The indentation data can be used to produce "wear maps", which provide a guideline for predicting the low-impact-velocity erosive wear resistance of brittle materials.

INTRODUCTION

The wear resistance of polycrystalline alumina has a strong dependence on grain size; in wear due to sliding [1-5] grinding [6,7], sawing [8], dry erosion [9] and wet erosion [10-13] the wear rate increases significantly with increasing grain size.

Franco et al.[11-13] also observed that the wet erosive wear rates of alumina varied strongly with grain size, even though Vickers hardness (Hv) and fracture toughness (K_{IC}) showed no significant grain size dependence.

The erosive wear tests, using large (~650 μm diameter) SiC grits at low velocities (~2.4 ms^{-1}) indicated that the principal erosion process is the linking of cracks from closely-spaced impacts to produce material loss, rather than by material loss from single impact events, as may happen as high impact velocities [14-15]. In this study, we use closely spaced Vickers indentations to model the process by which successive neighbouring impacts can lead to material loss.

EXPERIMENTAL PROCEDURE

Polycrystalline alumina materials (grain size, G=1.2, 3.8 and 14.1 μm) were fabricated using high purity, 99.9%, α-Al_2O_3 powder (Sumitomo AKP-50, Japan) of mean particle size ~180 nm. The details of sintering conditions and characteristics of the hot-pressed alumina specimens are described in Ref [11].

Grids of 3x3 Vickers indentations of varying loads and spacing were performed on surfaces of each material type to simulate damage produced by multiple impacts of sharp particles such as SiC grits.

Each series of indentations was performed using a microhardness tester (MHT1; Matsuzawa Seiki Co. Ltd. Japan) with a diamond pyramid indenter (Vickers type). Tests were performed on polished surfaces (final polishing: 1 μm) of each polycrystalline alumina material. The indentation load and indentation spacing were chosen to cover the regime in which cracks become linked to each other. The load range was 0.5 N to 3 N and the indentation spacings used were 40, 30, 20, 10 and 5 μm (in x and y directions). Indentations were performed in the order shown in figure 1. The damage produced by a given load-spacing combination was examined using SEM.

The amount of material removed for the 9-indentation grid, for the load/spacing combinations 3 N/10μm to 1 N/5 μm was measured for each material using an UBM non-contacting microfocus measurement system (UBM non-contacting Microfocus Measurement System, APT & T Advanced Products & Technologies Limited, Botley, Oxford 0X2 0JJ).

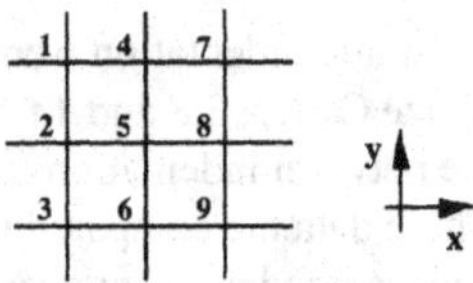

Figure 1. Schematic diagram of the Vickers indentation array. The indentation sequence is in numerical order.

EXPERIMENTAL RESULTS AND DISCUSSIONS

At the loads used, one indentation alone does not cause material loss by lateral fracture, but as subsequent indentations are produced nearby, damage may occur if they are closely enough spaced. The interlinking of microfractures present due to previous indentations with cracks from the new indentation leads to material removal from the surface. For a given load there is a critical indentation spacing below which such material removal occurs.

Figure 2 shows SEM micrographs of four square grids of 9 indentations of 3 N load at 40, 30, 20 and 10 μm spacings performed on each material type. For the 20 μm indentation spacing, severe damage is observed on the specimen with G=14.1 μm whereas less severe damage is observed on the specimen with G=3.8 μm. No damage is seen on the specimen with G = 1.2 μm. For the specimen with G = 1.2 μm severe damage is found for loads of 3 N only when the indentation spacing is ≤10 μm .

Table I shows the critical indentation spacing needed to produce severe material loss. For any indentation spacing greater than the critical value at a given load, while local microfracture may occur for individual indentations, no crack interlinking occurs, thus no significant amount of material is removed from the surface. For any spacing lower than the critical value for that load, material removal always occurs. Where the damage level changed from zero to full interlinking between two arrays of different spacing, (e.g. Figure 2(c)), then the critical spacing was taken as the mean spacing of the two arrays.

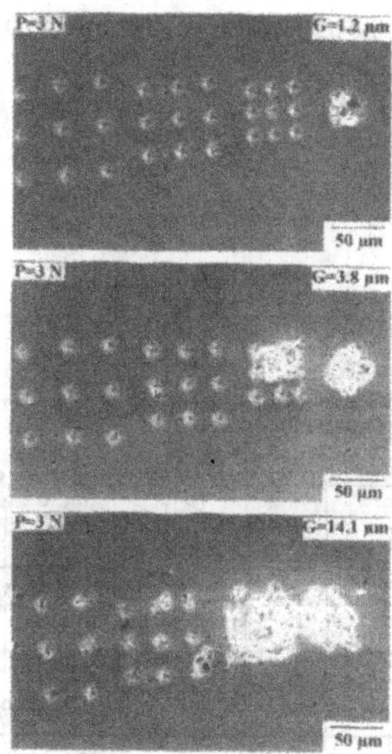

Figure 2. Typical damage produced by Vickers indentation arrays of various spacing. The indentation arrays shown were made with a 3 N load at 40, 30, 20 and 10 μm separations.

Table I. Critical indentation spacings for severe material loss.

Grain size (μm)	Indentation load (N)	Critical Indentation spacing (μm)	Volume removed per indentation ($x10^{-17}$ m^3/ind.)
1.2	3	10	11.34 ± 0.09
	2	7.5	6.33 ± 0.09
	1	5	3.72 ± 0.09
	0.5	2.5	-
3.8	3	17.5	33.13 ± 0.09
	2	12.5	15.56 ± 0.09
	1	7.5	4.28± 0.09
	0.5	3.5	-
14.1	3	20	76.34 ± 0.09
	2	15	25.28 ± 0.09
	1	10	5.45 ± 0.09
	0.5	5	-

Based on these observations a "wear map" for polycrystalline alumina was constructed as shown in figure 3.

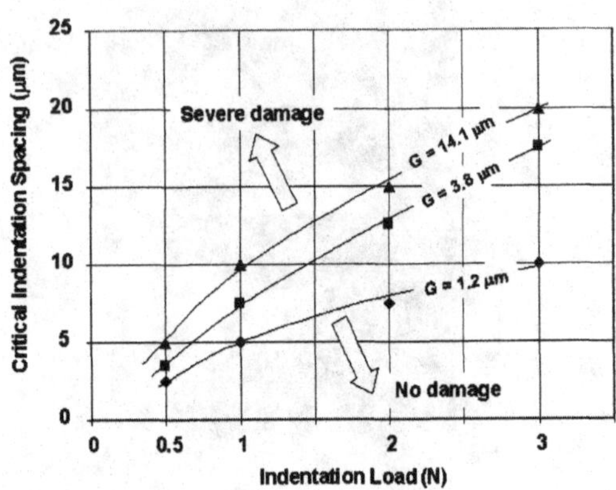

Figure 3. Wear map based on damage produced by grids of 9 indentations performed on polycrystalline alumina materials. Lines indicate critical conditions for producing material loss for the three materials tested.

For a given load the critical indentation spacing for crack linking giving substantial material loss is strongly grain size dependent; it is about two times greater for specimens of the coarsest grain size than for specimens of the finest grain size.

The wet erosion experiments of Franco and Roberts [11-13] used the same alumina materials as studied here. SiC particles of ~650 mm diameter travelling at ~2.4 ms^{-1} were used; comparison of the sizes of isolated impact craters with the sizes of indentations indicates that each impact event at 90° incidence roughly corresponds to a 2N load indentation [11].

Table II compares the "wear rate" for loads of 2 N and the wet erosion rate per particle [11]. Both the wet erosive and simulated wear rates per particle increase by about one order of magnitude with grain size over the range of grain size studied, with the indentation arrays giving loss rates per contact ~2.5 ±0.5 times those from wet erosive wear.

The difference in volume removed per indentation and per erodent particle could be attributed to several factors: (a) During impacts either the sharp or the round area of the SiC grits can strike the surface. Blunt (Hertzian) indentation requires higher loads to produce damage. This effect can be seen in the SEM micrographs shown by Franco et al. [11] for the early stage of erosion. (b) The grid spacings chosen for comparison with wet erosion are those just close enough for the cracks to coalesce. In wet erosion, impacts that produce damage will be on average closer and the volume removed per impact will thus be less. (c) The indentation tests were carried out in air, while the erosion tests were carried out submerged in water. The surface mechanical properties of alumina change with the moisture content of the test environment [16]; however, at the high loads used here, this is unlikely to be a significant effect.

Table II. Volume removed per indentation in grid tests and per particle in erosive wear.

Grain size (μm)	**Volume removed per indentation ($\times10^{-17}m^3$/ind.)**	**Erosion rate ($\times10^{-17}m^3$/particle) impacts at 90°**
1.2	6.33 ± 0.09	2.12 ± 0.08
3.8	15.56 ± 0.09	9.72 ± 0.08
14.1	25.28 ± 0.09	13.14 ± 0.06

Nonetheless, the "wear rates" from the two methods scale with each other well, implying that for low-speed erosive wear, the mechanism of material loss is by damage linkage between successive close impacts, and that the indentation grid method can be used to predict relative erosive wear rates of ceramic materials.

CONCLUSIONS

Arrays of closely spaced Vickers indentation can be used to produce "wear maps", which provide a guideline for predicting the low-impact-velocity erosive wear resistance of brittle materials.

ACKNOWLEDGEMENTS

A Franco Jr., would like to thank The Universidade Católica de Goiás for financial support. The experimental work was supported by CAPES (Brazilian government).

REFERENCES

1. C. Cm. Wu, R. W. Rice, D. Johnson, and B. A. Platt, *Ceram. Eng. Sci. Proc.* **7,** 995 (1985).
2. S-J. Cho, B. J. Hockey, B. R. Lawn, and S. J. Bennison, *J. Am. Ceram. Soc.* **72,** 1249 (1989).
3. D. E. Deckman, S. Jahanmir, and S. M. Hsu, *Wear,* **149,** 155 (1991).
4. H. Liu, and E. M. Fine, *J. Am. Ceram. Soc.* **76,** 2393 (1993).
5. R. W. Rice, *Ceram. Eng. Sci. Proc.* **6,** 940 (1985).
6. D. B. Marshall, B. R. Lawn, and R. F. Cook, *J. Am. Ceram. Soc.* **70,** C-139 (1987).
7. R. W. Rice, and B. K. Speronello, *J. Am. Ceram. Soc.* **59,** 330 (1976).
8. S. M. Wiederhorn, and B. J. Hockey, *J. Mater. Sci.* **18,** 766 (1983).
9. M. Miranda-Martinez, R. W. Davidge, and F. L. Riley, *Wear* **172,** 41 (1994).

10. M. Miranda-Martinez, R. W. Davidge, and F. L. Riley, in *Ceramics in Energy Applications*, Proceedings of the Institute of Energy's 2nd International Conference, London, 20-21 April 1994, The Institute of Energy London, pp.239-252. (1994).
11. A Franco,. *Erosive wear of Alumina*. D. Phil thesis, University of Oxford, Oxford, UK, 1996, pp 119.
12. A. Franco, and S. G. Roberts, *J. of Eur. Ceram. Soc.* **16,** 1365 (1996).
13. A. Franco, and S. G. Roberts, *J. of Eur. Ceram. Soc.* **18,** 269 (1998).
14. A. G. Evans, and T. R. Wilshaw, *Acta Metallurgica.* **24,** 939 (1976).
15. A. G. Evans, and T. R. Wilshaw, *J. Mater. Sci.* **12,** 97 (1977).
16. S. M. Wiederhorn, *Int. J. Fract. Mech.* **4,** 171 (1968).

Mater. Res. Soc. Symp. Proc. Vol. 843 © 2005 Materials Research Society T3.33

Adhesion Improvement of CVD Diamond Coatings on WC-Co Substrates for Machining Applications

Zhenqing Xu[1, 2], Ashok Kumar[1, 2], Leonid Lev[3], Michael Lukitsch[3], and Arun Sikder[2]
[1]Department of Mechanical Engineering, University of South Florida, Tampa, FL, 33620, USA
[2]Nanomaterials and Nanomanufacturing Research Center, University of South Florida, Tampa, FL, 33620, USA
[3]General Motors Corporation, Warren, MI 48090, USA

ABSTRACT

In order to improve the performance of WC-Co cutting tools, high quality microcrystalline diamond coatings were produced using microwave plasma enhanced chemical vapor deposition (MPECVD) method. The adhesion of the diamond film deposited on the substrate has been considered to play an important role in the performance of the cutting tools in machining applications. A thin layer of Cr was coated on the WC-Co substrate before the diamond deposition; 75 μm diamond powders were sandblasted on the surface at 40 Psi to increase the nucleation density. Diamond film has been successfully deposited on the substrate at temperature around 750°C with 1.5 % CH_4 in Hydrogen plasma. Scanning electron microscopy (SEM) has been used to study the surface morphology and Raman spectroscopy has been performed to characterize the quality of the diamond films and measure the residual stress. The adhesion of the diamond film has been evaluated by Rockwell indentation test. The results indicated that film grown on the Cr interlayer with diamond powder sandblasting has much better adhesion strength.

INTRODUCTION

Diamond Coating is an industrial solution to improve the surface properties in a wide variety of wear resistant applications because of its outstanding properties. Along with great wear resistance, the advantages of diamond coating include high surface hardness, high thermal conductivity, reduced friction, better corrosion protection and improved optical properties.

Cemented carbide is a material made by "cementing" very hard tungsten monocarbide (WC) grains in a binder matrix of tough cobalt metal by liquid phase sintering. The cemented tungsten carbide cutting tools are widely used all over the word in automotive industry, metal machining, mining and stone cutting industry. However, without a hard coating such as diamond, the cemented WC tools are found to wear rapidly when machining some particular materials such as green ceramics, abrasive composites or high silicon-filled aluminum. Successful development of diamond-coated tools will significantly benefit the performance of WC-Co cutting tools. As a result, diamond coated tools will have reduced down time, better cutting performance and less defective machined surface.

However, the poor adhesion of the diamond film on the substrate is the main cause of tool failures and becomes the main technical barriers for commercialization of diamond-coated tools [1-4]. The reason behind is that the deposition of diamond films on cemented carbides is strongly hindered by catalytic effect of cobalt. During the deposition, the Co leaching from the WC-Co substrate favors the formation of sp^2 carbon (non-diamond) phase instead of the sp^3 carbon [5]; this is considered the result why diamond films have poor adhesion to the WC-Co substrates. Various approaches have been reported to avoid or reduce the catalytic effect of cobalt on the growth of diamond films [6-10]. In our study, a thin Cr interlayer has been implanted between diamond film and substrate. Sandblasting with diamond powder on the Cr coated substrate was employed to improve the adhesion strength before the deposition.

EXPERIMENTAL

Cemented carbide coupon samples (Co 6%) were used as the substrates for diamond film deposition. In order to prevent the catalyst effect of Co, several surface pretreatments have been done prior to the deposition of diamond coating. Substrate surface was first cleaned using acetone and methanol, then a thin Cr layer (~5 μm) was coated on the WC-Co by unbalanced magnetron sputtering system. After that, the substrates were sandblasted by 75 μm diamond powder at pressure 40 Psi. During diamond powder sandblasting, half of the coupon sample was masked to compare the coating quality and morphology. This sandblasting procedure is believed to help the initial nucleation to get better adhesion strength. This treated substrate was then seeded by methanol solution containing 1μm diamond powder in an ultrasonic bath for 5 hours to further nucleate the surface. Diamond film deposition was carried out in the MPECVD system. Hydrogen and methane were used under the following conditions: microwave power: 800 W; substrate temperature, 750 °C; pressure, 80 Torr; gas composition, H_2:CH_4 (mass flow rate: 500/7.5 sccm); deposition time, 6 h. The surface morphology of diamond film was examined by scanning electron microscopy (SEM). The residual stress of the diamond film was estimated by Raman peak shift and curvature method. Rockwell indentation test was used to evaluate the adhesion of the diamond film. Pin-on-disk test was performed to investigate the wear resistance of the coating.

RESULTS AND DISCUSSION

Effect of Cr interlayer and diamond sandblasting

Cemented carbide substrate, polished to a mirror finish, was coated with a thin Cr interlayer (~5 μm) using sputtering system and sequentially coated with diamond film (~8 μm) at temperature around 750 °C. Figure 1 shows the SEM image of microcrystalline diamond films grown on bare WC-Co substrate. The crack of the film was observed right after the cooling down due to the catalyst effect of Co and the mismatch of coefficient of thermal expansion (CTE).

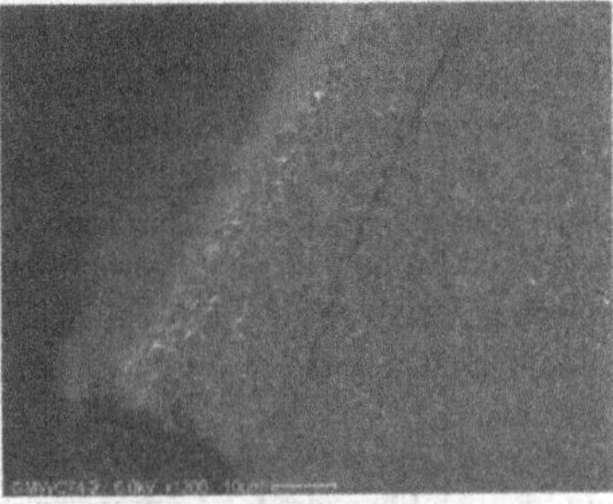

Figure 1. SEM image of diamond coating on bare WC-Co substrate.

Figure 2a shows the SEM image of diamond film on a Cr coated WC-Co substrate with two regions, the top part was masked during sandblasting and the bottom part was blasted by the diamond powder. Both samples have much better adhered coatings than the bare WC-Co substrate. The 5 μm Cr layer is good enough to inhibit the catalyst effect of Co and improves the adhesion strngth. Figure 2b shows the high resolution image of the coating in the blasted region. Much more diamond crystals grew in the blasted region and small diamond particles formed together like cauliflower in some places. This indicates that sandblasted surface has much higher nucleation rate. If diamond nucleation is low, the individual diamond crystals can increase in size before they grow together and form the layer. Voids between the crystals are formed at the interface, which makes the interface weaker. In the case of high diamond nucleation rate, the voids between the crystals remain much smaller and thus improve the adhesion. The adhesion strength of these two regions was evaluated by Rockwell indentation test.

Adhesion evaluation and residual stress measurement

Coating adhesion can be investigated qualitatively using Rockwell indent method [11, 12]. The diamond-coated surface was indented by a Rockwell indenter with a force of 150 kgf. Figure 3a and 3b show the SEM images of indentation area of two samples, with sandblasting and without sandblasting, respectively. Figure 3a shows the indentation area of the sample only with the Cr layer interlayer. Diamond delamination is observed and the average radius of the delamination area is around 300 μm. Figure 3b shows the SEM image of the indentation test done on the surface sandblasted with diamond powder. It is observed that the average delamination radius reduced to 180 μm. and the diamond film adhereed very well with the Cr interlayer along the edge where the coating delaminated. In the first sample, on the contrary, the diamond film peeled off along the edge, indicating that film deposited on the blasted WC-Co substrate has much stronger adhesion strength.

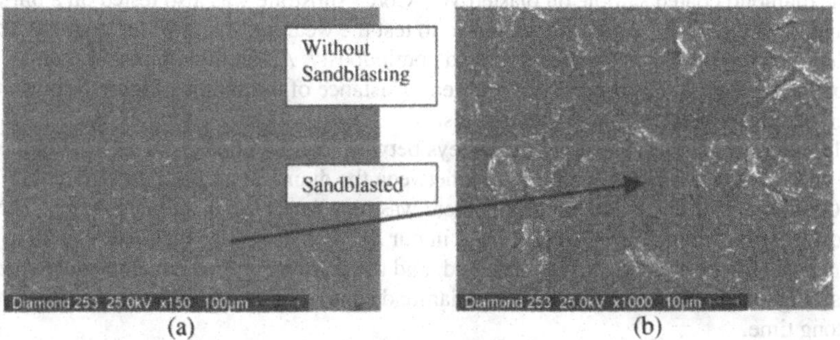

Figure 2. (a) SEM images of diamond coating on Cr coated WC-Co substrate with (bottom part) or without (top part) diamond powder sandblasting. (b) High resolution SEM image of diamond coating on the sandblasted region.

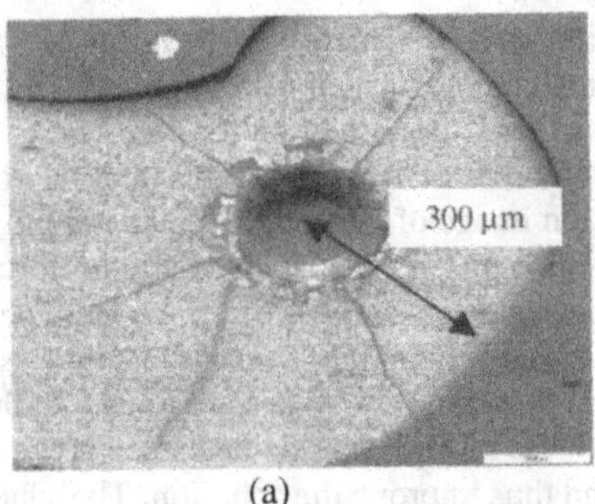

(a)

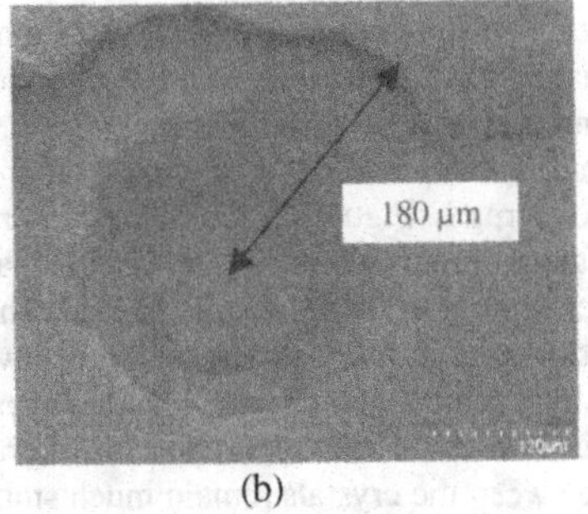

(b)

Figure 3. Effect of sandblasting with diamond powder on Cr coated WC-Co substrate. (a) delamination area without sandblasting; (b) delamination area with sandblasting.

The diamond coating delamination around the Rockwell indent indicates high level of residual stress in the coating. Residual stress of the coating could be measured by Raman shift. The Raman shift on diamond peak at 1332 cm^{-1} is proportional to the film stress according to Ralchenko et al. [13]. From Figure 4, we observed that the diamond peak 1337 cm^{-1} has 5 cm^{-1} shift, which is correlated to compression stress around 2.5 GPa. The residual stress has also been estimated by curvature method. Measured by WYKO optical profilometer, the curvature of the as deposited sample is about 11.5 μm. The curvature of the sliced sample, thickness is 250 μm, is about -0.43m. Using stoney's equation we estimated that the residual compression stress of the diamond coating is around 1.9 GPa.

Pin-on-disk test

The diamond coated sample on blasted WC-Co/Cr substrate was also tested on a pin-on-disk machine. This pin-on-disk machine was used to test the wear resistance of the coating. The sample was fixed on the machine and rotated continuously. An aluminum pin rode on a diamond-coated surface is used to test the wear resistance of the sample. Figure 5 shows the wear track covered by aluminum after the test.

Aluminum was observed filling the valleys between the diamond grain peaks (Figure 5a). Small chips and chunks of aluminum stuck between the diamond grains. After more than 100,000 rotations of pin-on-disk machine there was no wear of the diamond coating. The smallest wear that could have been detected in our SEM observations was less than 20 to 30 nanometers. However, no wear was observed, and the diamond grains remained very sharp after the test (Figure 5b). This suggests that the diamond coating on a machining tool is likely to last for a long time.

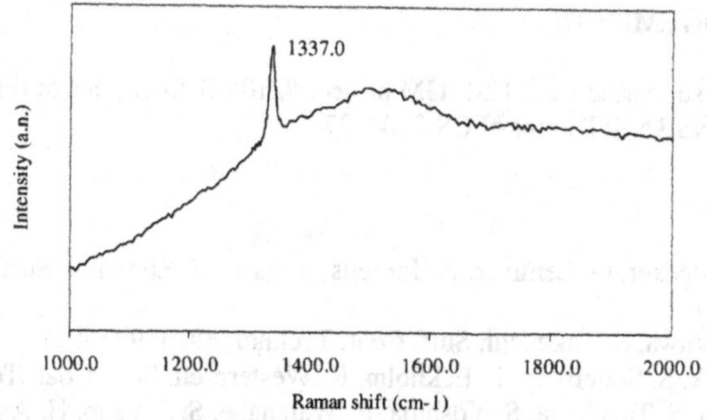

Figure 4. Raman spectrum of diamond film on blasted WC-Co/Cr substrate.

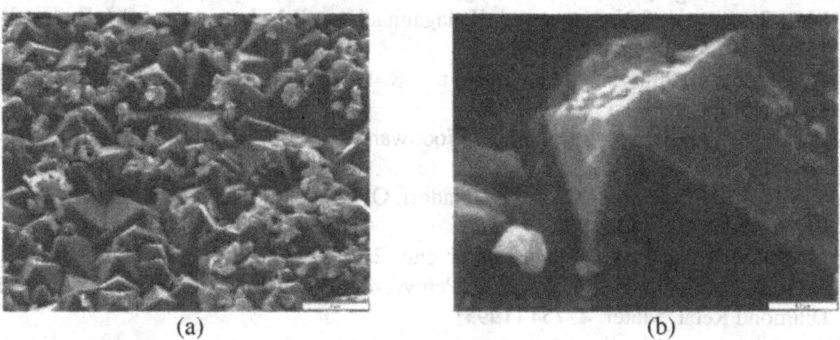

Figure 5. SEM micrographs of (a) diamond coating on cemented carbide substrate with a wear track; (b) diamond grain in the wear track after 100,000 rotations against an aluminum pin.

CONCLUSIONS

In our study, microcrystalline diamond films were deposited on WC-Co cemented carbide substrates using Cr interlayer to improve the adhesion strength. A thin Cr layer around 5 μm was coated on the substrate before diamond deposition, this layer were used as the diffusion barrier layer to inhibit the Co migration during CVD process. Continuous diamond film has been grown on the substrate and adheres well after the deposition. Sandblasting with 75 μm diamond powder has been performed to improve the adhesion and Rockwell indentation test shows that film grown on sandblasted substrate has much less delamination area and stronger adhesion strength. The diamond coating performs very well during the pin-on-disk test and no wear was observed. The residual stress was calculated based on the diamond peak shift by Raman spectroscopy measurement and curvature method. Compression stress around 2 GPa was obtained.

ACKNOWLEDGEMENTS

This work is supported from USF-GM project #2105-099LO. Part of this research is also supported from NSF NIRT grant #ECS 0404137.

REFERENCES

1. T. Leyendecker, O. Lemmer, A. Jurgens, S. Esser, J. Ebberink, Surf. Coat. Technol. **48,** 253 (1991).
2. M. Murakawa, S. Takeuchi, Surf. Coat. Technol. **49,** 359 (1991).
3. J. Reineck, S. Soderbery, P. Eckholm, K. Westergren, Surf. Coat. Technol. **57,** 47 (1993).
4. K. Kanda, S. Takehana, S. Yoshida, R. Watanabe, S. Takano, H. Ando, F. Shimakura, Surf. Coat. Technol. **73.** 115 (1995).
5. S. Amirhaghi, H.S. Reehal, R.J.K. Wood, D.W. Wheeler, Surf. Coat. Technol. **135,** 126 (2001).
6. R. Haubner, B. Lux, Int. J. Refract. Metal Hard Mater. **5,** 111 (1996).
7. Z. Ma, J. Wang, Q. Wu and C.Wang, Thin Solid Films, **390**, 104 (2001).
8. M.S. Raghuveer, S.N. Yoganand, K. Jagannadham, R.L. Lemaster and J. Bailey, Wear, **253**, 1194 (2002).
9. W. Ahmed, H. Sein, N. Ali, J. Gracio and R. Woodwards, Diamond Relat. Mater. **12**, 1300 (2003).
10. H Sein, W. Ahmed, M. Jackson, R. Woodwards, R. Polini, Thin Solid Films, **447**, 455 (2004).
11. S. Silva, V. P. Mammana, M. C. Salvadori, O. R. Monteiro and I. G. Brown, Diamond Relat. Mater. **8**, 1913 (1999).
12. M. D. Drory and J. W. Hutchinson, Science **263**, 1753 (1994).
13. V.G. Ralchenko, A.A. Smolin, V.G. Pereverzev, E.D. Obraztsova, K.G. Korotoushenko, Diamond Relat. Mater. **4**, 754 (1995)

Mater. Res. Soc. Symp. Proc. Vol. 843 © 2005 Materials Research Society T3.9

Comparison of Hardness Enhancement and Wear Mechanisms in Low Temperature Nitrided Austenitic and Martensitic Stainless Steel

S. Mändl, D. Manova, D. Hirsch, H. Neumann, and B. Rauschenbach,
Leibniz-Institut für Oberflächenmodifizierung,
Permoserstr. 15, 04303 Leipzig, Germany

ABSTRACT

Energetic nitrogen implantation into austenitic stainless steel or nickel alloys leads to the formation of a very hard and wear resistant surface layer with an expanded lattice, while similar results are reported for selected martensitic steels. A comparison of the phase formation under identical process conditions at 380 °C using an austenitic (304) and a martensitic (420) steel grade (in an annealed ferrite/cementite condition) shows that the initial microstructure is retained in both of them. As the formation of dislocations and stacking faults cannot account for the dramatic hardness increase, the build-up of compressive stress is proffered as an explanation covering these two steel types. Furthermore the energetic nitrogen implantation apparently stabilizes the expanded lattice with suppressing the chemical transition to iron nitrides and, for steel 420, the metallurgical transition towards austenite at high nitrogen contents.

INTRODUCTION

Austenitic stainless steels can be hardened at temperatures around 350 – 400 °C using energetic nitrogen ion implantation while retaining the corrosion resistance [1,2,3]. The so-called "expanded austenite" is characterized by a concentration dependent diffusion coefficient, strongly increasing at higher nitrogen concentrations [4] and an anisotropic lattice expansion [5]. Selected results on nitrogen implantation into martensitic stainless steels show a comparable lattice expansion [6,7] while others do not show this effect [8,9].

Following a previous investigation focused on the nitrogen diffusion and the lattice expansion in austenitic and martensitic stainless steel, as observed by X-ray diffraction (XRD) [10], a more detailed insight into the respective surface morphology is presented here. Additionally, possible origins of the increased hardness and wear resistance are discussed. The method of choice for nitrogen ion implantation is plasma immersion ion implantation (PIII) allowing a fast and cost efficient implantation for complex 3D-objects [11,12].

EXPERIMENT

This investigation focuses on austenitic stainless steel AISI 304 (X5CrNi18.10) and martensitic stainless steel 420 (X46Cr13), however additional experiments performed on 316Ti (X6CrNiMoTi17.12.2), 431 (X17CrNi16.2) and 430F (X14CrMoS17) did not yield significant differences. Samples prepared from 304 showed an austenitic structure with an average grain size between 10 and 15 μm whereas a ferritic microstructure with cementite inclusions was observed for 420 (see Fig. 1). Hence, an additional heat treatment step resulting in a change from the martensitic structure must have occurred prior to the receipt of the samples in the laboratory.

The experiments were performed in a HV chamber at a base pressure below 5×10^{-3} Pa and working pressure of 0.2 Pa. Negative high voltage pulses of 10 or 25 kV with a pulse length of 15 µs were applied to the samples, immersed in a nitrogen plasma with a density of 6×10^{9} cm^{-3}. The treatment time was between 1 and 2 hours resulting in an incident dose of 10^{18} nitrogen atoms/cm^2 and beyond. The process temperature was varied between 350 and 380 °C, adjusted by the repetition rate. No external heating or cooling was applied so the samples were heated only by ions.

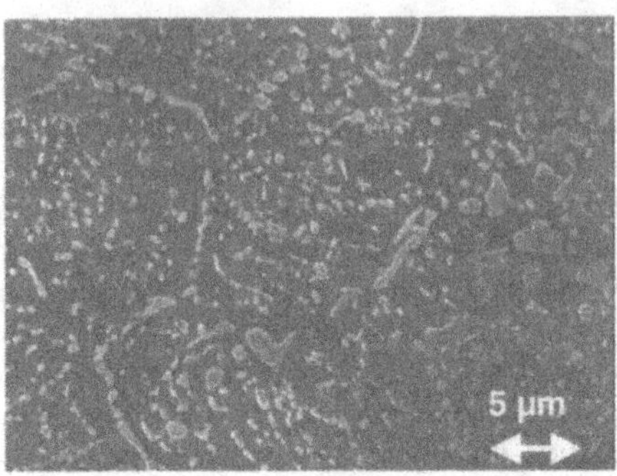

Figure 1. Plane view SEM viewgraph of AISI 420 before PIII treatment.

Elemental depth profiles were obtained from glow discharge optical spectroscopy (GDOS) while XRD was used to determine the phase composition. Metallographic cross-sections were prepared and analyzed using scanning electron microscopy (SEM). Additionally, dynamic hardness values were derived from Vickers nanoindentation measurements (load 1000 mN), together with specific wear data from an oscillating ball-on-disc test (WC ball, diameter 3 mm, normal load 3 N, contact pressure 1.0 GPa at an average velocity of 15 mm/s, track length 20 mm).

RESULTS

Fig. 2 shows GDOS and XRD data for samples from 304 and 420 steel implanted with 10 kV voltage pulses at 350 °C for 2 hours. A fast and deep nitrogen diffusion beyond the implantation range of 6.5 or 12 nm, respectively for molecular ions N_2^+ and atomic ions N^+, is observed for both steel grades. The nitrogen profile in 420 (Fig. 2.c) corresponds to the *erfc*-shape, whereas a kink is observed for 304 in Fig. 2a at a concentration of 20 at.%, indicative for the concentration dependent diffusivity [4]. With this increased nitrogen diffusion at higher concentrations, a similar diffusion rate, i.e. total layer thickness, is observed in both the fcc and bcc structures. Chromium enrichment, albeit uncorrelated with the nitrogen content is also present in these two steel grades.

The XRD spectra show satellites towards the left side of the base material reflections in Fig. 2.b&d. For steel 304, these broad structures are indicative of the formation of expanded austenite, with a larger lattice constant normal to the surface for the grains oriented with the (100) planes parallel to the surface than for the (111) oriented grains [5]. A similar effect, with an expansion larger for the (200) than for the (110) oriented grains is observed in 420. The relative X-ray intensity for the base material and the surface layer depends on the layer thickness, however an exact correspondence is not present as different point defect densities modify the absolute scattering intensities.

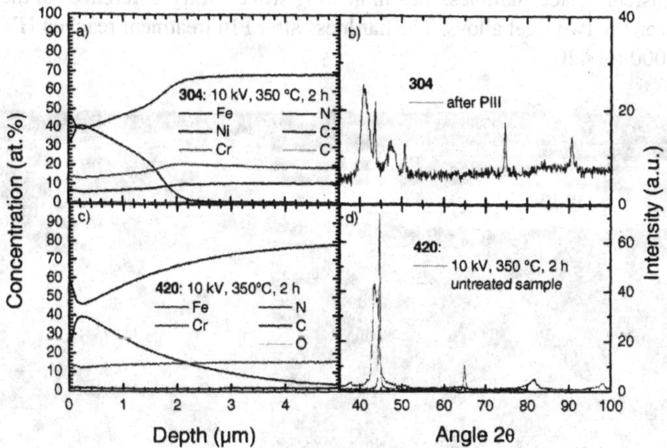

Figure 2. GDOS depth profiles and XRD spectra for 304 and 420 steel implanted with nitrogen PIII at 10 kV.

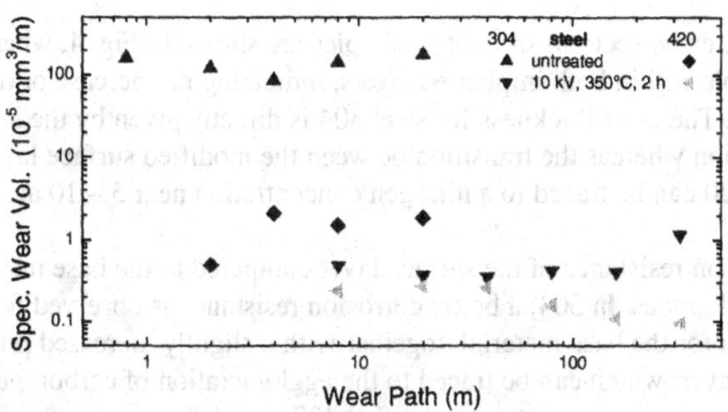

Figure 3. Specific wear volume as a function of the wear path for untreated and implanted 304 and 420 steel as determined in a ball-on-disc tribometer.

Fig. 3 presents the measured specific wear of 304 and 420 steel samples implanted with nitrogen at identical process parameters (i.e. 10 kV, 350 °C and 2 hours) together with the values for non-implanted samples as a function of the total wear length on a logarithmic scale. A similar wear resistance of both materials is measured after nitriding despite a completely different microstructure and a large difference in the wear rates of the base materials.

As the layer thickness of 304 steel grade sample is slightly thinner, a substrate influence leading to a minor increase in the wear rate at longer paths can be seen. In contrast, the wear of 420 decreases for longer wear paths, indicating some additional work hardening. For all four samples, adhesion of metallic particles onto the ceramic ball below accuracy of the used microbalance (20 μg) was observed. Albeit, less material is accumulated at the edge of the wear tracks on the austenitic steel samples, thus indicating still existing differences in the wear behavior between the two steel alloys. The hardness after PIII treatment reached HV 1200 for 304 and HV 1000 for 420.

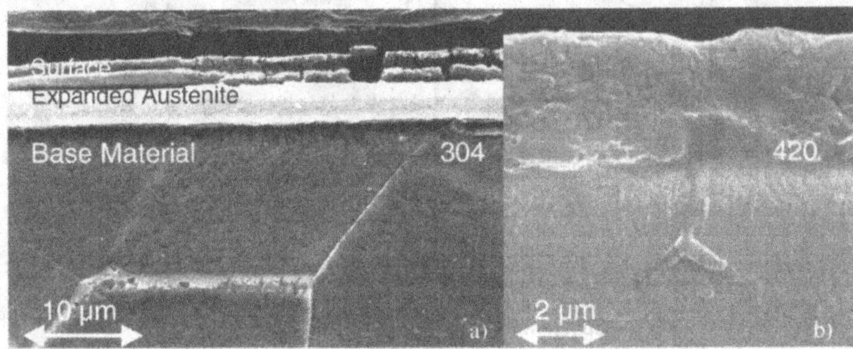

Figure 4. SEM cross-section after embedding and etching of nitrogen implanted 304 and 420 steel.

The metallographic cross-sections of treated samples are shown in Fig. 4, where no change of the grain size is visible within both implanted layers, indicating no increase or decrease of the grain boundary density. The layer thickness for steel 304 is directly given by the steep drop in the nitrogen concentration whereas the transition between the modified surface layer and the base material in steel 420 can be traced to a nitrogen concentration near 5 – 10 at.%.

The relative corrosion resistance of the surface layer compared to the base material is reversed in the two steel grades. In 304, a better corrosion resistance is observed for the expanded austenite than for the base material, together with a slightly increased pitting corrosion just below the surface layer, which can be traced to the agglomeration of carbon pushed ahead by the inward diffusing nitrogen. In contrast, modified 420 steel has a worse corrosion resistance than the base material. However, no absolute comparison is available from these two viewgraphs as different etchants and times were used.

DISCUSSION

No change in the crystal size is visible from the SEM viewgraphs, so that hardening via decrease of the grain size can be ruled out as an explanation for the hardness increase by factor of 3 – 4. At the same time, precipitation hardening – e.g. CrN – within the surface layer or hardening by an increased dislocation and stacking fault density can be excluded from both SEM images and XRD data. The formation of a very hard surface compound like Fe_3N or Fe_3C can be ruled out from GDOS and XRD data.

As the only difference between the modified surface layer and the base material is an expanded lattice, it can be speculated that compressive stresses inside the surface layers, reaching values of 1.5 GPa and more [13] are responsible for the very high hardness values. This could also explain wear and hardness data nearly independent of the actual microstructure of the material.

Beside a similar change in the mechanical surface properties, tendencies for a modification of the surface morphology after nitrogen PIII can be compared for the austenite and ferrite in the present investigation. It has to be pointed out again that a nominally martensitic steel was transformed into a ferrite/cementite compound due to prior heat treatment in the supply chain. Using XRD data alone, without a detailed morphological investigation, such changes remain unnoticed and may contribute to differing results in the literature on the effect of nitrogen implantation into "martensitic steel" [6,8].

An increased Cr and Ni content (respective the equivalent values after inclusion of the additional alloying elements [14]) leads to an enlargement of the area for austenite stabilization, hence the massive nitrogen incorporation should induce a transformation towards austenite from the ferritic and martensitic part of the phase diagram [15]. However no such transition was neither observed in steel 420 nor in 440 B nor in 630 [10], and thus it must be concluded that such a metallurgical transition towards the stable austenitic structure is thermodynamically not favoured for these investigated alloys in the temperature region near 350 – 380 °C.

However, at slightly higher temperatures near 420 – 450 °C, the onset of Cr mobility [16] leads to the formation of CrN precipitates for both austenitic and martensitic steels [17,10]. For low alloy steel grades, an additional reaction path exist, leading to ε-$Fe_{2\text{-}3}N$, predominantly at temperatures below 400 °C and γ'-Fe_4N at higher temperature [18]. The initial accommodation of implanted nitrogen (or carbon) on interstitial places, accompanied by a lattice expansion is a common observation in transition metals before the formation of nitrides [19]. However, why is this reaction suppressed in stainless steels with a high Cr and Ni content. One possible explanation could be the modification of interatomic potentials and the electronic structure by the alloying elements. Especially for the transition metals, a very strong influence of the number of d (or f-electrons) on the phonon dispersions is known [20,21], which may be the origin of the window of opportunity below 400 °C to stabilize the expanded lattice in austenite, ferrite and martensite and allow the favorable mechanical properties.

SUMMARY & CONCLUSIONS

Using nitrogen PIII at intermediate temperatures of 350 – 380 °C, an expanded lattice was formed in stainless steel 304 and 420 with nearly identical hardness and wear properties. An investigation of the microstructure revealed an ferrite/cementite composition instead of the nominally martensitic structure of steel 420. Hence, expanded ferrite was observed instead of expanded martensite, which cannot be distinguished by XRD alone. Compressive stress induced by the ion bombardment, together with a possible influence of the electronic structure of stainless steel may be responsible for the mechanical properties as well as the phase stabilisation.

ACKNOWLEDGEMENTS

E. Richter, Forschungszentrum Rossendorf, is acknowledged for providing wear data. This work was supported by EU/SMWA 6204/947.

REFERENCES

1. D. L. Williamson, J.A. Davis and P. J. Wilbur, *Surf. Coat. Technol.* **103-104**, 178 (1998).
2. E. Menthe and K.-T. Rie, *Surf. Coat. Technol.* **116-119**, 193 (1999).
3. S. Mändl and B. Rauschenbach, *Defect Diffus. Forum* **188-190**, 125 (2001).
4. S. Mändl and B. Rauschenbach, *J. Appl. Phys.* **91**, 9737 (2002).
5. S. Mändl and B. Rauschenbach, *J. Appl. Phys.* **88**, 3323 (2000).
6. S. K. Kim, J.S. Yoo, J. M. Priest and M. P. Fewell, *Surf .Coat. Technol.* **163-164**, 380 (2003).
7. K. Marchev, C. V. Cooper and B. C. Giessen, *Surf. Coat. Technol.* **99**, 229 (1998).
8. I. Alphonsa, A. Chainani, P. M. Raole, B. Ganguli and P. I. John, *Surf. Coat. Technol.* **150**, 263 (2002).
9. A. Leyland, D. B. Lewis, P. R. Stevenson and A. Matthews, *Surf. Coat.Technol.* **62**, 608 (1993).
10. S. Mändl and B. Rauschenbach, *Surf. Coat. Technol.* **186**, 277 (2004).
11. J. R. Conrad, J. L. Radtke, R. A. Dodd, F. J. Worzala and N. C. Tran, *J. Appl. Phys.* **62**, 4591 (1987).
12. G. A. Collins, R. Hutchings, J. Tendys and M. Samandi, *Surf. Coat. Technol.* **68-69**, 285 (1994).
13. S. Sienz, S. Mändl and B. Rauschenbach, *Surf. Coat. Technol.* **156**, 185 (2002).
14. A. L. Schäffler, *Metal Progress* **56**, 680 (1949).
15. V. G. Gavriljuk and H. Berns, *High Nitrogen Steels*, (Springer, Berlin, 1999) p. 236.
16. D. L. Williamson, O. Ozturk, R. Wie and J. P. Wilbur, *Surf. Coat. Technol.* **65**, 15 (1994).
17. X. Li, M. Samandi, D. Dunne, G. Collins, J. Tendys, K. Short and R. Hutchings, *Surf. Coat. Technol.* **85**, 28 (1996).
18. K. H. Jack, *Acta Crystallogr.* **13**, 392 (1950).
19. M. Nastasi, J. W. Mayer and J. K. Hirvonen, *Ion-solid interactions,* (Cambridge University Press, Cambridge, 1996) p 332.
20. F. Willaime, *Adv. Eng. Mat.* **3**, 283 (2001) 283.
21. J. Wong, M. Krisch, D. L. Farber, F. Occelli, A.J. Schwartz, T.-C. Chiang, M. Wall, C. Boro and R. Xu, *Science* **301**, 1078 (2001).

Mater. Res. Soc. Symp. Proc. Vol. 843 © 2005 Materials Research Society T3.38

Pseudomorphic stabilization on crystal structure and mechanical properties of nanocomposite Ti-Al-N thin films

A. Karimi, Th. Vasco, A. Santana
IPMC - Faculty of Basic Science
Swiss Federal Institute of Technology Lausanne (EPFL), Switzerland

Abstract

As the dimensions of materials are reduced to the nanometer scale, the stabilization of pseudomorphic crystal structures that differ from their bulk equilibrium phases can occur. The pseudomorphic growth could provide a substantially larger bulk modulus and greater hardness than the average of the constituent materials in nanolayered and nanocomposite thin films. To evaluate this effect in $Ti_{1-x}Al_xN$ system, a series of nanostructured films with x up to 0.7 were deposited onto WC-Co substrates using arc PVD, and characterised in terms of structure-property relations. Chemical composition by RBS together with HRTEM and XRD analysis showed that for the Al content below $x = 0.4$ a solid solution single-phase film is formed, while for x values beyond 0.5 mixed structures made of fcc-TiN and hcp-AlN, or nanocomposite films made of fcc-TiN, hcp-AlN, and fcc-AlN appeared depending on deposition conditions. Hardness of solid solution films was found to increase almost linearly with the Al content, while two opposite behaviours were distinguished for composite structures. Hardness rapidly decreased according to the rule of mixture as soon as solid solution phase began to separate into TiN and AlN growing in their natural structures with misfit dislocation at the interface. In contrast, further hardness enhancement was measured when nanocomposites with coherent interfaces were formed due to pseudomorphic stabilization of fcc-AlN on fcc-TiN crystallites.

Introduction

Titanium-aluminium nitride thin films have been recognised to provide effective protection to cutting and forming tools operating under high speed cutting conditions [1, 2]. The improved performance of these coatings is believed to arise from their ability to retain higher hardness at high operating temperatures, thereby hindering abrasive and dissolution wears of cutting edges. Further improvement of the mechanical properties of TiAlN coatings could be achieved by grain size refinement down to the nanometer length scale through alloying of TiAlN coatings, or through advanced deposition techniques to favour the formation of nanocomposites and compositionally modulated nanolayers [3,4]. Several experimental works were confirmed that the nanostructured TiAlN-based nitrides exhibit better functional properties and elevated thermal and chemical stabilities compared to monolithically grown films as reviewed in references [5-7]. The effects of nanostructuring on deformation mechanisms and mechanical strengthening of monolithic materials were relatively well described by Hall-Petch relation stipulating a dependence of hardness enhancement on the square root of grain size as explained by dislocation pile-up and work hardening models [8]. In contrast, the remarkable property of nanocomposites and nanolayers is associated with their ability to impart larger strain anisotropy, and the structure barrier strengthening of the interfaces and related shear stress enhancement that can be transformed into high yield strength and high hardness, significantly beyond that given by the rule of mixture. One of the experimental methods to produce nanostructured $Ti_{1-x}Al_xN$ films is the phase decomposition by exceeding the solubility limit of Al in TiN. The solid solution single phase TiAlN begins to decompose into mixed structure made of fcc-(Ti,Al)N and würtzite AlN when the concentration of Al exceeds $x \approx 0.4$ [9-11]. The present work deals with the effect of Al content, but contrary to previous studies devoted to conventional coarse-grained films, measurements are extended into nanograin films where the pseudomorphic stabilization [12] is expected to modify crystal structures and mechanical properties as compared to coarse structured films. Unlike hardness, fracture is less considered in previous studies on nanostructured thin films, even though the abrasive wear of hard coatings is dominated by the formation of submicron debris resulting from propagation of near surface nanocracks or extension of intercolumnar through thickness microcracks. The present study tempts to provide an improved understanding of the nanostructuring effects on hardening mechanisms and crack propagation in hard thin films

Experimental Details

The studied films were grown on the polished tungsten carbide substrates in a cathodic arc PVD unit using the procedure described in reference [13]. The microstructure of coatings was investigated by means of transmission electron microscopy including conventional (Philips CM20) and high resolution (Philips CM300). To prepare TEM foils for cross sectional observations, the coated samples were first cut to slices of about 500 μm thickness, then thinned by mechanical polishing on diamond pads down to ~ 40 μm before subjecting to Ar^+ ion-milling (PIPS) for final thinning to be transparent to the electron beam. The crystalline structures and phase formation were analysed based on X-ray diffraction spectra (XRD) collected using a Rigaku diffractometer working under grazing incidence ($\alpha = 4°$) and θ-2θ diffraction geometry and Cu K_α radiation. Overall mechanical properties including hardness were determined by means of nanoindentation measurement methods (XP-MTS). Rutherford Backscattering Spectroscopy (RBS) was carried out for elemental analysis based on Particle Induced X-ray Emission (PIXE) using a 2.5 MeV proton beam.

Results

a - Crystalline structure with the Al content

The XRD spectra shown in Fig. 1 provide a picture of the influence of the Al content on crystalline structure and phase formation in $Ti_{1-x}Al_xN$ thin films. The samples with the Al content below ($x < 0.4$) exhibit a single phase fcc-NaCl type structure as distinguished by its (200) peak located at $2\theta \approx 43.36°$. Increasing the Al content up to the solubility limit ($x \approx 0.5$ - 0.6) results in the formation of mixed structures made of fcc-(Ti,Al)N and fcc-AlN with their (002) peak located at $2\theta \approx 42.73°$ and 43.90°, respectively. Only when the Al concentration reaches to $x \approx 0.7$, a notable quantity of würtzite hcp-AlN can be observed in the as deposited films, identified by its typical $(11\bar{2}0)$ peak at $2\theta \approx 59.36°$. These diagrams show that the position of (002) peaks in mixed structures is dependent of the Al content, in particular that of TiN phase. By changing the Al concentration from 0.5 to 0.7, the AlN(002) peak shifts only 0.1° towards higher 2θ, while that of TiN(002) moves $\Delta\theta = 0.6$ °, being from 42.75° to 43.35°. The peak shift in TiN could mainly be associated with the substitution of titanium atoms by aluminium atoms with smaller diameter, resulting in a linear reduction of lattice parameter. In contrast, the solubility of Ti in AlN lattice being limited to a few per cent [14], no significant change in lattice parameter of fcc-AlN is expected. The preferred orientation of coatings is (002) for both single-phase films as well as for mixed structure films of (TiN+AlN). There is no systematic study of texture development in arc deposited TiAlN films and little is known on controlling mechanisms. Experimental results from literature, mostly magnetron sputtered films [9-11] seem to bear out the occurrence of (111), (200), or even (220) textures depending on deposition conditions. Meanwhile, more data are available on texture evolution in TiN layers, strong function of surface energy, adatom mobility and chemical potentials [10]. Higher temperatures and greater ion flux were suggested to acquire enhanced diffusion mobility and thus facilitate adatoms to move to the thermodynamically stable positions which are the (002) planes with the lowest surface energy. The results of Fig. 1 tempt to show the growth of Ti-Al-N films behaves very similar to TiN films. From the full width at half maximum (FWHM) values of the (002) peaks and using the Scherrer equation [21] the average grain size of the crystallites was calculated as a function of Al content. The grain size of both TiN and AlN decreases from ≈ 30 nm down to about10 nm when the Al content of the films increases from x ≈ 0.4 to x ≈ 0.7 (Fig. 1b).

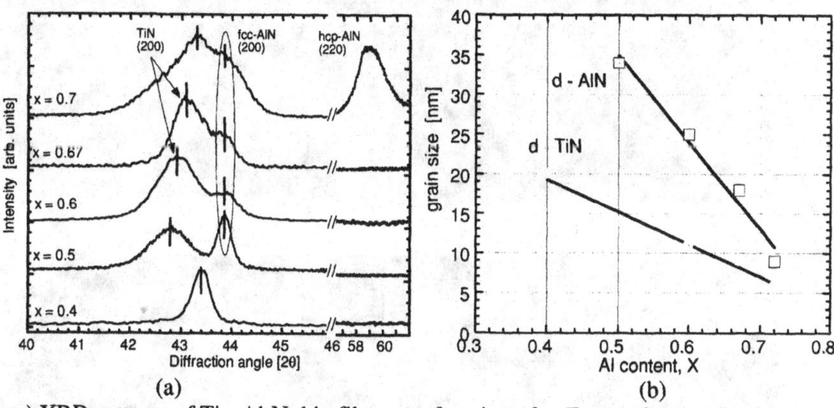

Fig. 1: a) XRD patterns of $Ti_{1-x}Al_xN$ thin films as a function of x. For $x < 0.4$, solid solution single-phase film are formed. For x = 0.5 - 0.6, two-phase films made of fcc-(Ti,Al)N and fcc-AlN are observed. For $x > 0.7$, nanocomposites made of fcc-(Ti,Al)N, fcc-AlN, and hcp-AlN are grown. b) The mean grain size decreases with the Al content of the films.

b - Structure refinement and hardness

Aluminium contributes to structure refinement of $Ti_{1-x}Al_xN$ films by two different ways. In solid solution single phase films it favours the secondary nucleation mechanisms. In composite films it supports phase decomposition beyond the solubility limit and the formation of polytype AlN structures, i.e., wurtzite-hcp, zincblend-fcc, and rocksalt-fcc. The refining role of Al is confirmed by transverse cross-sectional TEM observation of samples reported in Fig. 2. For solid solution films (Fig. 2a) the mean grain size decreases from about 100 nm to 30 - 35 nm when the Al content reaches to x = 0.4, while crystalline structure remains B1 TiN type as identified by its selected area diffraction pattern in Fig. 2d. Increasing the Al content up to x ≈ 0.5, favours the formation of AlN precipitates at grain boundaries of (Ti,Al)N (Fig. 2 b). From high resolution TEM micrograph the crystalline structure of this AlN precipitate cannot be identified, but according to XRD spectrum (Fig. 1) it could be zincblend cubic. The samples with the Al content

$x \approx 0.7$ develop a nanocomposite structure made of fcc-(Ti,Al)N, fcc-AlN, and hcp-AlN as distinguished by their respective SAD patterns in Fig. 2d right picture. Most of the interplanar spacing (d_{hkl}) of cubic AlN phases is very close to those of TiN and consequently the related diffraction rings are superposed. Regarding hcp-AlN, two diffraction rings (100) and (101) differ enough from (111) and (220) of (Ti,Al)N to be used for reliable indexation of different phases. The grain size of this sample 8-10 nm as estimated from HRTEM micrograph in Fig. 2c is comparable to that calculated from broadening of (002) XRD peaks in Fig. 1b.

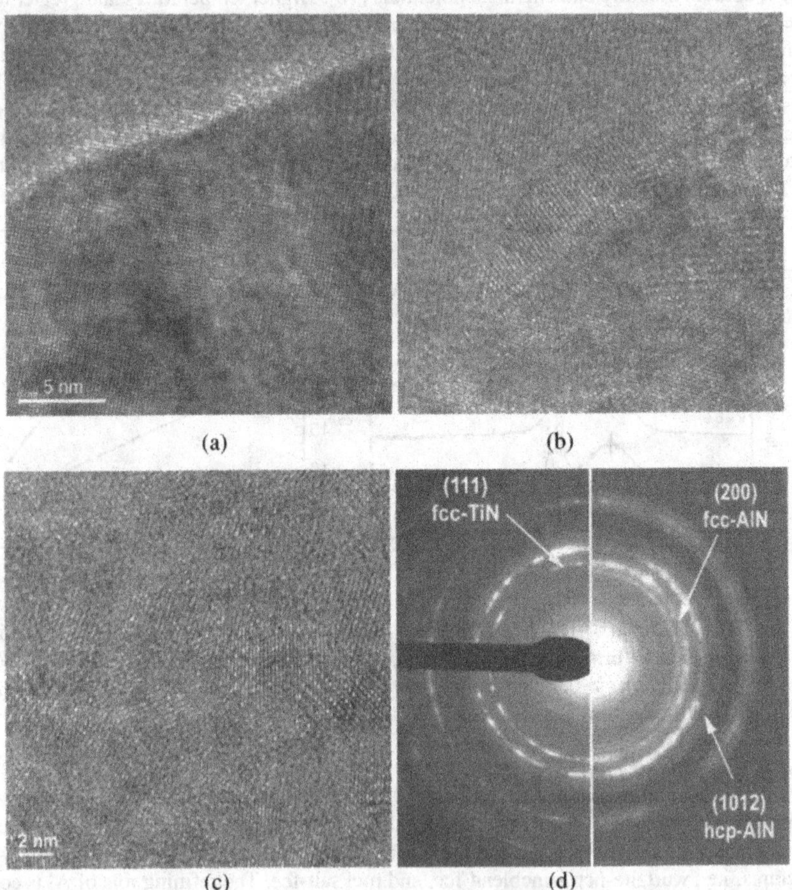

Fig. 2: HRTEM micrographs showing structure of films with different Al content, a) solid solution for lower Al, b) precipitation of AlN at grain boundaries for medium Al concentration, c) nanocomposite made of TiN, fcc-AlN, and hcp-AlN for higher Al, d) comparison of diffraction patterns from nanocomposite and solid solution films.

Previous publications on AlN claim that the equilibrium structure of AlN at atmospheric pressure is hexagonal-type wurtzite, which can be transformed into zincblend or rocksalt type under high pressure and epitaxial growth. According to XRD patterns, the lattice mismatch between cubic AlN (a = 0.412 nm) and (TiAl)N (a = 0.423 nm) being about 2.6% allows to expect the formation of epitaxial and coherent structures when coating grain size decreases to nm scale range. This could explain why in films with Al content between x = 0.4 - 0.7 the precipitation of fcc-AlN is favoured rather than the formation of more stable hcp-AlN. Such stabilization of pseudomorphic crystal structures that differ from their bulk equilibrium phases was reported to occur as the size of material is reduced to the nanoscale regime [12]. The variation of hardness presented in Fig. 3 exhibits two different behaviours depending on the film structure. In solid solution films, hardness increases almost linearly with the Al content. In contrast, for nano-composite coatings the measured hardness varies between an upper and a lower limit. The lower values correspond to two-phase films made of fcc-(Ti,Al)N and wurtzite-AlN obeying to the rule of mixture. The higher values correspond to nanocomposite films made of fcc-(Ti,Al)N, fcc-AlN, and hcp-AlN where coherency stresses compliment the intrinsic hardness. The formation of hcp-AlN, which is the softer phase, contributes to enhancement of fracture toughness (Fig. 3) estimated from the length of radial cracks.

Discussion

The crystalline structures of cathodic arc $Ti_{1-x}Al_xN$ films for various x values were identified using XRD (Fig. 1) and SAD (Fig. 2). Both methods confirmed the formation of single-phase B1 fcc-TiAlN for x < 0.4, mixed structures (B1 fcc-TiAlN and B3 zincblend fcc-AlN) for 0.4 < x < 0.6, and a composite film made of fcc-TiAlN, fcc-AlN and hcp wurtzite AlN for x > 0.7. These observations are somehow different from the results of other studies on phase formation and microstructures of Ti-Al-N films deposited using the cathodic arc ion plating [9], ultra high vacuum sputtering [10], and ion beam assisted deposition [11]. These investigations were commonly reported the formation of B1-NaCl structure for x < 0.4 - 0.5 and mixed phases of B1-cubic together with B4 wurtzite for higher values of x. The difference between the results of present work and previous investigations can be understood in terms of structure refinement and pseudomorphic stabilization. Pseudomorphysm occurs in materials with poltype crystal structures that slightly differ from their bulk equilibrium, and when the size of material is reduced to the nanoscale regime [12]. In fact, grain size of $Ti_{1-x}Al_xN$ films in previous works was within the submicron range being notably coarser as compared to the grain size of the present films being around 8 - 10 nm. At this scale, fcc-AlN can be preferentially stabilized by its isostructured fcc-TiN rather than the formation of incoherent wurtzite AlN [9-11].

Hardness of the nanocrystalline $Ti_{1-x}Al_xN$ films as a function of x (Fig. 3) presents a complex feature. It increases almost linearly with the Al content up to $x \approx 0.4$ and then two opposite behaviours can be distinguished for higher values of x; a rapid decrease when the solid solution phase begins to separate into fcc TiN-based and wurtzite AlN phases, or further hardness enhancement when nanocomposite structures with coherent interfaces are formed due to pseudomorphic stabilization. Solid solution hardening generated from substitution of Ti atoms by smaller Al atoms is understood by development of compressive stress

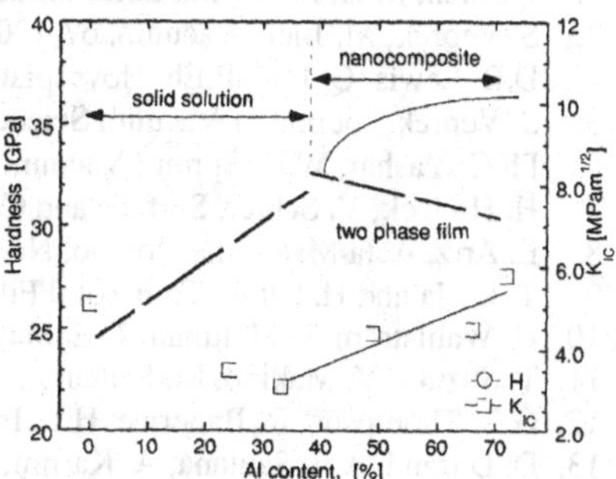

Fig. 3: Hardness and fracture toughness as a function of the Al content.

fields from local contractions and their interaction with the elastic stress field of dislocations. This hardening can be expressed by the Fleischer model [15] of substitutional solution hardening involving the atomic size misfit and elastic modulus mismatch. The increase of shear strength $\Delta\tau$ and thereby hardness H is proportional to the solution atom concentration (c),

$$\Delta H \propto \Delta\tau \propto \left|\varepsilon_G - \alpha\varepsilon_b\right|^{3/2} c^{1/2} \qquad (1)$$

where ε_G and ε_b are modulus mismatch and atomic size misfit, and α is a constant. In nanocomposite coatings further hardening mechanisms come into play because of phase separation; for example thermal and coherency stresses. Thermal stresses are generated since TiN and AlN have different thermal expansion coefficient α_1, α_2. A film deposited at temperature T_1 and then cooled or heated to a different temperature T_2 will experience a thermal stress estimated from following equation:

$$\sigma_{th} = E(\alpha_1 - \alpha_2)(T_2 - T_1)/(1-\nu) \qquad (2)$$

The elastic stress field due to the coherent growth of fcc-AlN on fcc-TiN is proportional to the lattice mismatch between the two crystallites.

$$\sigma_{coherency} \propto E/2(1+\nu)(a_{TiN} - a_{AlN})/a_{AlN} \qquad (3)$$

Thermal stresses evaluated to be less than 1 GPa in these films couldn't explain the hardness enhancement of 3-4 GPa in nanocomposites compared to solid solution films. In contrast, the coherency stresses estimated to 4-5 GPa could be considered a real source of hardening in these coatings. The role of coherency stresses on hardening of nanocomposites becomes still more evident when the results of previous works [9-11] are considered. In these studies no coherent structures were observed, and at the same time the hardness of mixed structures was found to remain always lower than that of solid solution single-phase films. In contrast, the pseudomorphic growth and related coherency stresses in the present work provide to nanocomposites further hardness, which is notably higher than that of solid solution thin films.

Acknowledgment

The Swiss National Science Foundation (SNSF) is acknowledged for the financial support of the project. We are grateful to the Centre Interdepartmental de Microscopie Electronique (CIME-EPFL) for access to SEM and TEM facilities.

References

1. S. PalDey, S.C. Deevi, Materials Science and Engineering, A342 (2003) 58 - 79.
2. J. Shieh, M.H. Hon, Thin Solid Films, 391 (2001) 101 - 108.
3. S. Veprek, M. Jilek, Vacuum, 67, (2002) 443 - 449.
4. D.B. Lewis, Q. Luo, P.Eh. Hovsepian, W.-D. Münz, Surf. Coat. Technol. 184 (2004) 225.
5. S. Veprek, Journal of Vacuum Science and Technology A17(5) (1999) 2401-2420.
6. Ph.C. Yashar, W.D. Sproul, Vacuum 55(1999) 179 - 190.
7. H. Holleck, V. Schier, Surface and Coating Technologies 76-77 (1995) 328 - 336.
8. E. Artz, Acta Materialia Vol. 46, No. 16 (1988) 5611 - 5626.
9. T. Ikeda and H. Satoh, Thin Solid Films, 195 (1991) 99 - 110.
10. U. Wahlström, L. Hultman, J.-E. Sundgren, J.E. Greene, Thin Solid Films, 235 (1993) 62.
11. T. Suzuki, Y. Makino, M. Samandi, S. Miyake, J. Mater. Sci. 35, (2000) 4193 - 4199.
12. G.B. Thompson, R. Banerjee, H.L. Fraser, Applied Physics Letters, 84 (2004) 1082 - 1084.
13. D. Durand, A.E. Santana, A. Karimi, Surf. Coat. Technol., 163-164 (2002) 260.
14. O. Ersen, M.-H. Tuilier, O. Thomas, P. Gergaud, Applied Physics Letters 83 (2003) 3659.
15. R.L. Fleischer, Substitutional solution hardening, Acta, Metallurgica 11 (1963) 203-209.

Ion Beam, Laser and Plasma Modification of Surfaces

Mater. Res. Soc. Symp. Proc. Vol. 843 © 2005 Materials Research Society T5.1

Simultaneous Effects on Topography, Composition and Texture in Ion Assisted Deposition of Thin Films

James M.E. Harper
Department of Physics, University of New Hampshire
Durham, NH 03824

ABSTRACT

Ion bombardment during deposition may simultaneously affect thin film topography, composition and crystallographic texture. Ion etching can produce periodic ripples that depend on the angle of ion incidence and surface temperature. When applied during deposition, ion bombardment can produce in-plane crystallographic orientation in polycrystalline materials for specific angles of incidence. In addition, ion bombardment changes the composition of multicomponent thin films according to the local angles of ion incidence and ion/atom ratios. Therefore, these three mechanisms may be linked under certain deposition conditions to generate novel topographically patterned materials with locally controlled composition and texture. Examples include metal alloys, oxides and nitrides, and recommendations for specific nanoscale structures are given.

INTRODUCTION

Ion bombardment (or other energetic particle bombardment) is commonly used during thin film deposition because it modifies materials in ways that may be essential to obtaining the desired properties. In plasma-based deposition systems, including magnetron sputtering, the degree of bombardment is typically not directly controlled. Ion beam assisted deposition systems, on the other hand, allow independent control of ion flux, energy and angle of incidence, providing means to identify the main components of bombardment that are responsible for property modification [1-4]. Three important effects of ion bombardment are modifications of topography, composition and texture (here, "texture" refers to crystallographic texture, or preferred orientation). In this paper, we explore the possibility of controlling simultaneously thin film topography, composition and texture. Using results from studies of ion etching and ion beam assisted deposition, we focus on the processes of periodic ripple formation, preferential sputtering and biaxial texture formation. By considering the angular dependence of these processes, we suggest that there are special conditions under which simultaneous control is possible. We propose a range of experimental conditions under which this control may be observed for several common types of crystal structure. Simultaneous control of properties allows the formation of self-patterned nanoscale heterogeneous structures and the growth of such structures on pre-patterned surfaces. The length scale of heterogeneity is in the range appropriate for nanotechnology applications. The ion beam direction also imposes anisotropy on the surface, giving directional material properties. Applications include directionally-sensitive detectors, anisotropic mechanical devices and electrically or optically anisotropic surfaces. While the methods described here are typically not used for single crystal materials synthesis, many properties of polycrystalline materials are improved by narrowing down the distribution of crystallite orientations into a small range, of less than 10-20°, which is feasible using the ion beam methods described here.

ION MODIFICATION OF TOPOGRAPHY (RIPPLE FORMATION)

The formation of periodic ripple structures on ion bombarded surfaces has been well documented. Ripples have been observed to form on Si [5,6], Cu [7], SiO_2 [8], yttria-stabilized ZrO_2 (YSZ) [9], CaF_2 [10] and other materials [11,12], using ion energies ranging from hundreds eV to tens keV. Typical wavelengths range from 0.02 μm to 1 μm, with amplitudes ranging from several nm to 10's nm. For large values of the angle of ion incidence (near-glancing), the ripple wave vector is

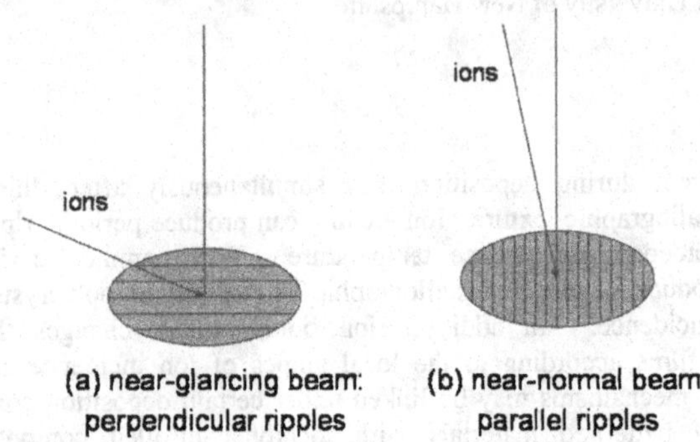

Figure 1(a) Perpendicular ripples formed by high ion incidence angles (near-glancing).
(b) Parallel ripples formed by low ion incidence angles (near-normal).

typically oriented perpendicular to the in-plane component of the ion beam direction [5] as shown in Figure 1(a). We shall refer to these as "perpendicular ripples". For small values of the angle of ion incidence (near-normal), the ripple wave vector is typically oriented parallel to the in-plane component of the ion beam direction [13], referred to as "parallel ripples" and shown in Figure 1(b). For normal ion incidence, surfaces may form a buckled topography or develop distinct dot structures [14]. At intermediate angles of incidence, transitional structures have been observed that evolve with time [15]. A significant feature of the oriented ripples is that perpendicular ripples are stationary, while parallel ripples are traveling waves moving in the projected ion beam direction. A linear theory based on a balance of ion-induced roughening and diffusion-mediated smoothing [16] accounts for the emergence of a dominant ripple wavelength with time, and also predicts the two main observed orientations. With increasing ripple amplitude, nonlinear effects become significant and slow the growth of features, which also become more complex in shape [5,12].

For the purposes of this paper, we simply observe that under off-normal ion bombardment, materials may develop either parallel or perpendicular ripple structures. We portray such a structure as a simple sine curve $h = h_o \sin(2\pi x/\lambda)$ with amplitude h_o and wavelength λ as shown in Figure 2. The difference between the overall angle of ion incidence (relative to the normal to the plane of the sample) and the local angles of ion incidence θ (relative to the local normal) is also shown. While most ripple formation studies have been made on ion etched materials, we note that ripples may also occur on materials during net deposition, as has been observed on YSZ [9]. Deposition under ion bombardment is typically slow enough that individual events of ion and atom arrival are completely

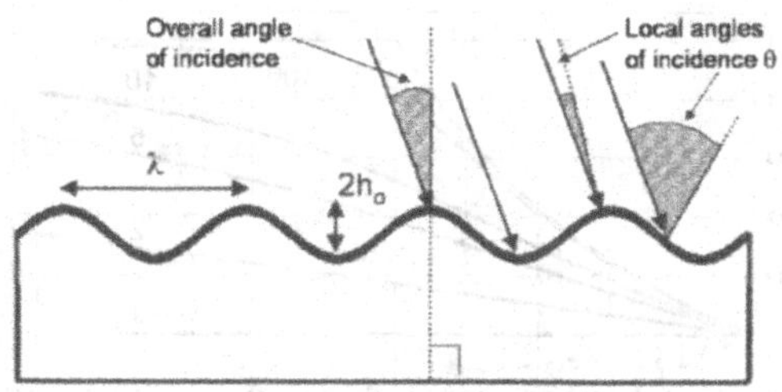

Figure 2. Sketch of sinusoidal ripple.

independent. That is, the arrival of new material uniformly on the rippled surface does not remove the mechanisms that cause the ripples to grow. Of course, if the arriving composition differs from the existing composition at and near the surface, then the ripple formation conditions may change and start evolving towards a new pattern or no pattern. Here, we will assume that the simultaneous processes of deposition and ion etching continue to develop ripples while the material undergoes net deposition. Below, we will consider the coupling of ripple shapes to the composition and texture.

ION MODIFICATION OF COMPOSITION (PREFERENTIAL SPUTTERING)

Ion bombardment during deposition commonly shifts the composition of multicomponent thin films because the sputtering yields of the components are seldom equal [17]. A convenient parameter is the resputtered fraction (z), which is the ratio of the flux of atoms resputtered from the growing film to the arrival flux of atoms. For $z = 0$ (no ion flux), the composition is unchanged, and for $z = 1$, there is zero net deposition rate. For intermediate values of z, the composition is shifted from the composition of the arrival flux according to the sputtering yield ratio (ε) of the components of the film and the value of z. The dependence of film composition on z is shown in Figure 3 for a 2-component material of A and B atoms, with an arrival composition $A_{0.5}B_{0.5}$ and values of ε ranging from 1 to ∞ [17]. The upper curves (solid) show the atomic fraction of B atoms, and the lower curves (dotted) show the atomic fraction of A atoms. The trends of diverging atomic fractions with increasing resputtered fraction, and increasing sputtering yield ratio, have been verified in binary metal alloys, e.g. Gd-Co, over a wide range of compositions [17].

A significant aspect of preferential resputtering that relates to surface topography is the dependence of the resputtered fraction z on the angle of ion incidence [18,19]. Even if the sputtering yield ratio ε remains unchanged with ion angle of incidence, the net sputtering yield increases with angle approximately according to:

$$s(\theta) = s(0)/\cos(\theta)$$

for angles of incidence up to about 50° for most materials. For higher angles, the sputtering yield increases more slowly up to a maximum at around 60-70°, after which it decreases to zero near 90° [20]. Here, we focus on angles of ion incidence up to about 50-60°, since at higher angles, redeposition may become significant on a rippled surface.

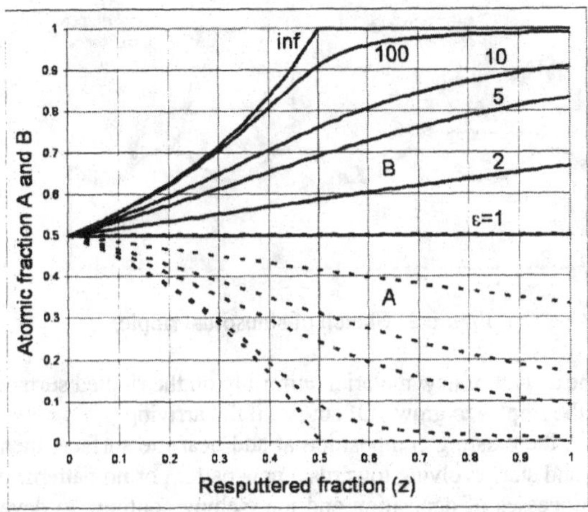

Figure 3. Atomic composition of deposit (A and B) vs. resputtered fraction [17].

In addition to z increasing with ion angle of incidence from the increase in net sputtering yield, we note that the local deposition rate and ion flux are both angular-dependent. Assuming that both fluxes have an overall normal angle of incidence, the local fluxes have a cos(θ)-dependence, which cancels out in the value of z. We therefore find that a first approximation to the angular dependence of z is:

$$z(\theta) = z(0)/\cos(\theta)$$

for values of θ between 0° and about 50-60°. If the ion flux or deposition flux have overall off-normal angles of incidence, the local angles will depend on both the overall angles and the local slope. The important result is that on a rippled (or otherwise non-planar) surface, ion bombardment during deposition produces a growing film composition that depends on the local angle of ion incidence. For example, an angle of incidence of 45° increases the value of resputtered fraction by a factor of 1.4 relative to normal incidence, which can substantially alter the local composition.

An example of a drastic shift in film composition on locally sloping features is found in the high-rate sputter deposition of Ni-La alloys under high-flux Kr^+ ion bombardment [21]. The substrate was a machined Cu tube, with machining grooves causing local steps in the surface. As the film was deposited onto this topographically-structured surface, the film in the flat areas had a composition close to that of the target material, with a significant fraction of incorporated sputtering gas (e.g. $Ni_{0.72}La_{0.18}Kr_{0.10}$), and was amorphous as expected for this alloy composition. In regions above the sharp surface steps, the growing film was found to be crystalline with a composition of essentially pure Ni ($Ni_{0.995}La_{0.005}Kr_{<0.01}$). These regions of Ni appeared as oriented precipitates, extending from the substrate surface up through the full thickness of the film (hundreds of μm) at an angle of 35-60° from the surface normal. The high ion bombardment flux had altered the composition on the sloping step edges by removing La relative to Ni. This direction of composition shift is consistent with studies of preferential sputtering in similar amorphous Gd-Co alloys [17]. The rejection of La was presumably assisted by the driving force of crystallization as the composition was enriched in Ni, since La has low solubility in Ni. Similarly, crystallization is expected to reject Kr from the Ni regions. Similar results were reported for Ni-Y alloys [21], including regularly-spaced Ni precipitates in films deposited onto substrates with pre-existing ripples resulting from machining and polishing of the surface. Even gentle ripple gradients were found to produce oriented precipitates. The precipitates and ripples remained clearly visible in films more than 100 μm thick.

Combining the angular dependence of preferential sputtering with the shapes of ripples formed under ion bombardment, we see that when local surface slope angles reach values of 30-50°, significant shifts in film composition can occur in these regions. For a sinusoidal surface morphology (Figure 2) exposed to an overall normal incidence ion bombardment, the composition shift will be highest in the mid-height region of the ripples, where the slope is highest. For off-normal overall ion incidence, the highest local angle of incidence will occur on the "downstream" side of the features, as indicated in Figure 2, leading to a greater composition shift than on the "upstream" side.

ION MODIFICATION OF TEXTURE (BIAXIAL TEXTURE FORMATION)

The third effect of ion bombardment during film growth being considered here is the development of biaxial (in-plane) texture under the influence of an off-normal ion flux. Strong in-plane texture has been obtained for materials in which there is strong alignment of one crystallographic axis with the surface normal, combined with an angle of ion incidence that coincides with a channeling direction. Some of the observed combinations are given in Table I, which lists the material, crystal type, out-of-plane orientation, in-plane orientation, ion channeling direction and ion incidence angle.

Table I. Examples of in-plane orientation induced by ion bombardment.

Material	Crystal type	Out-of-plane orientation	In-plane orientation	Channeling direction	Ion incidence angle (°)	Reference
Nb	bcc	110	100	111	35	22, 34
		110	110	100	45	23
		110	110 twinned	110	60	22
YSZ	cubic (CaF_2)	100	110	111	55	9, 29
MgO	cubic (NaCl)	100	100	110	45	26, 35
Al	fcc	110	110 twinned	110	60	33
AlN	hex (wurtzite)	001 (tilt 50°)	c-axis	near c-axis	75	28
		001 (tilt -15°)	c-axis	11*0	75	28

Examples of materials include Nb [22,23], Al [24,25], MgO [26], TiN [27], AlN [28] and YSZ [9,29]. The last example has been studied in great detail for its ability to serve as a template material for high-temperature superconducting Cu-oxides [29]. Mechanisms proposed for the alignment effect include the variation with crystal direction of sputtering yield [30] and of damage density and depth [25]. In addition, other mechanisms can promote in-plane anisotropy once the symmetry of the deposition process is broken, including shadowing, growth rate anisotropy and strain [31]. For the purpose of this paper, we note that there are special angles of ion incidence that promote in-plane orientation for certain materials, according to their crystal structure and symmetry. For example, body-centered-cubic materials typically grow with [110] directions normal to the substrate, and can be aligned using ion bombardment at 35°, 45° and 60°, corresponding to channeling directions of [111], [100] and [110], respectively. These combinations give in-plane orientations of [100], [110] and twinned [110], respectively. For face-centered-cubic materials, the usual texture is [111] normal to the surface, however, ion bombardment at an angle of 60° from normal promotes a [110] texture and [110] in-plane orientation. It has also been shown that the orienting effect of the substrate, caused by surface and interface energy minimization, can be overcome either by increasing the normal incidence ion intensity (flux and energy) [24,32] or by applying ion bombardment from two distinct angles [33].

PROSPECTS FOR SIMULTANEOUS CONTROL OF TOPOGRAPHY, COMPOSITION AND TEXTURE

Given the three effects of ion bombardment summarized above, it is likely that during deposition onto topographically-structured surfaces under ion bombardment, more than one effect may occur. To consider how these effects may couple with each other, we first comment on the evolution of periodic ripples with time. The linear model [16] predicts that a dominant ripple wavelength arises from the ion-roughened surface and does not change with time as the amplitude grows. This behavior is confirmed for small amplitudes, but nonlinear effects at higher amplitudes cause the ripple wavelength to increase with time [5]. Perpendicular ripples remain stationary as they increase in wavelength, however, parallel ripples move along the surface with a velocity that has been shown to decrease with time and increasing wavelength [13]. We note that a well-defined ripple structure can form either during net etching or net deposition, or even under conditions of zero net change in overall surface height. When the depositing composition contains more than one species, the possibility of angular-dependent preferential sputtering arises. The result is a slope-dependent film composition that can modify the ripple formation process and can increase or decrease the ripple amplitude and wavelength. Surface diffusion is an important component of these modification processes, and it is expected that the sample temperature will affect the balance of the three processes and the resulting morphology and microstructure.

In conjunction with the slope-dependent composition and evolving ripple structure, we also consider the local in-plane alignment mechanisms that can promote the growth of selected orientations. On a rippled surface under normal incidence ion bombardment, the local angle of ion incidence can vary from 0° to more than 45°, depending on the aspect ratio (h_o/λ). Many measured ripple structures have small aspect ratios, <0.1, which give a small range of slopes. However, there are examples of aspect ratios between 0.2 and 0.4 [6,9], giving profiles similar to Figure 1, with maximum slope angles of 30-40°. With these slope angles, ion bombardment at an overall normal incidence can produce significant local composition shifts and local alignment effects (see Table I). It should be noted, however, that in order to induce surface ripple structures, the overall ion flux must have an off-normal angle to the surface. Therefore, there is already a significant difference between local angles of incidence on the "upstream" and the "downstream" sides of the ripple features. Since the symmetry is broken by tilting the overall angle of incidence, it is clear that even ripples with small aspect ratios have significant local angles of incidence on the "downstream" side of the features. For example, an overall angle of incidence of 45°, commonly used in ripple formation, needs a local surface tilt of only 15° to cause a local angle of incidence of 60°. At this angle, the sputtering yield is enhanced by almost a factor of 2, causing significant shifts in local composition by preferential sputtering and potentially inducing local in-plane alignment of crystallites. On a periodic structure similar to that shown in Figure 2, the highest local angle of incidence will always occur on the downstream side of the features. We can therefore predict that the largest composition shift, and the first opportunity for in-plane alignment of crystallites, will occur on the downstream side of surface topographic features. The size of the resulting crystallites will depend on the material growth temperature relative to its melting temperature and the angular divergence of the ion beam. Given the small dimensions of the self-forming ripple structures, it is possible that single crystal precipitate phases may be formed in some materials.

To test these predictions, we suggest the following experiments:

(1) Deposit a fine-grained thin film on an existing rippled surface under off-normal ion bombardment and observe if in-plane alignment occurs at the downstream local angles predicted by the crystal symmetry.

(2) Use ion etching to develop a stable ripple structure, add simultaneous deposition of the same material and determine the stability of the ripples on a growing film. Examine the grown film for in-plane alignment on the downstream side of the features.

(3) Deposit a binary thin film material under ripple-generating conditions with a significant resputtered fraction (tens of percent) and examine the features for composition shifts as a function of local slope.

(4) Select a multicomponent thin film in which a major change in properties occurs at a certain composition (e.g. crystallization, metal-insulator transition, onset of magnetic ordering, etc.). Carry out deposition under off-normal ion bombardment and determine the locations of these regions with modified composition relative to the surface slopes. Establish whether local in-plane alignment can be induced at these angles for the structure of interest. Finally, determine whether a rippled surface remains stable under net growth, in which the regions of desired properties will occur at distinct locations on the topographic features.

SUMMARY

Three effects of ion bombardment were briefly reviewed, to consider whether one might expect to find simultaneous ripple formation, composition shifts and in-plane alignment during ion assisted deposition. The angles of incidence responsible for ripple formation are consistent with the angles needed for in-plane alignment based on crystal symmetry. In addition, the ability to shift film composition as a function of local slope strongly suggests that a combination of these three effects might be obtained on certain materials. The result is a self-forming, compositionally-defined, oriented crystalline phase with a regular periodic structure.

ACKNOWLEDGMENTS

Thanks are given to colleagues in thin film deposition and ion modification, including J.J. Cuomo, R.J. Gambino, H.R. Kaufman, R.M. Bradley, R.H. Hammond, S. Berg, M. Aziz, E. Chason and J. Erlebacher.

REFERENCES

1. H.R. Kaufman, J.J. Cuomo, and J.M.E. Harper, J. Vac. Sci. Technol., 21, 725 (1982).

2. J.M.E. Harper, J.J. Cuomo, and H.R. Kaufman, J. Vac. Sci. Technol., 21, 737 (1982).
3. J.M.E. Harper, J.J. Cuomo, R.J. Gambino, and H.R. Kaufman, pp. 127-162 in *Ion Bombardment Modification of Surfaces: Fundamentals and Applications* (O. Auciello and R. Kelly, eds.), Elsevier Science Publishers B.V., Amsterdam, 1984.
4. F.A. Smidt, International Materials Reviews, 35, 61 (1990).
5. J. Erlebacher, M.J. Aziz, E. Chason, M.B. Sinclair and J.A. Floro, Phys. Rev. Lett. 82, 2330 (1999).
6. T.K. Chini, F. Okuyama, M. Tanemura and K. Nordlund, Phys. Rev. B67, 205403 (2003).
7. W.L. Chan, N. Pavenayotin and E. Chason, Phys. Rev. B69, 245413 (2004).
8. T.M. Mayer, E. Chason and A.J. Howard, J. Appl. Phys. 76, 1633 (1994).
9. Y. Iijima, M. Hosaka, N. Tanabe, N. Sadakata, T. Saitoh, O. Kohno and K. Takeda, J. Mater. Res. 13, 3106 (1998).
10. M. Batzill, F. Bardou and K.J. Snowdon, J. Vac. Sci. Technol. A19, 1829 (2001).
11. G. Carter, J. Phys. D: Appl. Phys. 34, R1 (2001).
12. R. Cuerno and A.L. Barabasi, Phys. Rev. Lett. 74, 4746 (1995).
13. S. Habenicht, K.P. Lieb, J. Koch and A.D. Wieck, Phys. Rev. B65, 115327 (2002).
14. S. Facsko, T. Dekorsy, C. Koerdt, C. Trappe, H. Kurz, A. Vogt and H.L. Hartnagel, Science 285, 1551 (1999).
15. A.-D. Brown, H. Bola George, M.J. Aziz and J. Erlebacher, Proc. Mat. Res. Soc. 729 (2003).
16. R.M. Bradley and J.M.E. Harper, J. Vac. Sci. Technol. A6, 2390 (1988).
17. J.M.E. Harper and R.J. Gambino, J. Vac. Sci. Technol., 16, 1901 (1979).
18. P.J. Rudeck, J.M.E. Harper and P.M. Fryer, Appl. Phys. Lett. 53, 845 (1988).
19. J.M.E. Harper, S. Berg, C. Nender, I.V. Katardjiev and S. Motakef, J. Vac. Sci. Technol. A10, 1765 (1992).
20. H. Oechsner, Z. Phys. 261, 37 (1973).
21. R.W. Knoll, E.D. McClanahan and H.E. Kjarmo, Thin Solid Films 118, 93 (1984).
22. L.S. Yu, J.M.E. Harper, J.J. Cuomo, and D.A. Smith, Appl. Phys. Lett., 47, 932 (1985).
23. H. Ji, G.S. Was and J.W. Jones, Proc. Mat. Res. Soc. 434, 153 (1996).
24. Z. Ma and G.S. Was, J. Mater. Res. 14, 4051 (1999).
25. L. Dong and D.J. Srolovitz, Appl. Phys. Lett., 75, 584 (1999).
26. C.P. Wang, K.B. Do, M.R. Beasley, T.H. Geballe, and R.H. Hammond, Appl. Phys. Lett., 71, 2955 (1997).
27. I. Petrov, P.B. Barna, I. Hultman, and J.E. Greene, J. Vac. Sci. Technol., A21, S117 (2003).
28. A. Rodriguez-Navarro, W. Otano-Rivera, J.M. Garcia-Ruiz, R. Messier and L.J. Pilione, J. Mater. Res. 12, 1689 (1997).
29. Y. Iijima, K. Onabe, N. Futaki, N. Sadakata, O. Kohno, Y. Ikeno, J. Appl. Phys., 74, 1905 (1993).
30. R.M. Bradley, J.M.E. Harper, and D.A. Smith, J. Appl. Phys., 60, 4160 (1986).
31. O. Karpenko, J.C. Bilello and S.M.Yalisove, J. Appl. Phys. 76, 4610 (1994).
32. I. Petrov, P.B. Barna, L. Hultman and J.E.Greene, J. Vac. Sci. Technol. A21, S117 (2003).
33. L. Dong, D.J. Srolovitz, G.S. Was, Q. Zhao and A.D. Rollett, J. Mat. Res. 16, 210 (2001).
34. J.M.E. Harper, K.P. Rodbell, E.G. Colgan and R.H. Hammond, J. Appl. Phys. 82, 4319 (1997).
35. R.T. Brewer and H.A. Atwater, Appl. Phys. Lett. 80, 3388 (2002).

Mater. Res. Soc. Symp. Proc. Vol. 843 © 2005 Materials Research Society T5.4

On the Erosion of Material Surfaces caused by Electrical Plasma Discharging

Flavio A. Soldera and Frank Mücklich
Department for Materials Science - Functional Materials, Saarland University, P.O. Box 151150, D-66041 Saarbrücken, Germany.

ABSTRACT

The erosion of material surfaces produced by electrical discharges plays an important role on the degradation of many electrical devices, such as electrical contacts, switches or spark plugs. A discharge produces an extreme and concentrated flow of energy into the material that heats it and can even cause melting or vaporization. The plasma pressure may cause an even greater removal of material by the emission of droplets of molten material, producing craters in the surface of the material. In this contribution the microscopic erosion mechanisms on RuAl basis intermetallic compounds are compared with those for pure metals. Single discharge experiments at high pressure were done and the erosion structures were characterized with white light interferometry and scanning electron microscopy. The effects of microstructure on the surface erosion are discussed on the basis of different samples of RuAl. In certain high temperature applications, formation of oxide scales is an important process that may influence the discharge characteristics and erosion mechanisms. These effects are discussed on results in pre-oxidized samples. It was shown, that the surface can be additionally stabilized by controlling the protecting oxide coatings.

INTRODUCTION

The current transfer between the plasma of high-pressure electrical discharges and the cathode occurs in most cases in a constricted regime. The current is collected by a small area, which is called the cathode spot. The complex physical processes taking place in this area are associated with energy transfers between the plasma and the cathode. The cathode surface is primarily heated by the impact of ions accelerated in the cathode sheath, which deliver their kinetic, thermal, and recombination energy. In addition, resistive heating (also called Joule heating) by the discharge current within the cathode contributes to the enhancement of the surface temperature [1,2]. On the other side, emission of electrons, evaporation of cathode material and thermal conduction into the electrode produces a cooling effect on the hot spot [1,2]. Due to the extremely intense interaction between the cathode spot plasma and the solid surface, a crater is formed on this. Both very rapid, intense heating and strong mechanical forces are necessary for melting and excavation of craters and other surface deformations [1].

Previous investigations on material erosion by ignition discharges [3] showed that there is a strong relation between the eroded volume of material and the energy necessary for heating the material up to the melting point and to melt it. These results together with the presence of particles of metal on the cathode surface suggested that the electrode erosion occurs following the model of particle ejections proposed by Gray *et al.* [4] for low current, atmospheric-pressure arcs. According to this erosion mechanism the discharge produces a pool of molten material and the plasma exerts a high pressure on it. After the abrupt cessation of the discharge, molten metal is ejected from the pool due to the unbalanced recoil force from the material against the electrode gap: thus a crater forms on the electrode.

In applications at high temperature, like in spark plugs, the formation of an oxide layer in the electrode surface influences the crater formation and the material erosion. This influence was investigated on pre-oxidized samples of single phase RuAl as well as eutectic RuAl-Ru(Al). The oxide layers show different structures due to the different microstructures of the samples [5,6].

EXPERIMENTAL

Individual sparks were done on cylindrical samples (3 mm diameter) as cathode. A static electrode gap of 1 mm was used and the gap medium was air at 9 bar [7]. The discharge is composed of a breakdown, which takes several ns, and a second phase, which takes up to 1 ms. In the second phase, the discharge takes at the beginning an arc form and then it may change to a glow form. The eroded surfaces were measured using a "Zygo New View 200 3D Imaging Surface Structure Analyzer" [8] and examined using scanning electron microscopy (SEM). Cross sections of the craters were also done with a focus ion beam (FIB).

The RuAl samples were produced by three methods:

a) casting: Ru and Al powders with nominal composition $Ru_{70}Al_{30}$ (sample RAS-30) and $Ru_{50}Al_{50}$ (sample RAS-50) were melted under argon atmosphere in an arc oven with a cold copper plate. The microstructure was composed of primary RuAl dendrites and a surrounding eutectic phase RuAl-Ru(Al). The amount of the eutectic was obviously much higher in the sample RAS30 than in RAS50. The eutectic forms a continuous intergranular phase.

b) powder metallurgy (RAP): Ru and Al were mixed together in the relation 47,6 at. % Ru and 52,4 at. % Al, cold pressed at 990 MPa and axially hot pressed in high vacuum (10^{-5} mbar) in two steps: firstly at 500 °C and 264 MPa for 0,5 h and then at 700 °C and 264 MPa for 1 h. Finally they were annealed at 1600 °C for 12h in a vacuum atmosphere (10^{-5} mbar), which produced a single-phase structure, with a residual porosity of 1.9 % [9].

c) powder metallurgy + casting (RAC): After the powder metallurgy procedure described in (b), these samples were arc-melted like in case (a) to reduce the porosity (~ 0.6 %).

The samples were polished with diamond particles down to a grain size of 0.5 µm. The pre-oxidation was done on still air at 1000°C for different times (1 to 30 h).

RESULTS AND DISCUSSION

Crater formation in pure metals

Two extreme cases of erosion structures can be observed on pure metals: the *"normal craters"*, which present a typical depression in the middle and an external rim (Fig. 1 a), and very flat structures, which show a wavelike surface (Fig 1 c). The craters are formed due to the melting and displacement of material from the center of the crater to the crater rim and due to the ejection of molten particles. In the flat structures there is no proof of such material displacement. The material melts, oscillates and then re-solidifies [3].

The craters are typical for metals with low melting point and low surface tension in the molten state, like Al, Sn, Pb, and Cu. The flat structures are typical in materials with high melting point and high surface tension, like W, Ir and Ru. The material loss in the craters is higher than in the flat structures. In materials with intermediate values of melting point and surface tension like Pt and Ni, both kinds of craters are observed. Also some intermediate structures (Fig. 1 b) were observed in these metals as well as in Ag and Au. These structures present a flat part with only some waves and a deeper part, where material displacement occurred.

For high values of surface tension the molten pool oscillates without a loss of material, freezes during rapid cooling, and after solidification remains as a flat, wavelike structure. Apart from the main craters, also succession of micro-craters can be observed in some metals due to instabilities of the cathodes spot and its movement along the material surface [3,7].

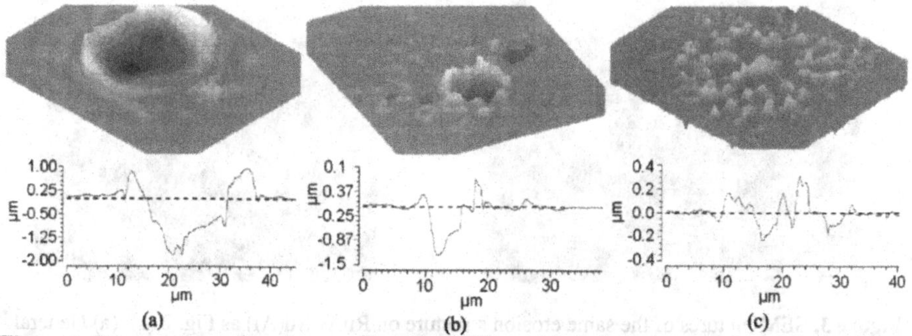

Figure 1. Topography and surface profile of the different kind of erosion structures obtained on platinum (WLI): (a) normal crater; (b) intermediate type; (c) flat, wavelike structure.

Crater formation in RuAl

The erosion structures in single phase RuAl are very flat (Fig. 2a and b). Although the shape of the crater of Fig. 2a is similar to the crater of Fig. 1a, it is much less deeper than that obtained on platinum. However, most of the craters on RuAl show the form described in Fig. b, with a wavelike surface structure and height differences between the peak and the valley of ca. 400 nm. The eroded volume was less than 5 µm³, what is much lower than for pure metals like platinum, nickel or aluminum (see also Fig. 5). The erosion structures in the multiphase RuAl samples (RAS-30) are always very flat with some waves in the surface (Fig. 2 c and e). After mechanical polishing, on the surface there is a height difference between the RuAl and Ru(Al) phases of approximately 60 nm (see Fig. 2d). This difference is much larger after the effects of

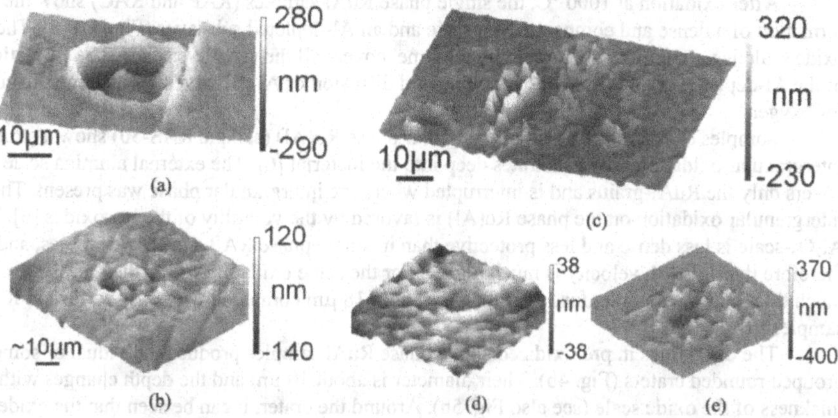

Figure 2. 3D representation of the topography of erosion structures in RuAl samples measured with WLI. (a), (b) Single phase RuAl sample (RAC). (c) Two-phase RuAl-Ru(Al) sample (RAS-30). (d) Topography of the eutectic region of the two-phase RuAl-Ru(Al) sample before erosion test and (e) erosion structure in the same sample. Notice the different scales between (d) and (e).

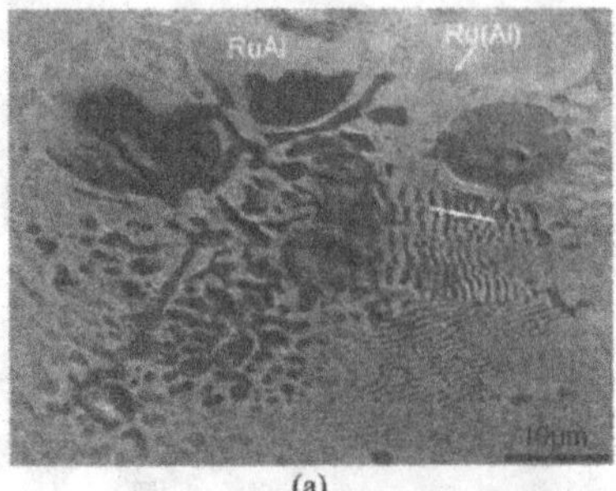

(a)

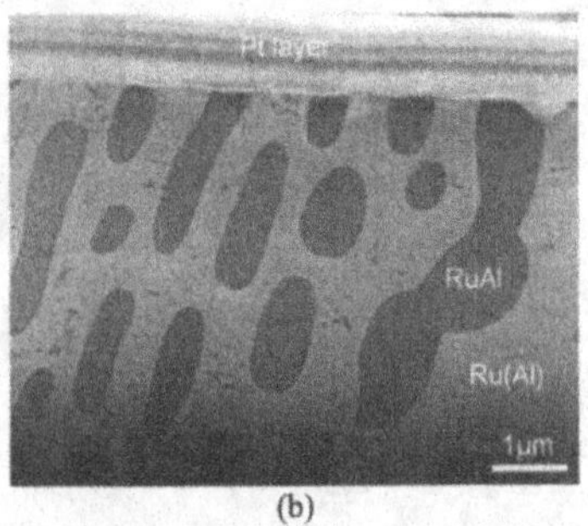

(b)

Figure 3. SEM pictures of the same erosion structure on RuAl-Ru(Al) as Fig. 2 (c). (a) General view, 52° tilted. (b) Detail of the cross section through the white line of Fig. (a).

the discharge (Fig. 2e). The plasma attacks preferably the RuAl phase, which is much darker in the SEM-image after the effects of the discharge (Fig. 3a). The Ru(Al) phase seems to remain unaffected. In Fig. 3b, it can be seen at the edge of the sample (below the Pt layer deposited with the FIB to protect the sample surface during the preparation of the cross section) that there is a height difference between both phases. The phase RuAl is deeper than the phase Ru(Al), what is believed to be due to the lower melting point (~2050°C vs. ~2300°C). The volume of material loss (less than 2µm³, Fig. 5a) is lower than in RuAl. The increased erosion resistance is attributed to the presence of a second phase, which, being more stable, makes difficult the displacement of molten material. This effect diminishes the erosion by the ejection of molten particles that is typical for single phase materials.

Influence of the oxide layer on the crater formation

After oxidation at 1000 °C, the single phase RuAl samples (RAP and RAC) show the formation of a dense and compact Al_2O_3 scale and an Al-depleted sub-layer (Fig. 4a) [5]. The oxide scale is very regular and, after enough time, covers all the sample-surface. The formation of the Al-depleted layer is caused by the outward diffusion of Al [5] driven by its larger affinity for oxygen.

Samples of RuAl having an intergranular phase Ru(Al) (sample RAS-50) show a strong intergranular oxidation, that penetrates deep into the material [6]. The external alumina scale covers only the RuAl-grains and is interrupted where the intergranular phase was present. The intergranular oxidation on the phase Ru(Al) is favored by the volatility of the Ru-oxides [6]. The Al_2O_3-scale is less dense and less protective than in single phase RAP and RAC samples, and therefore their growth velocity is much higher. For the same oxidation times, these scales are much thicker. For example, for 100 h they are about 18 µm compared with only 2,5 µm in RAP samples [5].

The discharges in pre-oxidized single phase RuAl samples produce individual or some grouped rounded craters (Fig. 4b). Their diameter is about 10 µm and the depth changes with the thickness of the oxide scale (see also Fig. 5b). Around the crater, it can be seen that the oxide layer is densified due to the melting and re-solidification produced by the thermal energy of the discharge. The discharge removes in each case the oxide layer until it reaches the metallic substrate (Fig. 4c). EDX analysis in the middle of the crater showed a large enrichment in Ru, which is due to the presence of the Al-depleted layer. In the cross section of the crater (Fig. 4c) it can be observed a deepening in the middle, where some metal was displaced to the sides. The

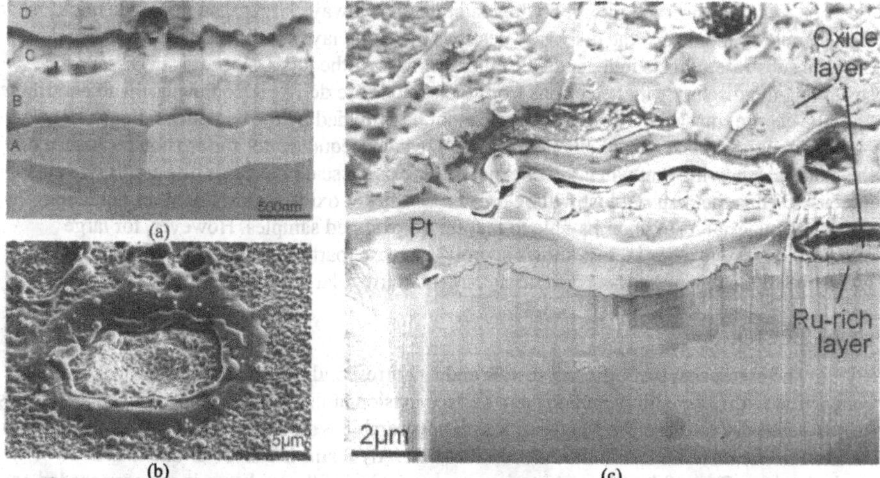

(a) (b) (c)

Figure 4. Oxidized single phase RuAl sample. The oxidation time was 30 h. (a) Detail of a cross section of the oxide scale: A- Ru-rich layer, B-intern part of the Al_2O_3 layer, C-external part of Al_2O_3 layer, Pt layer for the FIB preparation. (b) Typical crater (SEM secondary electrons image - tilt 52°). (c) Cross section of the crater.

microstructure beneath the crater reveals a small layer that may indicate the amount of material that was melted by the discharge and re-solidified with a different microstructure.

In pre-oxidized RAS-50 samples, the craters are always found in the intergranular phase and they have a very irregular form. Many small craters can be seen distributed in a large eroded region, of about 60 µm. The material erosion for comparable oxidation times is much larger than in single phase samples (Fig. 5a) due to the thicker oxide layers.

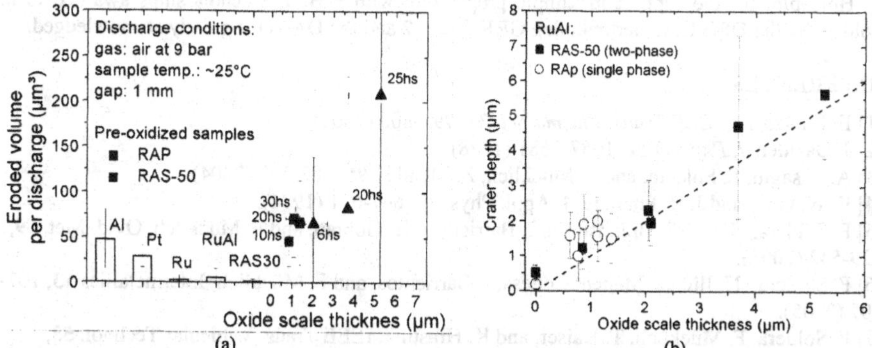

(a) (b)

Figure 5. (a) Volume of eroded material per discharge for the different samples. The times give the duration of the pre-oxidation treatment in RuAl-samples at 1000°C. (b) Depth of the craters as a function of the oxide scale thickness in pre-oxidized RuAl samples. The depth of the crater is larger than the thickness of the oxide scale.

As can be observed in Fig. 5b the crater depth is always larger than the oxide thickness in each sample. That means, the discharge destroys the oxide layer until the plasma reaches the metallic substrate. This is independent of the thickness of the layer, since it is necessary for sustaining the discharge that electrons from the cathode are delivered to the plasma to contribute with the charge transport. These electrons cannot be provided by the ceramic and must be provided by the metal. This phenomenon has a great consequence for the erosion resistance of the material, showing that when the thickness of the oxide scale increases, the crater size also increases. Thus the material lost will be greater for thicker oxide scales. For short oxidation times, the material loss is comparable to that for no oxidized samples. However, for large oxidation times the material loss shows a strong increase, particularly for two phase samples, where the oxide thickness was particularly large and irregular in its form (Fig. 5a).

CONCLUSIONS

The crater shape in different metals and its corresponding material loss show a strong connection with the melting enthalpy and surface tension at the melting point. Erosion structures in RuAl are very flat, and their material loss is comparable with that of Ru and much lower than Al and Pt. The material lost in the eutectic RuAl-Ru(Al) is still much lower due to the strengthening effect of the two-phase microstructure. One of these phases is stronger eroded and the other is stable against the discharge. The structure and thickness of the oxide scale in RuAl depends strongly on the previous microstructure. Single phase RuAl have a continuous Al_2O_3-scale. In two-phase RuAl there is a strong intergranular oxidation, which interrupts the oxide scale. Moreover, the scale is much thicker than in single phase material. The sparks destroy in each case the oxide scale, since they need the contribution of the metallic substrate for the emission of new electrons. In this way, for thicker scales, like in two-phase RuAl, the craters and consequently the material loss is much larger than for thinner scales.

ACKNOWLEDGMENTS

The authors would like to thank A. Lasagni and J. Fiscina for the fruitful discussions and C. Holzapfel for the support in sample preparation with FIB. The scholarships awarded to F. Soldera by the DFG Graduiertenkolleg GRK 232/3-2 and the DAAD is greatly acknowledged.

REFERENCES

[1] E. Hantzsche, *IEEE Trans. Plasma Sci.* **31**, 799-808 (2003).

[2] J. Daalder, *J. Phys. D* **11**, 1667-1681 (1978).

[3] A. Lasagni, F. Soldera, and F. Mücklich, Z. Metallk. **95**, 102-108 (2004).

[4] E. W. Gray and J. R. Pharney: J. Appl. Phys. **45**, 667-671 (1974).

[5] F. Soldera, N. Ilić, S. Brännström, I. Barrientos, H. Gobran, and F. Mücklich, Oxid. Met. **59**, 529-542 (2003).

[6] F. Soldera, N. Ilić, N. Manent Conesa, I. Barrientos, and F. Mücklich, Intermetallics **13**, 101-107 (2005).

[7] F. Soldera, F. Mücklich, T. Kaiser, and K. Hrastnik, IEEE Trans. Vehicular Technol. **53**, 1257-1265 (2004).

[8] F. Soldera, M. Sierra Rota, N. Ilić, and F. Mücklich, Prakt. Metallog. **37**, 477-486 (2000).

[9] H. Gobran, N. Ilić, F. Mücklich, Intermetallics, **12**, 555-562 (2004).

Mater. Res. Soc. Symp. Proc. Vol. 843 © 2005 Materials Research Society T5.5

Low Damage Smoothing of Magnetic Materials using Oblique Irradiation of Gas Cluster Ion Beam

Shigeru Kakuta[1, 3], Shinji Sasaki[1, 3], Kenji Furusawa[1, 3], Toshio Seki[3], Takaaki Aoki[3] and Jiro Matsuo[2]
[1]Storage Technology Research Center, Hitachi, Ltd., 292 Yoshida-cho Totsuka-ku Yokohama, 244-0817, Japan
[2]Quantum Science and Engineering Center, Kyoto Univ., Gokasyo Uji, 611-0011, Japan
[3]Collaborative Research Center of Nano-scale Machining with Advanced Quantum Beam Technology, Japan

ABSTRACT

Oblique irradiation using a gas cluster ion beam (GCIB) has been studied in order to achieve low-damage smoothing of magnetic materials. We investigated how the surface morphology and surface roughness depended on the angle of incidence. Quite smooth surfaces could be obtained using both normal and grazing-incidence irradiation. At incident angles larger than 45°, periodic ripples were formed. The orientation of the ripples changed from perpendicular to parallel with respect to the GCIB when the incident angle exceeded a critical value. Surface roughening resulting from the formation of ripples was observed at incident angles between 45° and 65°. Fluctuations in the Ni/Fe component ratio and the intermixing of oxygen from the native oxide were evaluated. As the angle of incidence increased, both the thickness of the layer in which the component ratio was fluctuating and the depth of oxygen intermixing decreased. As a result, it was determined that low-damage smoothing of magnetic materials could be performed by using grazing-incidence irradiation from a GCIB.

INTRODUCTION

As the areal density of hard disk drives (HDDs) has increased, the flying height of the magnetic recording head over the disk has been decreasing. In order to further reduce the flying height, a surface smoothing technology that induces minimal damage in magnetic materials is extremely important, since surface roughness and damage to surface layers decreases the sensitivity of the sensor. The goal of this study is to develop a method that enables damage-free and ultra-smooth surface processing of magnetic recording heads, which is a requirement for those wishing to manufacture such storage systems with high reliability.

Traditionally, surface smoothing of magnetic recording heads has been performed by a polishing process using diamond abrasives. However, this polishing process is, by nature, liable to leave scratches on the surface, and these can cause severe damage to magnetic recording heads.

Recently, a gas cluster ion beam (GCIB) technology has been proposed as a new surface smoothing technique [1]. Gas cluster ions are aggregations of a few to several thousands of atoms or molecules. Extremely low energy ion irradiation can easily be achieved, since atoms or molecules consisting of cluster ions share their kinetic energy. Therefore, GCIB technology can be utilized to obtain a low-damage and ultra-smooth surface finish on magnetic materials.

EXPERIMENTAL

Figure 1 shows a schematic diagram of the GCIB apparatus. The GCIB was generated by adiabatic expansion of Ar gas at high pressure through a nozzle into a vacuum and subsequent ionization by electron impact. The acceleration voltage was fixed at 20 kV. Monomer ions, which are generated simultaneously with the cluster ions, were eliminated by a magnetic field perpendicular to the beam. Beam-current measurement and dose control were effected with a Faraday cup.

The cluster size (i.e. the number of aggregated atoms) is one of the most important parameters in the GCIB process. The cluster size distribution can be controlled by the source pressure, the ionization voltage and the ionization electron current [2]. The cluster size distribution was measured by the time of flight (TOF) method in a separate experiment. In this paper, a GCIB whose size distribution has a maximum at around 3,000 was used.

In order to investigate the effects of GCIB irradiation on magnetic materials, 50 nm-thick NiFe thin films were deposited by Ar sputtering onto Si wafers for use as test samples. The film thicknesses before and after irradiation were measured by the X-ray fluorescence (XRF) method. Thickness calibration was performed by the X-ray reflection (XRR) method.

In this study, both normal and off-normal GCIB irradiation of the NiFe films was carried out. The angle of incidence (as measured from the normal) was varied from 0° to 90°. The surface morphology of the GCIB-irradiated surfaces was observed by scanning electron microscopy (SEM) and atomic force microscopy (AFM). Depth profiles of Ni, Fe and O were obtained using secondary ion mass spectrometry (SIMS).

RESULTS AND DISCUSSION

Figure 2 shows the surface morphology of NiFe films after GCIB irradiation at incident angles of 0°, 25°, 45°, 65° and 85°, as observed by SEM. The GCIB was irradiated

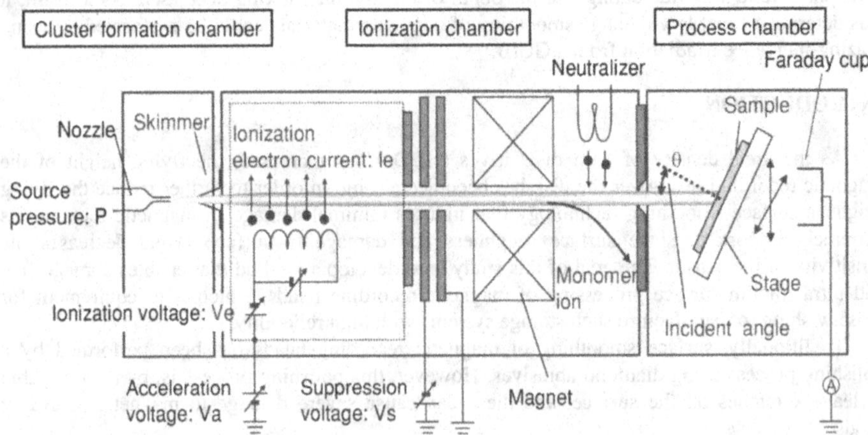

Figure 1. Schematic diagram of the GCIB apparatus.

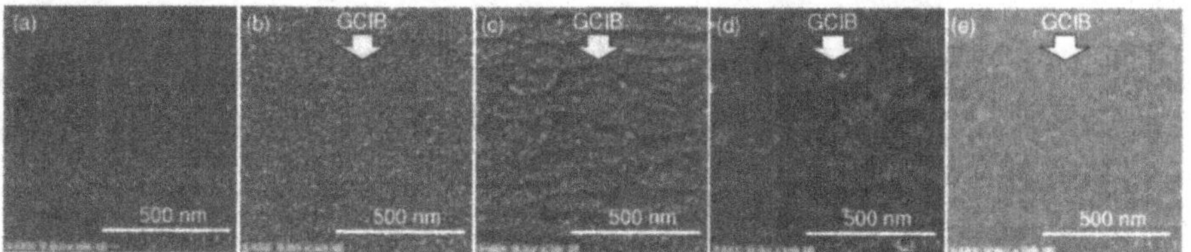

Figure 2. SEM images of NiFe film surfaces after GCIB irradiation at an acceleration voltage of 20 kV and a dose of 1×10^{16} cm-^{2}. (a) Normal and oblique irradiation at incident angles of (b) 25°, (c) 45°, (d) 65° and (e) 85°. The arrows on the pictures denote the direction of the GCIB for oblique irradiation.

at an acceleration voltage of 20 kV and at a dose of 1×10^{16} cm^{-2}. In the case of the samples that underwent oblique irradiation, the arrows shown in the pictures represent the beam direction. It was found that the surface morphology varied drastically with the incident angle. After irradiation at an incident angle of 25°, many small hillocks that were several tens of nm in size were observed on the surface, as shown in Fig.2(b). This result is quite similar to that for normal irradiation (Fig.2(a)). At an incident angle of 45°, periodic ripples perpendicular to the GCIB were formed on the surface. As the angle of incidence increased, the ripples perpendicular to the GCIB disappear and those parallel to the GCIB become more obvious. A critical angle where the ripple orientation changes from perpendicular to parallel to the GCIB should exist at about 65°. The wavelengths of the ripples perpendicular and parallel to the GCIB are about 120 and 20 nm, respectively. The wavelength of the ripples perpendicular to the GCIB is similar to that observed in previous work [1]. Ripples with a wavelength of 200 nm were produced on a Cu surface by off-normal irradiation with incident angles ranging from 45° to 60°.

Note that the wavelength of the ripples perpendicular to the beam is six times larger than that of the parallel ripples. The formation of ripples by monomer ions arriving at a glancing angle of incidence is a well-known phenomenon that has been explained by thermal surface diffusion and the dependence of the sputtering yield on curvature [3]. Koponen et al. simulated ripple formation by off-normal irradiation with a monomer ion beam, and they determined that the wavelengths of the ripples perpendicular and parallel to the beam produced on a C surface were 14 and 19 nm, respectively [4]. Chan et al. demonstrated ripple patterns produced on a Cu surface perpendicular and parallel to a monomer ion beam, and their wavelengths were 250 and 400 nm, respectively [5]. These results indicate that the wavelengths of ripples that lie both perpendicular and parallel to a monomer ion beam are similar. The large wavelength of the ripples perpendicular to the GCIB might result from the fact that multiple collisions occur along the direction of travel of the cluster ions.

Figure 3 shows the average roughness of NiFe films after GCIB irradiation as a function of incident angle. The initial roughness of the NiFe films was about 0.5 nm. At normal irradiation, quite a smooth surface with an average roughness of 1.0 nm was obtained. An average roughness of 1.0 nm was also obtained by GCIB irradiation normal to a NiFe film with an initial roughness of 9.0 nm. The average roughness at an incident angle of 25° increased slightly to 1.5 nm. This increase in average roughness with oblique irradiation can be explained by anisotropic energy deposition, as discussed in Ref. [1]. At an incident angle of 45°, the surface roughness drastically increased because of ripple formation. As the incident angles increases beyond 45°, the surface

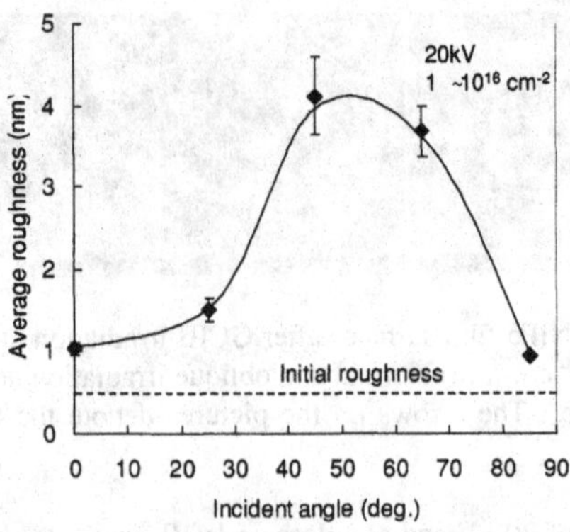

Figure 3. Average roughness of NiFe films after GCIB irradiation with an acceleration voltage of 20 kV and a dose of 1×10^{16} cm^{-2} as a function of incident angle

roughness decreases. Finally, quite smooth surfaces with an average roughness of 0.9 nm were obtained at an incident angle of 85°. This result indicates that the use of grazing incidence is useful for surface smoothing when utilizing a GCIB, as well as when using a monomer ion beam.

To realize a practical smoothing process for magnetic materials, irradiation damage needs to be studied. Depth profiles of Ni, Fe and O in a NiFe film before and after off-normal GCIB irradiation are demonstrated in Fig. 4. The GCIB irradiation was performed with an acceleration voltage of 20 kV and a dose of 1×10^{16} cm^{-2}. The high oxygen density near the surface of the NiFe film before irradiation (Fig. 4(a)) denotes the existence of a native oxide layer. The Ni/Fe component ratio near the surface tends to fluctuate because of the native oxide layer. At an incident angle of 25°, it was proved that the oxygen distribution extended into the interior of the film. The native oxide layer and any adsorbed water molecules on the NiFe surface are considered to be potential sources of oxygen. A more serious problem is that a 13 nm-thick Ni/Fe

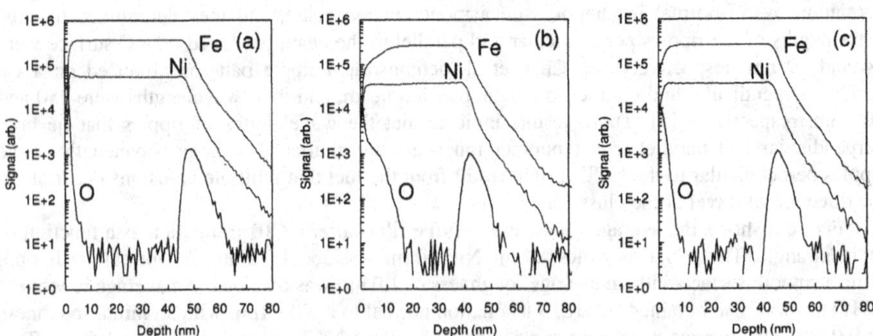

Figure 4. Depth profiles of Ni, Fe and O in NiFe film. (a) Before irradiation and after off-normal irradiation at incident angles of 25° and 85°. The GCIB irradiation was performed at an acceleration voltage of 20 kV and a dose of 1×10^{16} cm^{-2}.

layer in which the component ratio changes exists near the surface. The selective etching of Fe with respect to Ni can be caused by the fact that the heat of evaporation of Fe (354 kJ/mol) is somewhat less than that of Ni (381 kJ/mol). On the other hand, no selective etching or oxygen mixing was observed at an incident angle of 85°, as shown in Fig. 4(c). The high oxygen density is considered to be due to a native oxide, since the NiFe films were exposed to air after GCIB irradiation.

Figure 5 indicates how the thicknesses of the layer with the changing Ni/Fe composition ratio altered as a function of incident angle. The GCIB was irradiated at an acceleration voltage of 20 kV and a dose of 1×10^{15} cm^{-2}. The dashed line denotes how the thickness of the layer with the changing composition was affected by the native oxide. The fluctuation of the component ratio is thought to result from kinetic energy deposition of cluster ions. Consequently, the thickness of the layer with the modified component ratio should be proportional to $\cos^2 \theta$, where θ denotes the incident angle. At low incident angles, the thickness of the layer in which the component ratio was changing decreased in line with $\cos^2 \theta$, as shown in Fig. 5. At high incident angle, the thickness of the modified layer is dominated by the thickness of the native oxide.

To summarize the above discussion, grazing irradiation from a GCIB can be applied to achieve low-damage smoothing of magnetic materials. Quite smooth surfaces can be obtained with grazing irradiation, and fluctuation of the Ni/Fe component ratio can be suppressed.

CONCLUSIONS

In order to achieve low-damage smoothing of magnetic materials, oblique irradiation using a GCIB has been studied. The dependence of surface morphology and surface roughness on the incident angle were observed by SEM and AFM. Quite smooth surfaces could be obtained using both normal- and grazing-irradiation. At incident angles larger than 45°, periodic ripples were formed. The orientation of the ripples changed from perpendicular to parallel to the GCIB when the incident angle exceeded a critical angle. Surface roughening resulting from the formation of ripples was observed at incident angles between 45° and 65°. Fluctuations in the Ni/Fe component ratio and intermixing of oxygen from the native oxide were evaluated. Depth profiles of Ni, Fe and O in GCIB-irradiated NiFe films were obtained using SIMS. Both fluctuations in the component ratio and the intermixing of oxygen were observed. As the incident angle

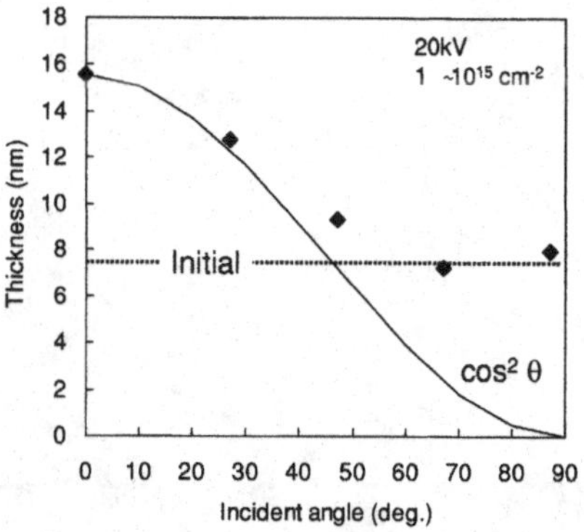

Figure 5. Incident angle dependence of Ni/Fe component ratio changed layer thickness by GCIB irradiation at an acceleration voltage of 20 kV and a dose of 1×10^{15} cm^{-2}.

increased, both the thickness of the layer in which the component ratio fluctuated decreased and the depth of the oxygen intermixing decreased. The thickness of the layer in which the component ratio was fluctuating was proposed as being proportional to the square cosine of the incident angle. As a result, low-damage smoothing of magnetic materials can by performed by grazing-incidence irradiation from a GCIB.

ACKNOWLEDGMENTS

This work is supported by the New Energy and Industrial Technology Development Organization (NEDO) of Japan.

REFERENCES

[1] I. Yamada, J. Matsuo, N. Toyoda and A. Kirkpatrick, Mater.Sci.Eng. R.34 (2001) 231.
[2] T. Seki, J. Matsuo, G. H. Takaoka and I. Yamada, Nucl. Instr. and Meth. in Phys. Res. B 206 (2003) 903.
[3] R.M. Bradley and J.M.E. Harper, J. Vac. Sci. Technol. A 6 (1988) 2390.
[4] I. Koponen, M. Hautara and O.-P. Sievänen, Phys. Rev. Lett. 78 (1997) 2612.
[5] W.L. Chanm N. Pavenayotin and E. Chason, Phys. Rev. B 69 (2004) 245413.

Mater. Res. Soc. Symp. Proc. Vol. 843 © 2005 Materials Research Society T5.7

MOLECULAR DYNAMICS STUDY OF SUFACE STRUCTURE AND SPUTTERING PROCESS BY SEQUENCIAL FLUORINE CLUSTER IMPACTS

Takaaki Aoki[1,2] and Jiro Matsuo[1]
[1]Quantum Science and Engineering Center, Kyoto University,
Gokasho, Uji, 611-0011, JAPAN
[2]Osaka Science and Technology Center,
Utsubo-Honmachi, Nishi-ku, Osaka, 550-0004, JAPAN

ABSTRACT

In order to study the surface reaction process under cluster ion impact, molecular dynamics simulations of sequential cluster impacts of fluorine and neon clusters were performed. $(F_2)_{300}$ and Ne_{600} clusters were accelerated with totally 6keV and irradiated on bare Si(100) target. By iterating impact simulation sequentially, change of surface morphology and composition of desorbed materials were studied. In the case of fluorine cluster impact, the rough surface structure was made compared with neon cluster impact because fluorine atoms adsorb on the target, which work to keep pillar structure on the surface, whereas Ne atoms evaporate immediately leaving spherical mound structure on the surface. From the study of desorption process, it was observed that large number of Si atoms are desorbed in the form of silicon fluorides. The major sputtered material was SiF_2, but various types of silicon-fluorides, including the molecules consists of several tens to hundreds silicon and fluorine atoms, were observed. The distribution of clusters in desorbed materials obeyed the classical model of cluster emission from quasi-liquid phase excited with ion bombardment. From these results, the irradiation effects of reactive cluster ions were discussed.

INTRODUCTION

The impact processes of gas clusters on solid surfaces are of great interest for the new surface modification techniques [1,2]. Cluster is an aggregated material, which consists of several tens to thousands atoms. When a cluster is accelerated and impacts on a solid target, high-density atomic collision occurs and the kinetic energy of the cluster is deposited on very shallow area of the target. This collisional process is different from that of monomer ions, and is expected to develop new surface modification technologies. The study of gas cluster ion impact has been started with argon gas cluster mainly, which is inert material and is suitable to understand the physical effect of multiple collision effect of clusters. Recently, it has been reported that, gas cluster ion beam can be generated with various gas-phase materials such as SF_6, CF_4, etc [3,4]. Expansion of source materials implies that gas cluster ion beam will be available for various surface modification process by combining the non-linear effects inherits in cluster impact and chemical properties of source gas materials. Fot the example in the field of sputtering, it has been demonstrated that, when a $(SF_6)_{2000}$ cluster is irradiated on Si and W substrate, the sputtering yield is 10 times higher than that of Ar_{2000} cluster, which in turn is 10 times higher than monomer ions with the same total acceleration energy of 20keV [4,5]. The high etching ratio of reactive cluster ions is now studied to apply to the industrial use of precise nano-scale patterning process [6]. So, it becomes more important to understand the mechanism of collisional and etching process of reactive clusters in order to optimize the surface process depending on industrial requirements. In this study, molecular dynamics (MD) simulations of fluorine cluster impacting on bare silicon surface were performed. By the iteration of the MD simulations, change of surface morphology and accumulation of fluorine atoms in the target were studied. Additionally, desorbed materials during cluster impact were analyzed from the viewpoint of chemical component in order to discuss the sputtering process by reactive cluster irradiation.

SIMULATION MODEL

In this work, change of surface morphology and sputtering process under sequential cluster ion impact were studied using molecular dynamics method. In order to describe the interactions between Si-Si, Si-F and F-F, the Stillinger and Weber (SW) type potential model was applied. The SW type potentials were developed by Stillinger et al. [7-9] and modified by Weaklime et al. [10]. $(F_2)_{300}$ cluster was used as a projectile, which was annealed in the simulation at 10K using the Langevine dynamics method before impact. After annealing, the fluorine cluster has a spherical and amorphous structure, which avoids artifacts due to the structure of the cluster. On the other hand, Ne_{600} cluster was applied in order to compare reactive and non-reactive cluster impact process. The interactions between Ne-Ne and Si-Ne are described only in Ziegler, Biersack and Littmark (ZBL) universal potential model [11] in which no attractive force is applied because the binding energies of Ne-Ne and Si-Ne are much lower than those of Si-Si, Si-F, F-F and also less than the incident energy of projectile atoms used in this work. The initial structure of Ne_{600} was given as spherical solid with icosahedral structure.

Si(100) substrates with the dimension of 173Å×173Å×173Å were prepared as the target. This target consists of about quarter million Si atoms. In this case, each Si atom receives 0.02eV of kinetic energy at each impact. So, thermal bath region was implemented at the bottom of target, which controls target temperature at 300K. The MD simulation of sequential cluster impact was performed by iterating following process; I) Select impact point randomly II) Perform MD simulation of $(F_2)_{300}$ or Ne_{600} cluster impact in 6keV for 16ps, after the impact simulation molecules/atom desorbed into the vacuum are removed. III) Cool down the target to 300K for 4ps and remove desorbed materials after during cooling down process. In this study, 160 impacts were simulated for each cluster species, which corresponds to $5.5\times10^{13}/cm^2$ impacts.

RESULTS AND DISCUSSION

Figure 1 shows the snapshots of Si surfaces irradiated with $(F_2)_{300}$ and Ne_{600} clusters. Si atoms are indicated as light-gray spheres, while F and Ne atoms are described as dark spheres. Both clusters consist of 600 atoms and accelerated with 6keV, so each constituent atom carries very low energy of 10eV/atom. However, the left pictures shows that very large hole and crater structure is formed with only one impact. This is the result of the multiple collision process caused by high-density particle irradiation, which is specific phenomenon for cluster impact [12]. The shape of cluster is very similar to each other, except for that some fluorine atoms remain on the Si target. The difference of cluster species becomes more distinct as the number of impact increases. After the 160 sequential impacts of clusters (it corresponds to about $5.5\times10^{13}/cm^2$), the Si surface irradiated with Ne_{600} cluster shows several spherical mounds on the surface. However, after $(F_2)_{300}$ cluster irradiation, some pillars of Si densely covered with F atoms, are found on the surface. This pillar structure is also found at the MD simulations of the F monomer impact [13].

Figure 2 shows the evolution of surface roughness by sequential impacts of F and Ne clusters. The surface roughness is calculated as the root mean square (RMS) of the set of the highest atoms in every 5×5Å^2 mesh. For both cases, the surfaces become rough as number of impacts increases. At the impact of Ne cluster, the surface roughness converges to about 10Å, which is similar dimension of the crater caused by single impact. On the other hand, the surface

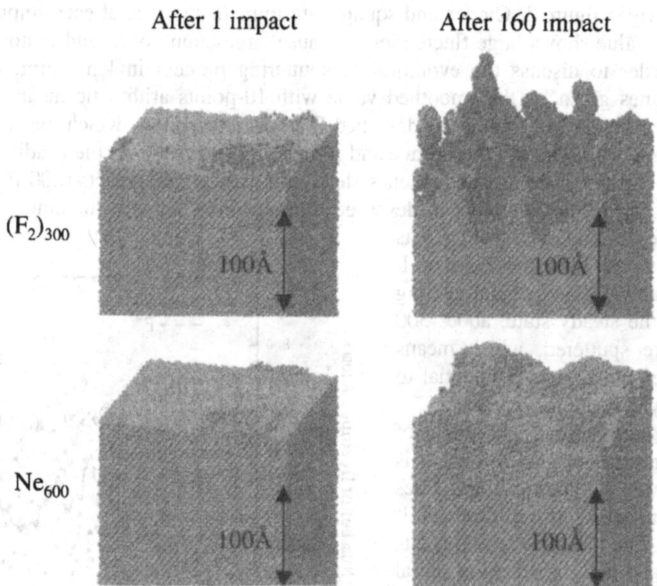

Figure 1. Snapshots of $(F_2)_{300}$ and Ne_{600} impacting on Si(100) surface, after 1 and 160 iterations.

roughness under F cluster irradiation takes much number of impacts to converge and the final roughness is very rough compared with the Ne cluster case. This difference is due to that the F cluster would be accelerated near the target because of attractive potential between Si-F, which results in deep damage and rough surface. Another reason is considered that, the pillar structure could be stabilized by fluorine atoms. The pillar structure has small cross section against normal incident of cluster, so that it could not be deformed or removed by cluster impact and remains for long time.

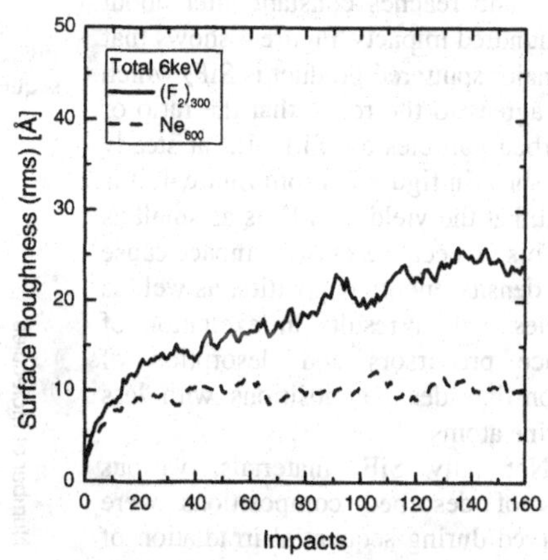

Figure 2. Evolution of surface roughness by $(F_2)_{300}$ and Ne_{600} sequential impacts.

The difference between reactive and non-reactive cluster impact was also shown in desorption and sputtering processes. As for Ne cluster impact, almost of all Ne atoms were immediately evaporate to vacuum at the impact (278 Ne atoms remained after 160 Ne_{600} impacts) and very few Si atoms were sputtered (133 Si atoms for 160 Ne_{600} impacts). However, much number of sputtered particles was observed at F cluster impact, which enables statistical analysis. The evolution of number of desorbed atoms under F cluster impact is shown in figure 3. Circles and squares are immediate value at each impact. However, the immediate value shows large fluctuations because large chunk of Si and F atoms eventually desorbs. In order to discuss the evolution of sputtering process in long term, figure 3 also includes the lines given by the smoothed value with 10-points arithmetic mean. At the initial stage of cluster irradiation, number of desorbed F atoms is very low, which means that most of incident F atoms are stick onto the surface and form Si-F precursors. As the irradiation proceeds, desorption of F atoms increases and reaches steady state after 100 impacts (600 F atoms desorb at one $(F_2)_{300}$ impact). The number of desorbed Si atoms also increases in similar way to the F atoms. This result implies that Si atoms could be sputtered with the help of F atoms, in other words, chemical etching proceeds. At the steady state, about 300 Si atoms were sputtered, which means reactive cluster impact has a potential to achieve high-yield etching process.

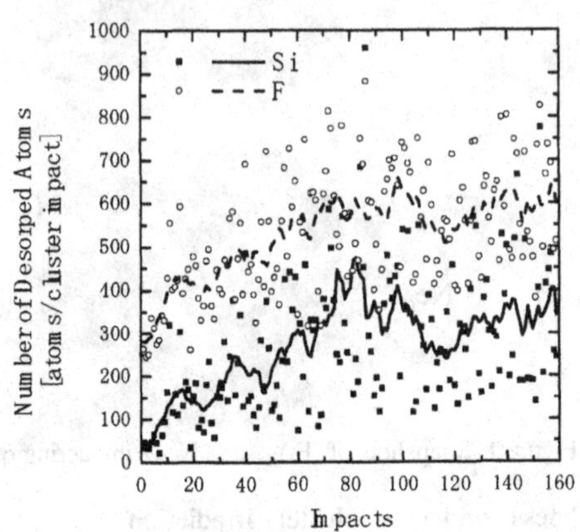

Figure 3. Evolution of number of desorbed atoms by sequential $(F_2)_{300}$ impacts

The evolution of number of sputtered particles of $SiF_y(y=0\sim4)$ is shown in figure 4. The dots and lines indicate as similar to those of figure 3. It can be seen that fluoride components such as SiF_2 and SiF_3 increases at initial stage of impacts, which corresponds to the accumulation of fluorine atoms in the target, and reaches constant after about 100 hundred impacts. Figure 4 shows that the major sputtered product is SiF_2 which well agrees to the result that the ratio of desorbed particles are Si:F=1:2 at steady state show in figure 3. From figure 4, it is noted that the yield of SiF_4 is as small as Si. This is because cluster impact cause high-density energy deposition as well as paticles, which results in excitation of surface precursors and desorption of silicon-fluorides compositions with less fluorine atoms.

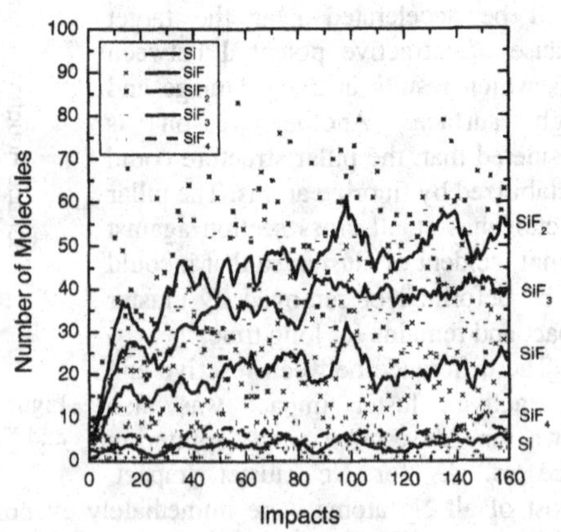

Figure 4. Evolution of desorption yield of SiF_y molecules

Not only SiF_y materials, various types of desorbed compositions were observed during sequential irradiation of fluorine atom. Figure 5 shows the distribution of Si_xF_y clusters within desorption materials. The distribution was described as the number of Si atoms (x in Si_xF_y) in the compositions. It is found that the distribution function well matches the $x^{-(7/3)}$, which is the classical model of cluster emission from quasi-liquid phase excited with ion bombardment [14]. In the case of sequential impact of reactive cluster ions, the irradiate surface forms the mixture of silicon and fluorine atoms, which is considered as very fragile material and as intermediate state of solid and liquid. In figure 5, very large Si-F clusters were also found as desorption materials. The structure of these large Si-F clusters consists of pure Si as a core material and its surface is covered by F atoms. This structure is very similar to that of pillar resides on the surface as shown in figure 1. It is concluded that these large pillar structure could be desorbed directory by reactive cluster ion impact.

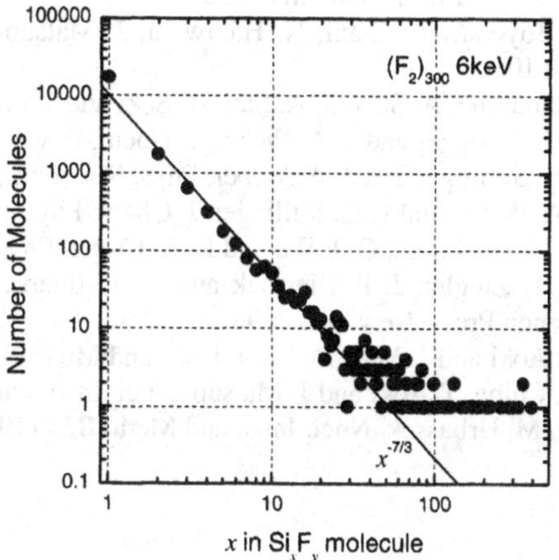

Figure 5. Distribution of molecules in desorbed particles discriminated by the number of Si atoms in molecules.

SUMMARY

Molecular dynamics simulations of sequential cluster impact of fluorine and neon clusters were performed in order to study the surface reaction process under cluster ion impact. As the number of cluster impact increases, large difference in surface morphology was observed between F and Ne clusters. In the case of fluorine cluster impact, the rough surface structure was made compared with neon cluster impact. This is because that fluorine cluster accelerated by attractive potential to cause deep damage and fluorine atoms adsorb on the target, which work to keep pillar structure on the surface. From the study of desorption process, it was observed that large number of Si atoms are desorbed in the form of silicon-fluoride composition. At the case of $(F_2)_{300}$ cluster impact on Si target at 6keV, the major sputtered product was SiF_2, so that 300 of Si sputtering yield was achieved at steady state. In addition to mono-silicon fluorides, various types of silicon-fluorides were observed as sputtered materials. The distribution of clusters in desorped materials obeyed the classical model of cluster emission from quasi-liquid phase excited with ion bombardment. From these results, the irradiation effect of reactive cluster ions were studied and it was shown that reactive cluster ion process would be available for high-yield and precise surface etching processes.

ACKNOWLEDGEMENT

This study is supported by New Energy and Industrial Development Organization (NEDO) in Japan.

REFERENCES

1. I. Yamada, J. Matsuo, N. Toyoda, T. Aoki, E. Jones and Z. Insepov, Mater. Sci. and Eng., **A253** (1998) 249
2. I. Yamada, J. Matsuo, Z. Insepov, T. Aoki, T. Seki and N. Toyoda, Nucl. Instr. and Meth. B, **164-165** (2000) 949
3. A. Kirkpatrick, Nucl. Instr. and Meth. **B 206** (2003) 830.

4. T. Seki and J. Matsuo, this issue.
5. N. Toyoda, H. Kitani, N. Hagiwara, J. Matsuo and I. Yamada, Mater. Chem. and Phys., **54** (1998) 106.
6. E. Bourelle, A. Suzuki, A. Sato, T. Seki and J. Matsuo, Jpn. J. Appl. Phys., **43** (2004), L1253.
7. F. H. Stillinger and T. A. Weber, J. Chem. Phys., **88** (1988) 5123
8. F. H. Stillinger and T. A. Weber, Phys. Rev. Lett., **62** (1989) 2144
9. T. A. Weber and F. H. Stillinger, J. Chem. Phys., **92** (1990) 6239
10. P. C. Weakliem, C. J. Wu and E. A. Carter, Phys. Rev. Lett., **69** (1992) 200
11. J. F. Ziegler, J. P. Biersack and U. Littmark, "The stopping and range of ions in solids", Pergamon Press, New York, 1985.
12. T. Aoki and J. Matsuo, Nucl. Instr. and Meth. **B216** (2003) 185.
13. S. Chiba, T. Aoki and J. Matsuo, Nucl. Instr. and Meth. **B180** (2001) 317.
14. H. M. Urbassek, Nucl. Instr. and Meth. **B31** (1988) 541.

Mater. Res. Soc. Symp. Proc. Vol. 843 © 2005 Materials Research Society T5.8

Laser induced hierarchical nano-composites in metallic multi-films: structural characterization

C. Daniel[1] and F. Mücklich
Functional Materials, Department for Materials Science, Saarland University, Saarbruecken, GERMANY

ABSTRACT

Biological solutions to enhance strength and stability often use lateral and hierarchical composite structures from nano- to micro-scale. The effect does not consist of a large chemical variety but it is realized by structural composites (namely phase changes and orientations). A new developed bio-mimetic laser interference metallurgical technique uses this biological approach to optimize mechanical properties of surfaces and thin films. A hierarchical nano-composite is realized by both a vertical nm-scaled period of PVD-processed multi-films and a lateral µm-scaled period of laser interference metallurgy.

In past, laser interference irradiated Ni/Al multi-films showed periodical properties in the range of interference period. The hardness and modulus could be varied periodically and the texture and stress situation could be significantly changed.

In this work, the micro-structural evolution of irradiated Ni/Al multi-films is analyzed by TEM measurements to justify the properties change. The grain size can be obtained to be laterally oscillating between 5 to 10 nm and up to 100 nm and the layer interface to be semicoherent. Up to a certain depth, intermetallic compounds are found in the layer interface.

Keywords: thin multi-film, laser interference metallurgy, structural composite, transmission electron microscopy

INTRODUCTION

Biological hierarchical composite structures show much higher relative strength and stability than technical systems because of the optimized and adaptive structure in nano- and micro-meter scale. These effects can be ascribed to both the interface bonding between the compositional phases [1] and the hierarchical and small dimensions [2].

In technical composite systems, the interface between the phases is not reticulated and the adherence is week. This might be resolved by growing of one phase in another instead of mixing two phases.

The hierarchical and small dimensions and a long range order can be also realized by an interference pattern of two or more coherent laser beams. The interference pattern of two beams shows an intensity distribution dependent on the used wavelength λ and the angle in between the beams θ to be [3]:

$$d = \frac{\lambda}{2\sin\left(\frac{\theta}{2}\right)} \quad (2)$$

This interference pattern of a high-power pulsed laser allows a direct, periodical, and local heating of metallic surfaces through the photo-thermal interaction between laser and metal [5,6]. Such a heating with specified threshold energies leads to both a direct structuring of the surface by topography effects [4,7] and the phase-microstructure effects like re-solidification, re-crystallization [7], and phase transformation [3]. Thus, the laser structuring

[1] Corresponding author: Phone: +49 681 302 3048, Fax: +49 681 302 4876, Email: C.Daniel@matsci.uni-sb.de

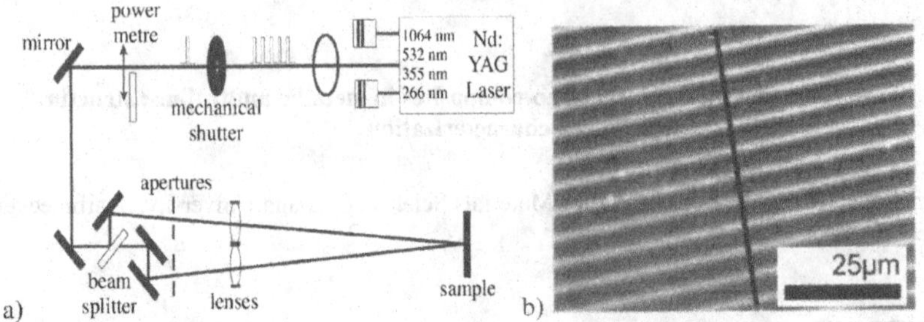

Figure 1: a) Experimental setup, laser interference system with the laser and the optical elements to guide the beams; b) Resulting line like structure on a sample (from [3,9])

can produce periodically new phases or controlled defect structures to realize a bio-mimetic composite effect with much more reticulated interfaces than common structures.

In past, Ni-Al-multi film systems are irradiated by the laser interference technique. In the ref. [9] and [10] the local and periodic mechanical effect of this structuring technique is demonstrated. The nano-hardness shows a lateral oscillation between 4.27 GPa and 5.74 GPa during the residual stresses in the individual layers reaches from unstructured 790 MPa and 130 MPa up to 1,300 MPa and 340 MPa for Ni and Al, repectivelly. The microstructural texture changes from a Ni (111) and an Al (100) fibre texture to a spiral fibre texture with a tilt of 15° of Ni (111) and 35° for Al (100). In past x-ray diffraction measurements, a phase formation to intermetallic phases could not be obtained.

In this work, we focus on the microstructure (namely grain size distribution, local phases and texture) in and the morphology of the thin film system by transmission electron microscopy study to understand the changes of mechanical properties, texture and stresses. TEM observations are compared to these published mechanical properties and x-ray diffraction measurements in ref. [9, 10].

EXPERIMENTAL SETUP

Film preparation

The multi-layered Ni/Al film was produced by physical vapor deposition with an Ar-ion-gun sputtering facility on a glass substrate under a vacuum of 10^{-5} Pa. Twenty Ni/Al bi-layers with a total thickness of around 1 µm were deposited. The thickness of the individual layer was monitored in-situ by a micro balance based on a quartz oscillating crystal to be 20 nm for Ni and 30.3 nm for Al. The resulting atomic ratio was expected to be 1:1 [8] and confirmed by EDX.

Laser interference metallurgy

The third harmonic of a q-switched Nd:YAG high-power laser with a wavelength of 355 nm, a pulse duration of 10 ns, and a frequency of 10 Hz was employed for the laser interference irradiation experiments. The primary laser beam was split into two beams and guided by an optical system to interfere with each other on the sample surface (see fig. 1 and ref. [3]). An angle of 4.04° in between the beams generates a laterally line like interference pattern with a period of 5.04 µm (see fig. 1b). Due to the fraction of molten material and stability of the film during irradiation [9], a laser fluence of 245 mJ/cm^2 and a number of 10 and 20 pulses were chosen respectively for each interference micro-structuring experiment.

Target preparation by lift out technique

The target preparation of electron transmissible cross-sectional foils of the thin films was managed using a focused ion beam dual beam workstation. The workstation contains of

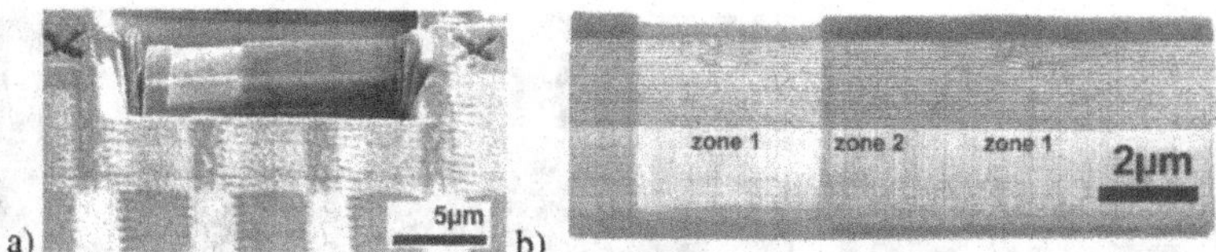

a) b)

Figure 2: Target preparation by focused ion beam: Milling process imaged in SEM mode (left hand side); prepared TEM foil imaged in STEM mode with a total thickness of 50 nm and a thinner area of 20 nm for high resolution imaging (right hand side). The foil shows two laser interference maxima (zone 1) separating by a laser interference minimum (zone 2) to analyze the periodical microstructure.

an integrated focused ion beam microscope in a scanning electron microscope. Focused Ga-ions are used to manipulate materials down to 7 nm resolution while the electrons give the possibility for an in-situ observation of the manipulation process. Manipulation means ablation (milling) or deposition.

Fig. 2 shows the TEM foil standing alone in the thin film system and the foil in STEM mode after lifting out. Two laser interference maxima are included in the TEM foil. Interesting zones for investigations are just under one (zone 1) and in between two laser maxima (zone 2) (see fig. 2b).

RESULTS

The layer morphology, grain size distribution, phase formation, and texture can be demonstrated in a TEM cross-section of the irradiated films.

Layer morphology

The individual layers in the multi-film show a strong plastic deformation in the area of laser intensity maxima (fig. 3). There is a significant difference in the morphology of the individual Ni and Al layers. The Al layers have changed the thickness and layer morphology

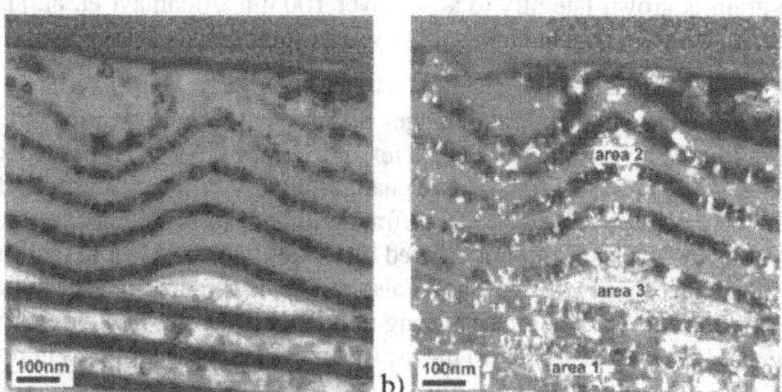

a) b)

Figure 3: Cross-sectional TEM image on a laser interference maximum: a) Bright field with dark Ni and bright Al sublayers; b) dark field. The microstructure in the individual Al sublayers appears to be fine grained in the area 2 (5 to 10 nm), one large grain in the area 3 (>100 nm), and unchanged nano-grained structure (due to the depth) in the rest of the sample named area 1 (20 to 30 nm).

a) 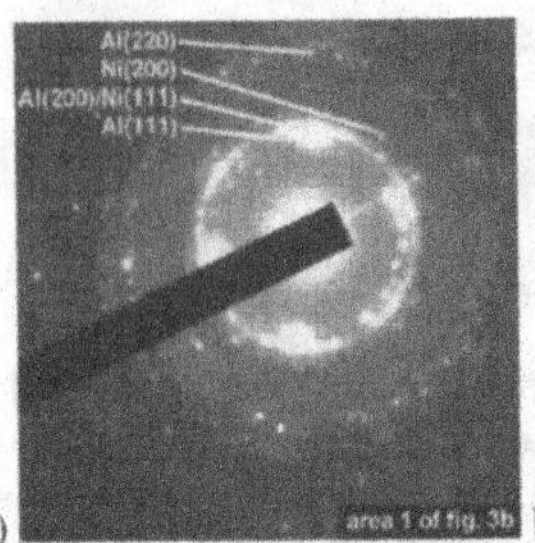

b) 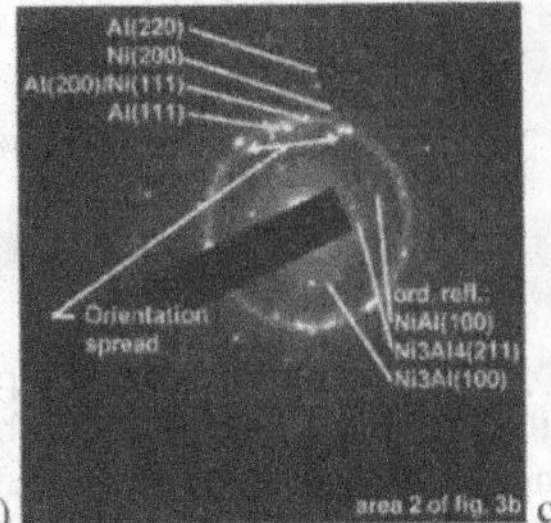

c) 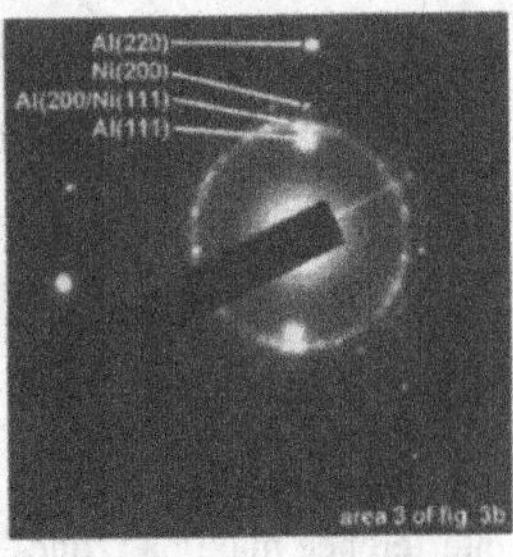

Figure 4: TEM diffraction images: a) Un-irradiated layers without ordered intermetallics' reflexes and a strong fiber texture (fig. 3b, area 1); b) Strongly irradiated layers with ordered intermetallics' reflexes and a texture change with an orientation spread (fig. 3b, area 2), c) One weakly irradiated and laterally grown Al grain (fig. 3b, area 3).

while the Ni layers are only deformed without any significant thickness change.

In the upper two bi-layers, the layer structure is strongly deformed and it can be found a zone of inter-diffusion between this layers (see also section "Texture and phase formation").

Grain size

Fig. 3 shows a TEM micrograph in the area of an intensity maximum (zone 1 of fig. 2b). In dark field TEM images (fig. 3b), the grain sizes are determined by analyzing the size of the bright grains in dark field images generated by different diffracting reflexes. The grain size distribution is statistical and uniform but it is grouped in local regions of laser structuring. The unchanged sample regions show grain sizes of about 20 to 30 nm due to the individual layer thickness (see fig. 3b area 1). This region is unchanged due to the depth. The same situation can be found in zone 2 (fig. 2b) where the laser intensity was very low. The average grain size on surface (see area 2) can be determined to be less than 10 nm which was also found by Liu et. al. [12] in similar experiments. In the last manipulated Al layer (see area 3) exactly one grain is grown laterally to a size over 100 nm. Aichmayr et. al. [11] have found similar results in structuring semiconductors by laser interference irradiation.

Texture and phase formation

Fig. 4 shows three electron diffraction images on different sample positions. It can be shown a strong texture in the layer (fig. 4a). Such a texture is normal for PVD produced films [8].

The texture and phase-microstructure changes after laser interference irradiation in dependence on the depth (see fig. 4). The diffraction image of unchanged layers (fig. 4a) shows only Ni and Al reflexes but not ordered intermetallics' reflexes. In the first two bi-layers, the diffraction image (fig. 4b) shows also the higher ordered intermetallics' reflexes of NiAl, Ni_3Al_4, and Ni_3Al (indication according to ref. [17]). In fig. 4b, a orientation spread of the Al(200)/Ni(111) reflexes similar to the x-ray diffraction measurements in ref. [9] can be obtained.

High resolution images

In high resolution images, the interface behavior of the individual layers in the multi-film can be obtained. The layers show semi-coherent interfaces in netplanes Al(200) and Ni(111) due to the very similar plane spaces of 0.2024 nm and 0.2034 nm, respectively (fig. 5a). On the sample surface, the nanocrystalline structure of the indiviual Al sublayers can be proofed in fig. 5b.

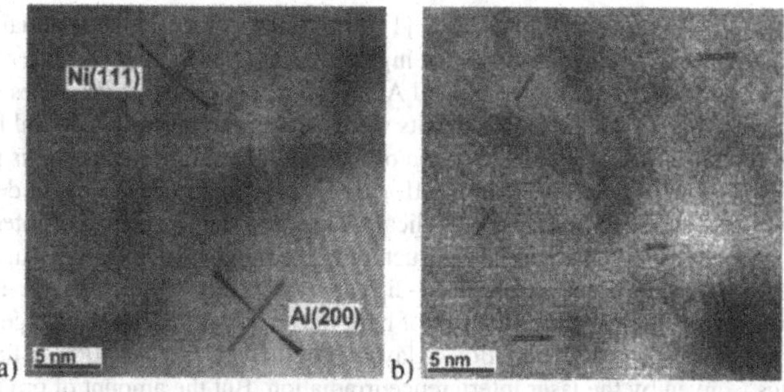

Figure 5: Lattice plane images: a) Interface between Ni (left upper corner) and Al (right bottom corner) in the 5th Al layer, the interface is semi-coherent; b) Grain structure in the 2nd Al layer with its nano-crystalline structure.

DISCUSSION

The laser interference irradiation introduces a very fast periodical heat treatment to the material. Due to this treatment, the laser interference irradiation changes generally three micro-structural parameters to manipulate the resulting defect structure: The morphology, the texture and the grain size distribution.

The periodical hardening process described in the introduction can be partially explained by the obtained morphology of the indiviual Ni sublayers, the texture evolution of the film, and the grain size distribution of the individual Al sublayers.

Morphology: The periodical heat treatment reaches to a periodical melting of the Al layers but leaves the Ni layers in the solid phase. A thermal simulation of the process is obtained in ref. [9]. Because of the resulting compressive stress by periodically and partially molten Al layers during the fast thermal treatment, the Ni layers receive a periodic deformation (see fig. 2 and 3) with high velocity which result in a work hardening process of the Ni films. Such kind of hardening results are shown in literature with "high-speed deformation" experiments of Titanium alloys by Tamirisakandala et. al. in ref. [13].

Texture: The described deformation of Ni layers tilt the crystals in preferred directions perpendicular to the line like interference structure and realize a orientation spread in the electron diffraction image (fig. 4b). This can also be a reason for hardness change because the resulting Ni texture is mechanically more stable than before irradiation. Mendik et. al. [15] have shown that the mechanical properties of single-crystalline or strongly textured Ni is dependent on the mechanically applied direction. Because of the melting and re-solidification process of the Al layers, the as sputtered fiber texture of Al changes with similar results for mechanical properties.

Grain size distribution: The cooling rate after melting is calculated to be in the order of 10^{10} K/s on the surface (see ref. [9]), but it is not constant and independent on the lateral and vertical coordinates. The grain size is changing accordingly to the relative position to the laser interference pattern. Due to the lateral periodicity of the thermal treatment, the local grain size monitors the local cooling rate. In the upper layer in the area of laser intensity maxima, the structure is fine grained to a size of 5 to 10 nm (area 2 of fig. 3b). In the last deformed Al layer the lower cooling rate is performed to induce a lateral grain growth to a size of more than 100 nm (area 3 of fig. 3b) while the grain sizes in the unchanged regions of sample rest to be about 20 to 30 nm fixed by the individual film thickness (area 1 of fig. 3b). Similar grain structures are found in crystallization and re-cristallization experiments of a-Si and c-Si wafers by Aichmayr et. al. (see ref. [11]). In this way introduced residual compressive stresses can reach values much higher than the yield stress because of the small thickness of 20 nm and 30 nm for the Ni and Al layers, respectively. Thicknesses smaller than 100 nm constrain dislocation movements which are important for a material flow [14].

Phase formation: The volume fraction of intermetallics all over the film is too small to be detected by x-ray diffraction. Electron diffraction images make it possible to detect small and local volume fractions of phases. Periodically local and small amounts of intermetallic phases can be found. The separated layer structure in the multi-film is changed in the zone of laser maximum. In the upper layers, an inter-diffusion between Ni and Al can be found. This interdiffusion can be understood as a result of both the temperature-drived change of the diffusion coefficient and the deformation of layers. In this way, the initiation of formation of intermetallics is given by the laser interference irradiation. But the amount of ordered intermetallics in this experiments might be to less to account for the change in mechanical properties. In other experiments, Liu et. al. [16] have shown a hardness change up to 300% if a phases transformation becomes important.

CONCLUSION

Laser interference irradiation gives the possibility for periodical modulation of the microstructure. This microstructural change (namely the morphology, grain size distribution, texture, and phase microstructure) is responsible for a micrometer-scaled periodical change of mechanical properties. If the structure is realized on a nanometer-scaled multi-film, a two step hierarchical bio-mimetic composite structure can be obtained.

ACKNOWLEDGMENT

The authors would like to thank R. Geurts from the FEI Company Lab Eindhoven for his help in TEM preparation and J. Schmauch from the Saarland University for his help in TEM observations. The research work of Prof. F. Mücklich was supported by the Alfried Krupp Prize for Young University Teachers awarded by the Krupp Foundation.

REFERENCES

[1] S. Weiner,W. Traub, H.D. Wagner, J. Struct. Biolog. **126**, 241 (1999).
[2] I. Jäger, P. Fratzl, Biophys. J. **79**, 1737 (2000).
[3] C. Daniel, F. Mücklich, Z. Liu, Appl. Surf. Sci. **208-209**, 317 (2003).
[4] M.K. Kelly, B. Dahlheimer, Phys. Stat. Sol. A **156**, K13 (1996).
[5] M. von Allmen, A. Blatter, *Laser- Beam Interactions with Materials*, 2nd ed. (Springer, Berlin, Germany, 1995).
[6] D. Bäuerle, *Laser Processing and Chemistry*, 2nd ed. (Springer, Berlin, Germany, 1996).
[7] M.K. Kelly, J. Rogg, C.E. Nebel, M. Stutzmann, Sz. Kátai, Phys. Status Solidi A **166**, 651 (1998).
[8] U. Rothhaar, H. Oechsner, M. Scheib, R. Müller, Phys. Rev. B **61,** 974 (2000).
[9] C. Daniel, A. Lasagni, F. Mücklich, Surf. Coat. Technol. **180-181,** 477 (2004).
[10] C. Daniel, A. Lasagni, F. Mücklich, Mat. Res. Soc. Symp. Vol. **795**, U10.4 (2004).
[11] G. Aichmayr, D. Toet, M. Mulato, P.V. Santos, A. Spangenberg, S. Christiansen, M. Albrecht, H.P. Strunk, Phys. Stat. Sol. A **166**, 659 (1998).
[12] K.W. Liu, F. Mücklich, Mat. Res. Soc. Symp. Vol. **753**, BB6.5 (2003).
[13] S. Tamirisakandala, P.V.R.K. Yellapregada, S.C. Medeiros, W.G. Frazier, J. C. Malas, B. Dutta, Adv. Eng. Mater. **5**, 667 (2003).
[14] O. Kraft, C.A. Volkert, Adv. Eng. Mater. **3**, 99 (2001).

[15] M. Mendik, S. Sathish, A. Kulik, G. Gremaud, P. Wachter, J. Appl. Phys. **71**, 2830 (1992).
[16] Z. Liu, X.K. Meng, T. Recktenwald, F. Mücklich, Mater. Sci. Eng. A **342**, 101(2003).
[17] International Center for Diffraction Data, PDF-No. Al [04-0787], Ni [04-0850], NiAl [44-1188], Ni_3Al [09-0097], Ni_3Al_4 [46-1037] (Powder Diffraction File, 1998).

Mater. Res. Soc. Symp. Proc. Vol. 843 © 2005 Materials Research Society T5.9

Si(100) Nitridation by Remote Plasmas for Enhanced High- κ Gate Performance

P. Chen and T. M. Klein
Department of Chemical and Biological Engineering, The University of Alabama,
Tuscaloosa, AL 35487-0203, USA

ABSTRACT

Hafnium oxide is the leading high-κ candidate for next generation CMOS devices, however, it is plagued by problems such as a propensity to react with the Si(100) substrate, to crystallize, and to have fixed and trapped charges and low transistor mobilities. A remote N_2/He plasma was used as a reactant during MOCVD with Hf (IV) t-butoxide which incorporates 6 at.% nitrogen located at the film-substrate interface and reduces the interdiffusion problem upon anneal [1]. A new process for pretreating the silicon wafer with a N_2/He plasma was developed to improve nitrogen concentrations at the interface and reduce interdiffusion further. Films deposited with N_2/He plasma and the pretreatment method were compared to O_2/He plasma deposited films and N_2/He plasma deposited films without the nitridation step. It is shown that a 16Å interface SiN_x layer is sufficient to prevent reaction during a 1000ºC Ar/O_2 anneal at atmosphere and an intermediate annealing step is crucial for desired reduction in interdiffion. Thick films show some crystalline peaks by XRD which are suppressed using the N_2 process. Electrical measurements on thick films show the pretreatment process results in the lowest leakage current density and the highest dielectric constant of 21.5.

INTRODUCTION

Different dielectric materials are currently being studied to replace SiO_2 as the gate dielectric in the future complementary metal-oxide-semiconductor (CMOS) transistor [2-4]. Among them, HfO_2 is a promising one with high dielectric constant κ=30 [5], good conduction band offset [6] and thermodynamic stablility [7]. However, HfO_2 suffers from interfacial reaction with the substrate [8], crystallization and dopant penetration through from a polysilicon gate [9]. Nitrogen incorporation into high-k dielectrics decreases the interfacial reaction, crystallization [9], and impurity and dopant penetration [10, 11].

Nitrogen is mainly incorporated into the oxide and silicate interfacial layers as a pseudo ternary compound $[HfO_2]_x[(SiO_2)_{1-y}(Si_3N_4)_y]_{1-x}$ [12, 13]. However, silicon oxynitride may suffer from degraded mobility [14], and high fixed charges density [15], and the nitrogen can be lost during anneal in an oxidizing environment [16]. The nitrogen composition decrease causes interfacial structure deterioration, crystallization of the amorphous film and lowers the overall capacitance of the gate stack. Our previous work confirms that the nitrogen incorporation decreased the interfacial reaction significantly [1], however, we also found that the nitrogen composition decreased after annealing which can potentially cause interfacial instability, crystallization and low capacitance. In order to increase the nitrogen concentration and at the same time decrease the nitrogen loss in the film after annealing, we redesigned the original film deposition process by using nitrogen/helium plasma pretreatment of wafer before the hafnium layer deposition.

EXPERIMENTAL DETAILS

The sample substrates used in the experiments were cut from n-type 4-6 Ω/cm Si (100) 4-inch diameter wafers. Resistance of the wafer used for the I-V and C-V has been measured by the four-point probe method. Before the experiment, silicon wafers were cleaned by using a diluted HF (10%) solution and rinsed with de-ionized water for 1 minute. Remote, inductively coupled plasmas of N_2 and He, or O_2 and He mixtures with flow rates of each gas maintained at 100 sccm and plasma power at 50W were used to generate active oxidant species. The reactor pressure was maintained at 1 torr while the deposition temperature was 390°C. Hf (IV) t-butoxide was delivered downstream from the plasma through a gas dispersion ring. Vapor was produced in a bubbler heated by a 65°C oil bath without a carrier gas. Deposition rates were obtained by thickness by a J. A. Woollam VASE ellipsometer and were targeted to 4 nm for interface stability experiments using XPS, 17 nm – 25 nm for the I-V and C-V measurements, and 30 nm for XRD. Different annealing methods have been used. The samples used in the electric measurement were annealed inside the reaction chamber with the temperature 600°C, 1 torr with argon for 2 minutes. Capacitors were made by sputtering platinum electrodes and then were annealed at 200°C in a forming gas of N_2:H_2 = 9:1 for 1 hour. Interface stability studies were done using a furnace anneal at 1000°C, in argon with small amount of oxygen (from backdiffusion) at 1 atm for 2 minutes. XPS data was collected with a Physical Electronics APEX surface analysis system and an Omicron EA125 hemispherical analyzer, XPS measurements were done with an Al Kα source at take-off angles of 25°, and peaks were shifted to an adventitious C(1*s*) peak at 284.8 eV to compensate for charging. XRD analysis as done with a Philips X'pert system, while a Radiant precision workstation system and HP 4284A were used for capacitor measurements.

RESULTS AND DISCUSSION

Composition of a thin 2.5 nm hafnium oxide sample deposited using a N_2/He plasma was measured by an Auger depth profile and is shown in Figure 1. Two trends are apparent which show the nitrogen concentration increasing with the silicon concentration, and concurrently, the hafnium atomic concentration decreasing. This indicates at these low nitrogen concentrations SiO_xN_y is likely to form at the interface rather than HfO_xN_y in the bulk which is desirable for interface quality and stability.

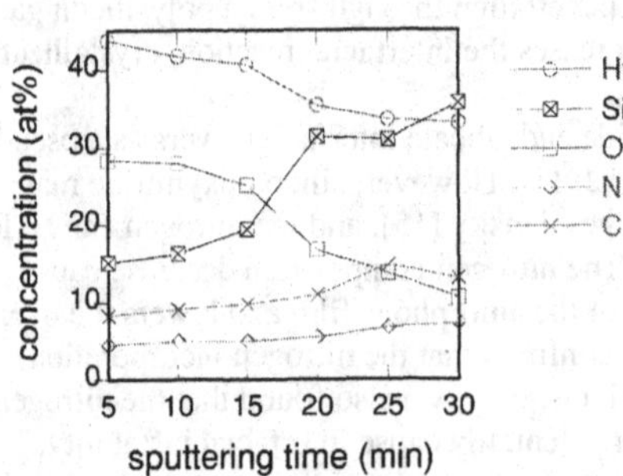

Figure 1. Auger depth profile N_2/He plasma deposited hafnium oxide film.

To improve N incorporation at the interface, the Si substrate was exposed to a 50W N_2/He plasma for 5 min before MOCVD with the Hf precursor. Figure 2 shows the XPS Si 2p region of the pretreated substrate before and after 1000°C anneal in Ar (with residual oxygen) at 1 atm for different pretreatment time and power. It is useful to define a fraction (Si-O,N%) of the of the Si 2p peak area corresponding to Si-O,N versus total peak area of the Si-O,N at high binding energy and the Si° peak at low binding energy from the substrate as a measure of the extent of reaction after anneal as the Si° peak disappears with increasing

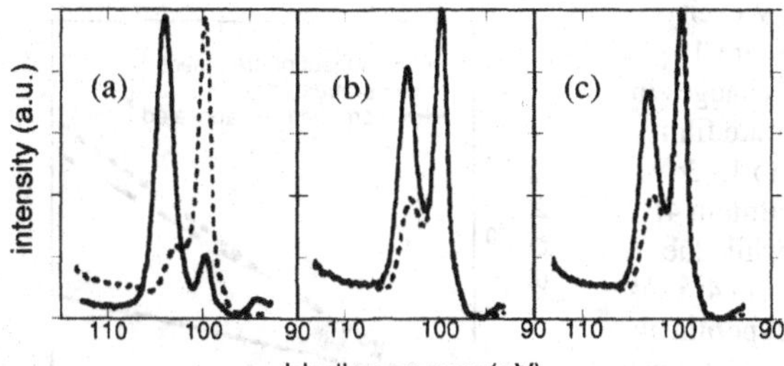

Figure 2. Si 2*p* XPS peak of the plasma pretreated substrate. (a) 5min 25W (b) 3min 50W and (c) 5min 50W. Dashed line: before anneal, Solid line: after anneal.

thickness due to the finite escape depth of the photoelectrons and the Si-O,N peak grows with reaction at the interface. For an annealed pretreated sample exposed to a 25W plasma for 5 min the thickness changes from 12Å to 32Å upon anneal while the Si-O,N% changes from 21.87% to 86.59%. The 50W, 3 min pretreatment results in a 14Å to 17Å thickness change and a Si-O,N% change from 30.99% to 50.07%. The 50W, 5 min pretreatment resulted in 16Å with no thickness change by ellipsometry upon anneal, but with a Si-O,N% change from 33.19% to 46.34%. All samples lost nitrogen during the anneal as observed by the N 1s XPS peak (not shown). The 25W, 5min N concentration changed from 6.36 at. % before annealing to 0.62 at. % after annealing; the 50W, 3 min sample changed from 6.47 at. % to 4.94 at. %; while the 50W, 5 min changed from 11.07 at. % to 7.48 at. %. It is clear that more nitrogen at a higher plasma power results in a smaller thickness change and a reduced Si-O,N% increase. During anneal, trace oxygen replaces nitrogen which desorbs resulting in a peak shift to higher binding energies as shown in XPS data in Figure 2. The 50W, 5 min pretreatment was used for subsequent hafnium dielectric deposition, as the 16Å interface layer seems sufficient for film interface stability, although further experiments show, an intermediate anneal between the pretreatment and dielectric deposition step is required for optimal physical and electrical properties.

Figure 3 shows XPS Si 2p data from three, approximately 4 nm thick, hafnium oxide films deposited with a N_2/He plasma on an untreated Si substrate, a pretreated substrate, and a pretreated substrate with an intermediate 600ºC, 2 min Ar anneal at 1 torr inside the deposition chamber. All three samples were annealed after deposition to 1000ºC for 1 min. at 1 atm. Ar with trace O_2 and the XPS traces shown are for the samples before and after this anneal. It is seen that the pretreatment with an intermediate anneal has the lowest Si-O,N% change from 48.8% to 62.3%. This sample was 45Å by ellipsometry before and after the anneal indicating minimal interface reaction. In comparison, the pretreated sample without the intermediate anneal increased in thickness from 41Å to 45Å and the Si-O,N% increased from 47.5% to 82.6%. The sample with no pretreatment increased from 39Å to 44Å and the Si-O,N% increased from 44.3% to 86.7%. A similar Si 2p shift is observed as in Figure 2 for the untreated and pretreated substrates without anneal, however the intermediate

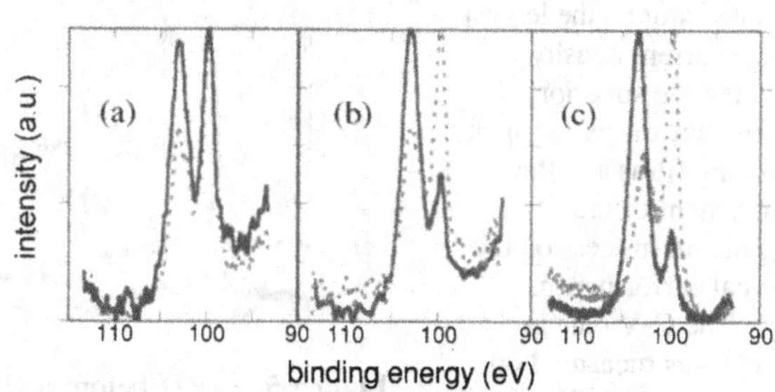

Figure 3. Si 2*p* XPS peak of hafnium oxide deposited on (a) pretreat/anneal, (b) pretreat w/o anneal, and (c) untreated substrates. Dashed line: before anneal, solid line: after anneal.

anneal fixes the Si 2p peak at 102.9eV even after the 1000°C annealing. The hafnium 4f peak (data not shown) also indicates changes in oxidation state as the untreated substrate film hafnium 4f peak shifts from 17.8eV to 18.2eV, while the pretreated substrate film hafnium 4f peak shifts from 17.8eV to 18.1eV, while the intermediate anneal results in a hafnium 4f peak fixed at 17.8 eV. The result of this experiment shows the pretreatment decreases Si and oxygen diffusion during the high temperature anneal to some degree, however, the intermediate 600°C anneal has the largest effect on stability. This is due to the solidification of loosely bonded nitrogen introduced by the plasma to create a dense and relaxed network impermeable to atom diffusion. Figure 4 shows the extent of change in the Si-O,N% for the three samples.

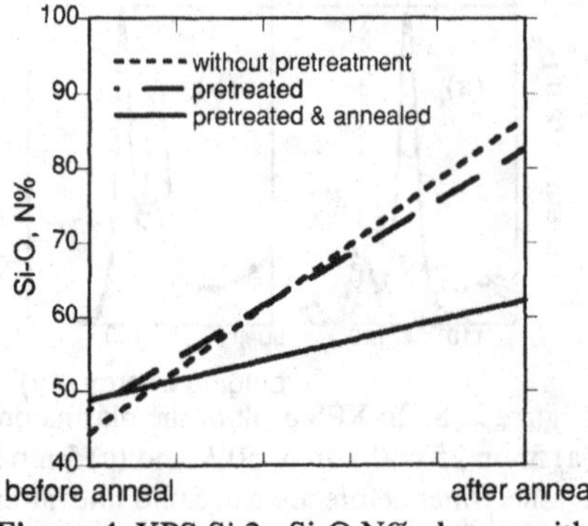

Figure 4. XPS Si 2*p* Si-O,N% change with anneal for different pretreatment methods.

Crystallization of the films were studied by XRD on 30 nm thick films before and after annealing to 1000ºC in 1 atm of Ar (with trace oxygen). The results are shown in Figure 5 for a O_2/He plasma deposited film without pretreatment, a N_2/He plasma deposited film without pretreatment and a N_2/He plasma deposited film with pretreatment and anneal. The peaks were calibrated by the Si (400) substrate 2θ peak located at 69.13°. Two significant HfO_2 peaks appear at 28.4° and 31.6° corresponding to HfO_2 monoclinic (111). The O_2/He plasma deposited film (a) shows severe crystallization before and after annealing, while the sample produced by the N_2/He plasma (b) shows only limited crystallization although annealed at high temperature and only containing 6 at. % nitrogen. The N_2/He plasma deposited film on the pretreated substrate (c) shows little difference than the film deposited on the untreated substrate (b), except a small crystallization peak at 31.6° observed after annealing.

The electrical properties of the O_2/He and N_2/He deposited hafnium oxide films on untreated Si(100) were compared to N_2/He films deposited on a pretreated substrate. I-V and C-V results are shown in Figures 6 and 7 for capacitors 20.2 nm, 24.2 nm, and 17.4 nm thick respectively with thickness measured after a 2 min., 600ºC anneal in 1 torr of Ar inside the deposition chamber. Of these samples, the O_2/He plasma film shows the largest leakage current density, while the pretreated substrate exhibits the lowest leakage current density indicating the superior performance of the nitrogen containing films and the improvement of the pretreatment process on the electrical performance.

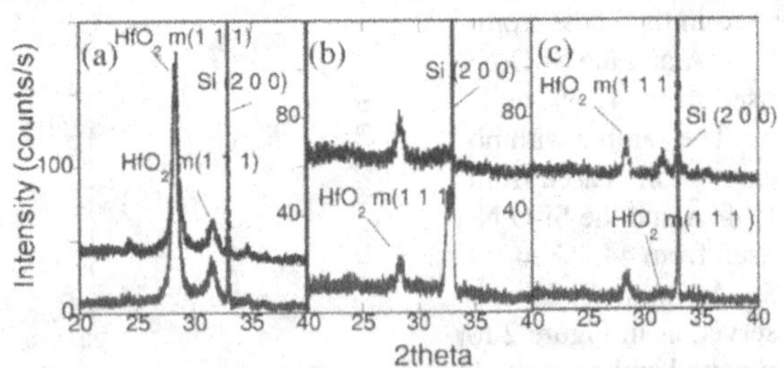

Figure 5. XRD before and after anneal for hafnium oxide films deposited with (a) O_2/He plasma and (b) N_2/He plasma on untreated substrates and (c) N_2/He on pretreated substrate.

The C-V data in Figure 7 was measured on capacitors with O_2/He plasma deposited and N_2/He plasma

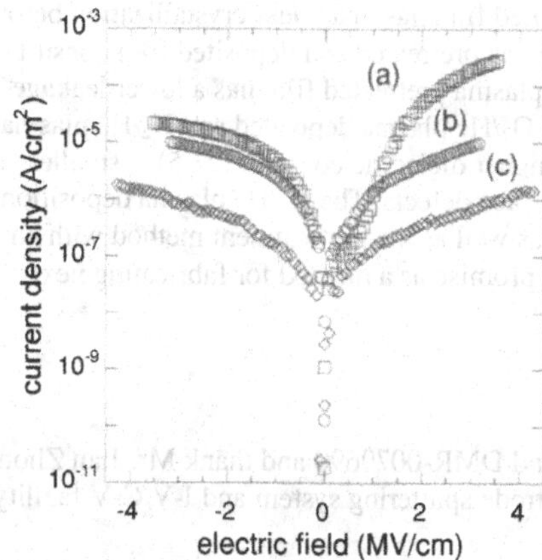

Figure 6. I-V for hafnium oxide films deposited with (a) O_2/He plasma and (b) N_2/He plasma on untreated substrates and (c) N_2/He on pretreated substrate.

deposited films on untreated Si substrates and with a N_2/He plasma deposited film on a pretreated sample with areas of 7.9×10^{-4} cm^2, 9.1×10^{-4} cm^2, and 4.9×10^{-4} cm^2 respectively. Some frequency dispersion is observed which is greatest for the O2/He deposited film. Using Hauser's model for C-V data, the dielectric constant and hysteresis at flat band of the three films were 15.1 and 80 mV, 18.7 and 90 mV, and 21.5 and 10 mV for the O_2/He and N_2/He plasma films on untreated substrates and on the N_2/He plasma deposited film on the N_2/He pretreated substrate. The ΔV_{fb} for the O_2/He plasma sample is -0.1eV, while the value is 0.1eV for the other two samples which can be expect for N_2/He treated films as nitrogen is shown to produce positive fixed charges [17]. In addition, the stretch out in the C-V curve of the O_2/He film is visibly greater than the N_2/He deposited films, indicating a higher density of interface defects. The pretreatment and annealing process can be optimized to form a minimal interfacial layer for hafnium oxide gates resulting in t_{ox} equivalent less than 1 nm.

CONCLUSIONS

In summary, N_2/He remote plasma deposited hafnium oxide films contain nitrogen mainly at the interface in the form of SiO_xN_y. In order to achieve strong nitrogen bonds at the interface and also increase the nitrogen concentration in the interface layer, a N_2/He plasma pretreatment of the silicon wafer was developed. A 16Å SiN_x layer is needed before the high-κ dielectric deposition to minimize possible interfacial reaction of the dielectric and substate. Furthermore, annealing after the pretreatment and before high-κ deposition can further significantly decrease the interface oxidation during post deposition anneals. Compared to the high-κ dielectric film

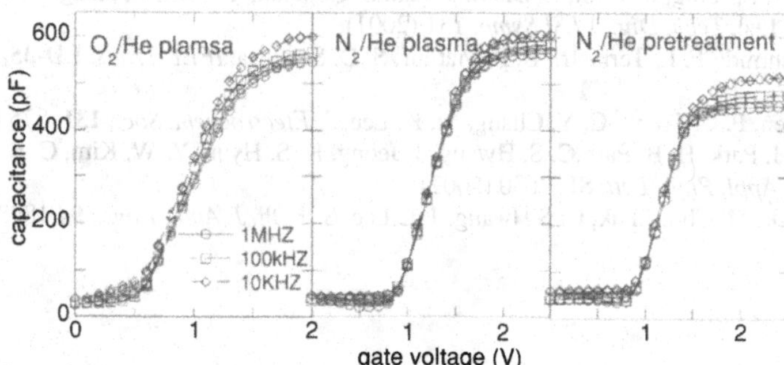

Figure 7. C-V of hafnium oxide film capacitors deposited with O_2/He and N_2/He plasmas on untreated and N_2/He plasma treated substrates.

deposited by O_2/He plasma, the N_2/He plasma deposited film has much less crystallization before and after annealing by XRD, however, the N_2/He plasma pretreated and deposited film doesn't show a completely amorphous structure. The N_2/He plasma pretreated film has a lower leakage current density from I-V measurements compared to O_2/He plasma deposited and N_2/He plasma deposited films on untreated substrates as well as a higher dielectric constant (21.5), a smaller trapped charge density, and a smaller number of interface defects. The N_2/He plasma deposition method for nitrogen incorporation in hafnium oxide as well as the pretreatment method with an intermediate anneal for nitriding the interface shows promise as a method for fabricating next generation CMOS gates.

ACKNOWLEDGEMENTS

We acknowledge NSF for award #s ECS-0084703 and DMR-0079690 and thank Mr. Jian Zhong and Prof. R. K. Pandey for assistance with their electrode sputtering system and I-V/C-V facility.

REFERENCES

1. P. Chen, H. Bhandari, T. M. Klein, *Appl. Phys. Lett.* **84**, 1574 (2004).
2. *The International Technology Roadmap for Semiconductors* (Semiconductor Industry Association, San Jose CA, 2001).
3. D. A. Buchanan, *IBM. J. Res. Dev.* **43,** 245 (1999).
4. G. D. Wilk, R. M. Wallace and J. M. Anthony, *J. Appl. Phys.* **89**,5243 (2001).
5. Y. S. Lin, R. Puthenkovilakam, J. P. Chang, *Appl. Phys. Lett.* **81**, 2041 (2002).
6. J. Robertson, *J. Vac. Sci. Technol.* **B18**, 1785 (2000).
7. K. J. Hubbard, D. G. Schlom, *J. Mater. Res.* **11**, 2757 (1996).
8. M. A. Quevedo-Lopez, M. El-Bouanani, B. E. Gnade, R. M. Wallace, M. R. Visokay, M. Douglas, M. J. Bevan and L. Colombo, J. Appl. Phys. **92**, 3540 (2002).
9. Y. Morisaki, T. Aoyama, Y. Sugita, K. Irino, T. Sugii, T. nakamura, Tech. Dig. – Int. Electron Devices Meet. **2002**, 861 (2002)
10. S. V. Hattangady, H. Niimi, G. Lucovsky, *Appl. Phys. Lett.*, **66**, 3495 (1995).
11. G. Lucovsky, *J. Vac. Sci. Technol.* **A17**, 1340 (1999).
12. M. R. Visokay, J. J. Chambers, A. L. P. Rotondaro, A. Shaneware, L. Colombo, *Appl. Phys. Lett.* **80**, 3183 (2002).
13. K. Onishi, L. Kang, R. Choi, E. Dharmarajan, S. Gopalan, Y. Joen, C. Kang, B. Lee, R. Nieh and J. C. Lee, *Tech. Dig., VLSI Symp.* 131 (2001).
14. M. A. Schmidt, F. L. Terry Jr., B. P. Mathur, S. D. Senturia, *IEEE Trans.* **ED-35**, 1627 (1988).
15. C. T. Chen, F. C. Tseng, C. Y. Chang, M. K. Lee, *J. Electrochem. Soc.*, **131**, 875 (1984).
16. M. Cho, J. Park, H. B. Park,C. S. Hwang, J. Jeong, K. S. Hyun, Y. W. Kim, C. B. Oh and H. S. Kang, *Appl. Phys. Lett.* **81**, 3630 (2002).
17. H. B. Park, M. Cho, J.Pak, C. S.Hwang, J. C.Lee, S. J.Oh, *J. Appl. Phys.*, **94**,1898 (2003)

Mater. Res. Soc. Symp. Proc. Vol. 843 © 2005 Materials Research Society T3.1

Surface Nitridation of Amorphous Carbon by Nitrogen Ion Beam Irradiation

Chihiro Iwasaki, Masami Aono, Nobuaki Kitazawa, and Yoshihisa Watanabe
Department of Materials Science and Engineering, National Defense Academy,
Kanagawa 239-8686, Japan

ABSTRACT

Amorphous carbon was irradiated with a nitrogen ion beam and changes in composition and tribological properties after surface nitridation have been studied. The nitrogen ion beam energy was varied from 0.1 to 2.0 keV under the constant ion current density. Composition and chemical bonding states near the surface were analyzed by X-ray photoelectron spectroscopy (XPS). Tribological properties were studied by a pin-on-disk type tribotester. XPS studies show that the nitrogen concentration near the surface increases after nitrogen ion irradiation. The depth profiles of nitrogen in the irradiated specimens display that the nitrogen concentration near the surface is maximized after 0.2 to 0.7 keV ion irradiation. From the pin-on-disk testing, it is found that the friction coefficient of amorphous carbon increases after irradiation by 1.5 keV nitrogen ions, while the friction coefficient does not change drastically after irradiation by 0.2 keV ions.

INTRODUCTION

Since it was predicted that crystalline carbon nitride (β-C_3N_4) would be comparable with or harder than diamond [1], various attempts have been made to synthesize this hypothesized material using different techniques, such as laser ablation deposition [2], ion beam assisted deposition [3], reactive sputtering [4], hot-filament chemical vapor deposition. [5] In addition to the deposition methods, nitrogen ion irradiation into carbon materials has been proposed as an effective technique for preparation of carbon nitride (CN_x). [6]

Hoffman *et al* irradiated graphite and diamond by 500 eV nitrogen ions in the temperature range from room temperature to 800 K and studied effects of irradiation temperature, post-irradiation annealing and ion dose. According to their results, it is found that the nitrogen concentration increases and implanted nitrogen is present in three different bonding states of nitrogen and carbon. [6-8] Galán *et al* studied X-ray photoelectron spectroscopy of CN_x films grown by nitrogen ion implantation on graphite and hydrogenated amorphous carbon in the ion energy region from 250 to 5000 eV. [9] They found that higher nitrogen concentration is obtained by consecutive implanting from higher to lower energies and the N_{1s} line shape is composed of four chemical species. Husein *et al* prepared CN_x films by implanting amorphous carbon films with nitrogen using plasma immersion ion implantation and conventional ion beam implantation. [10] Raman spectra show that the structure of implanted CN_x films becomes more amorphous compared with the as-deposited films.

Recently, the present authors irradiated bulk amorphous carbon using a nitrogen ion beam to form CN_x layers. [11] Amorphous carbon shows unique properties, such as isotropic mechanical properties and non-cleavage, compared with graphite. In the previous work, the effect of nitrogen ion beam irradiation on formation of CN_x was mainly studied with the constant ion energy of 0.2 and 1.5 keV and it is found that 0.2 keV ion irradiation is more effective for

CN_x formation than 1.5 keV ion irradiation. [11] Since the nitrogen ion energy was limited in the previous work, it is interesting to extend the ion energy region and study how nitrogen ion beam energies affect on nitridation of amorphous carbon.

In the present paper, bulk amorphous carbon is irradiated by a nitrogen ion beam with different ion energies, and the effect of ion beam energy on the composition and tribological properties of the irradiated amorphous carbon is examined.

EXPERIMENTAL

Amorphous carbon (a-C) substrates of 10 x 10 x 0.7 mm, produced by Mitsubishi Pencil Co, were used as a specimen. A specimen was introduced into a vacuum chamber after ultrasonic cleaning by ethanol and then irradiated using a nitrogen ion beam. The nitrogen ion beam energy was varied from 0.1 to 2.0 keV and the nitrogen ion beam current density was kept at constant, about 70 $\mu A/cm^2$. After the vacuum chamber was evacuated to 1.3 x 10^{-4} Pa, nitrogen gas was introduced and then nitrogen ions were generated by arc discharge. The pressure was kept at 4.7 x 10^{-3} Pa during ion beam irradiation. Ion irradiation was performed at the right angle against the a-C substrate and continued for 10 min at room temperature.

Composition and chemical bonding states between carbon and nitrogen were analyzed by X-ray photoelectron spectroscopy (XPS) (ESCA 1600, Physical Electronics) with a non-monochromatic Mg K_α X-ray source. Depth profiles of carbon and nitrogen in the a-C substrates after ion irradiation were obtained by sputtering them by 2 keV argon ions. Tribological properties were studied by a pin-on-disk type tribotester using a steel ball. The normal load to the ball was fixed at 730 mN and the specimen was rotated with 60 rpm. Carbon structure is examined by Raman spectroscopy using an argon laser (λ = 514.5 nm).

RESULTS AND DISCUSSION

Similarly to previous work, carbon, oxygen and nitrogen peaks are clearly observed in the wide XPS spectra from the a-C substrate irradiated by nitrogen ions. However, after the surface was sputtered by 2 keV argon ions for 30 sec, the oxygen peak disappears, while the nitrogen peak is still observed. From these results, it is concluded that oxygen is adsorbed on the surface and nitrogen is absorbed into the a-C substrates.

Typical depth profiles of nitrogen and carbon in the a-C substrates are shown in Fig. 1 (a) to (d) for the a-C substrates after irradiation by 0.1, 0.5, 1.0 and 2.0 keV nitrogen ions for 10 min, respectively. In each figure, it can be seen that the nitrogen concentration is found to be less than 20 % near the surface, and the carbon concentration gradually increases with the sputtering time and the nitrogen concentration decreases correspondingly. Compared Fig. 1 (a) with (b), it is found that the nitrogen concentration near the surface increases with increasing the nitrogen ion energy. However, the nitrogen concentration near the surface decreases slightly when the nitrogen ion energy increases to 1.0 keV, as shown in Fig. 1 (c). When the nitrogen ion energy increases further to 2.0 keV, the nitrogen concentration near the surface increases slightly again, as shown in Fig. 1 (d). In addition, Fig. 1 (d) shows that the nitrogen is absorbed in the a-C substrate more deeply than any other specimens when the ion beam energy is 2.0 keV.

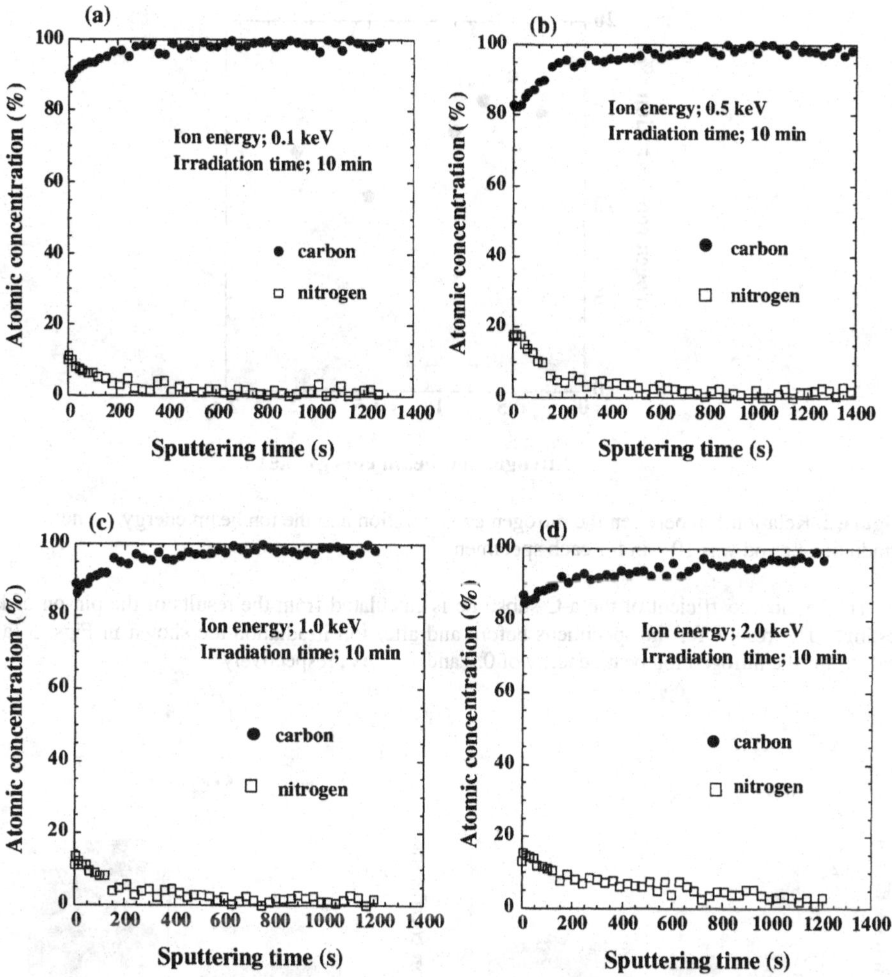

Figure 1. Depth profiles of carbon and nitrogen in the amorphous carbon substrate after irradiation by a nitrogen ion beam with different ion energies; (a) 0.1 keV, (b) 0.5 keV, (c) 1.0 keV and (d) 2.0 keV. Irradiation period was 10 min for each specimen.

The nitrogen concentration is plotted as a function of the ion beam energy and shown in Fig. 2. From this figure, it is evident that the nitrogen concentration is maximized from 0.2 to 0.7 keV. This result can be explained by assuming that, when the nitrogen ion energy is low like 0.1 keV, nitrogen ions are not absorbed in the substrate because of their insufficient energy, and the amount of absorbed nitrogen increases with increasing the ion energy, and back-scattering of nitrogen ions becomes dominant when the nitrogen ion energy increases above 1.0 keV.

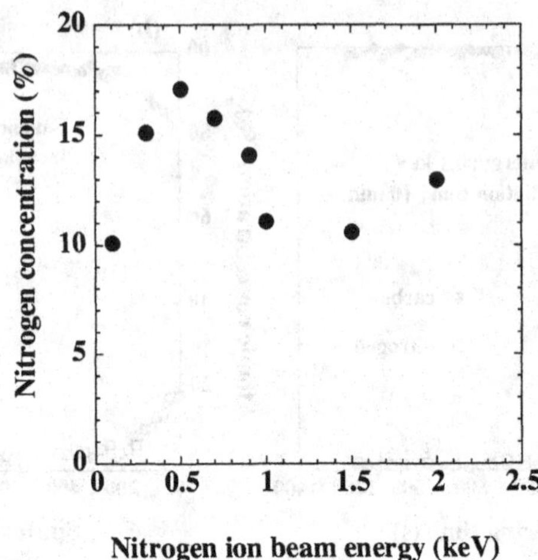

Figure 2. Relationship between the nitrogen concentration and the ion beam energy. The irradiation period was 10 min for each specimen.

The friction coefficient of the a-C substrate is calculated from the results of the pin-on-disk testing. The results for the specimens before and after ion irradiation are shown in Figs. 3 (a) and (b) for the nitrogen ion beam energy of 0.2 and 1.5 keV, respectively.

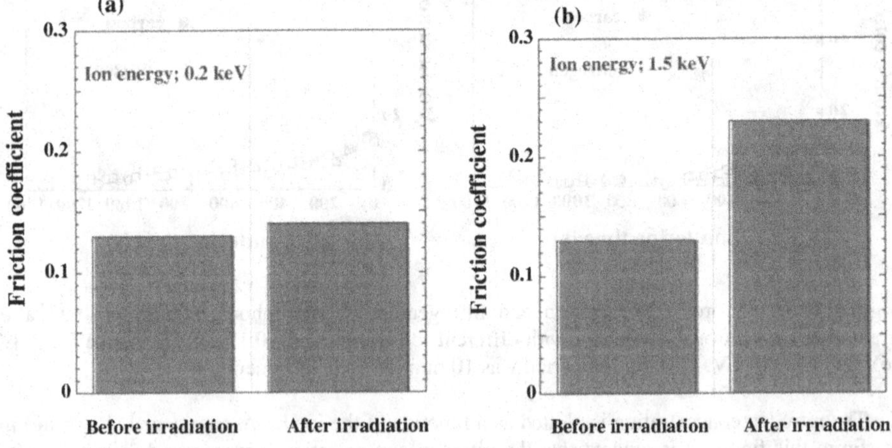

Figure 3. Changes in the friction coefficient of amorphous carbon substrates between before and after ion irradiation.

From these figures, it is evident that the friction coefficient of amorphous carbon increases after irradiation by 1.5 keV nitrogen ions, while the friction coefficient does not change drastically after irradiation by 0.2 keV ions.

We have reported that the root-mean square roughness (R_{MS}) of the a-C substrate increases from approximately 2.0 nm to 2.5 nm after nitrogen ion irradiation and the considerable difference in the value of R_{MS} is not discerned between 0.2 and 1.5 keV ion irradiation. [11] These results lead us to the conclusion that the difference in the friction coefficient after the ion irradiation is not result of the change in the surface morphology.

The friction coefficient of CN_x films prepared by ion beam assisted deposition is reported to be between 0.15 and 0.2 by Ohta et al. [12] According to their results, the friction coefficient depends on Ar/N_2 ratio during deposition and it is found that the higher the nitrogen concentration, the higher the friction coefficient. Although the friction coefficient obtained in the present work is in agreement with their results, the trend is opposite to them, the higher the nitrogen concentration, the lower the friction coefficient. This difference can be explained by assuming that the friction coefficient is sensitive to surface conditions and they depend on preparation methods.

The typical Raman spectra are shown in Fig. 4 (a) and (b) for the a-C substrates irradiated by 0.2 and 1.5 keV ions. In both spectra, *D*- and *G*-peaks are clearly seen around the wave number of 1350 and 1580 cm^{-1}. According to Tuinstra *et al*, the intensity ratio of the *D* to *G*-peak, I_D/I_G, is related to the carbon cluster size. [13] The value of the I_D/I_G is found to be approximately 1.5 for the as-received a-C substrate, while the ratio is changed to approximately 1.2 and 0.65 after irradiation by 0.2 and 1.5 keV nitrogen ions. This suggests that the carbon cluster size does not change drastically after 0.2 keV ion irradiation and it changes after 1.5 keV ion irradiation. This change in the carbon cluster size may be responsible to the change in the friction coefficient. To confirm the relation between the friction coefficient and nitrogen concentration, friction tests for samples with the maximum nitrogen concentration, for example after 0.5 keV ion irradiation, are necessary as future work that will be performed.

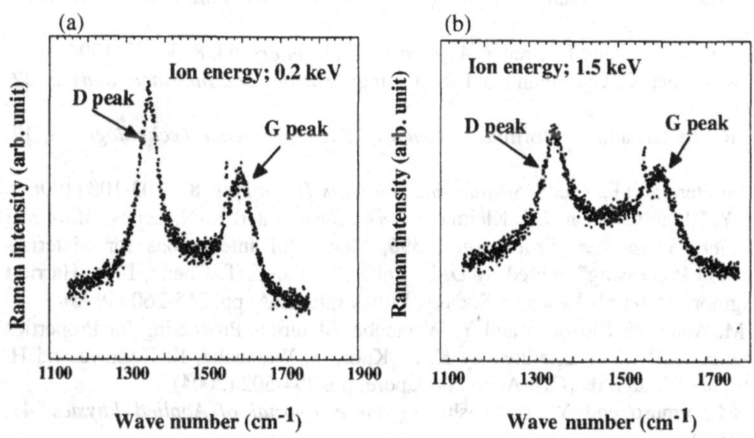

Figure 4. Typical Raman spectra of amorphous carbon substrates irradiated by (a) 0.2 keV and (b) 1.5 keV nitrogen ions.

CONCLUSIONS

Amorphous carbon (a-C) substrates were irradiated with the nitrogen ion beam with different ion energies from 0.1 to 2.0 keV for 10 min. The depth profiles of nitrogen in the irradiated specimens display that the nitrogen concentration near the surface is maximized after 0.2 to 0.7 keV ion irradiation. From the pin-on-disk testing, it is found that the friction coefficient of amorphous carbon increases after irradiation by 1.5 keV nitrogen ions, while the friction coefficient does not change drastically after irradiation by 0.2 keV ions. The change in the friction coefficient may be related to the change in the carbon cluster size.

ACKNOWLEDGEMENT

The authors would like to thank Prof. A. Matsumuro and Mr. T. Suzuki, Nagoya University, for their kind help in measurements of tribological properties. The nitrogen ion beam irradiation was performed using the ion vapor deposition apparatus in the Advanced Materials Laboratory of National Defense Academy.

REFERENCES

1. A. Y. Liu and M. L. Cohen, *Science,* **245**, 841-842 (1989).
2. M. Okoshi, H. Kumagai and K. Toyoda, *Journal of Materials Research*, **12**, 3376-3379 (1997).
3. D. Marton, K. J. Boyd, A. H. Al-Bayati, S. S. Todorov and J. W. Rabalais, *Physical Review Letters,* **73**, 118-121 (1994).
4. H. Sjöström, W. Lungford, B. Hjörvarson, K. Xing and J. E. Sundgren, *Journal of Materials Research*, **11**, 981-988 (1996).
5. Y. Chen, L. Guo and E. G. Wang, *Journal of Physics: Condensed Matters* **8**, L685-L690 (1996).
6. A. Hoffman, I. Gousman, and R. Brener, *Applied Physics Letters,* **64**, 845- 847(1994)
7. I. Gousman, R. Brener, C. Cytermann and A. Hoffman, *Surface and Interface Analysis,* **22**, 524-527 (1994).
8. I. Gousman, R. Brener, and A. Hoffman, *Journal of Vacuum Science Technologies A,* **17**, 411-420 (1999).
9. L. Galán, I. Montero and F. Rueda, *Surface and Coatings Techinique*, **83**, 103-108 (1996).
10. I. F. Husein, Y. Zhou, C. Chan, J. I. Kleiman, Y. Gudimenko and K.-N. Leung, *Materials Research Society Symposium Proceedings*, **396**, "Ion-Solid Interactions for Materials Modification and Processing", edited by D. B. Poker, D. Ila, Y.-T. Cheng, L. R. Harriott and T. W. Sigmon, Materials Research Society, Pittsburgh, USA, pp. 255-260 (1996).
11. C. Iwasaki, M. Aono, N. Kitazawa and Y. Watanabe, Materials Processing for Properties and Performance (MP^3) Vol. 2, edited by K. A. Khor, Y. Watanabe, K. Komeya and H. Kimura, Institute of Materials (East Asia), Singapore, pp. 494-502 (2004).
12. H.Ohta, A. Matsumuro and Y. Takahashi, *Japanese Journal of Applied Physics*, **41**, 7455-7461 (2002).
13. F. Tuinstra and J. L. Koenig, *Journal of Chemical Physics*, **53**, 1126-1130 (1970).

Mater. Res. Soc. Symp. Proc. Vol. 843 © 2005 Materials Research Society T3.2

C, Si and Sn Implantation of CVD Diamond as a means of Enhancing Subsequent Etch Rate

P.W. Leech[1], T. Perova[2], R.A. Moore[2], G.K. Reeves[3], A.S. Holland[3] and M. Ridgway[4],
[1]CSIRO Manufacturing and Infrastructure Technology, Clayton, Victoria, Australia.
[2]Dept. of Electronic and Electrical Engineering, University of Dublin, Trinity College, Ireland.
[3]RMIT University, School of Computer Systems and Elect. Eng., Melbourne, Australia.
[4]Dept. of Electronic Materials Engineering, ANU, Canberra, Australia.

ABSTRACT

Diamond films were implanted with C^+, Si^+ or Sn^+ ions at multiple energies in order to generate a uniform region of implantation-induced disorder. Analysis of the C^+ implanted surfaces by micro-Raman spectroscopy has shown only minor increase in the proportion of non-diamond or sp^2-bonded carbon at doses of 5×10^{13} - 5×10^{15} ions/cm^2. In comparison, an amorphization of the structure was evident after implantation with either Si^+ ions at a dose of 5×10^{15} ions/cm^2 or with Sn^+ ions at $\geq 5 \times 10^{14}$ ions/cm^2. At a given implantation dose, the etch rate of the diamond film in a CF_4/O_2 plasma increased with the mass of the implanted species in the order of C^+, Si^+ and Sn^+. For a given implant species, the etch rate was directly proportional to vacancy concentration as controlled by the dose or the implantation-induced disorder.

INTRODUCTION

Diamond films manufactured by chemical vapour deposition (CVD) have shown important potential in applications such as inert biosensors and ultra-low friction electromechanical devices. A critical step in the realisation of many of these devices has been the development of techniques for the patterning and selective etching of the diamond films. Several techniques have been successfully reported in the etching of diamond as reviewed in ref. [1]. One of the most widely applied of these techniques has been reactive ion etching (RIE) in an oxygen plasma. But while oxygen RIE has allowed the definition of fine structures in diamond, the etching has been characterised by surface roughening. The alternative use of SF_6 [2] or CF_4 [3] gases has produced a significantly smoother surface than with oxygen plasmas, although also with the limitation of a low etch rate.

In this paper, we have examined for the first time the use of implant-induced damage as a means of increasing the rate of subsequent reactive ion etching of diamond in CF_4/O_2 plasmas. The etch rate of the films has been examined as a function of the ion dose and the mass of the implant species (C^+, Si^+ or Sn^+). Previous work has identified a correlation between implant dose and the extent of damage in both CVD films [4,5] and single crystal diamond [5]. Ion implantation at low doses resulted in the formation of clusters of point defects around each track [4,5]. At intermediate doses, the implantation produced a partial graphitisation while a full amorphisation was evident above a critical level of dose, D_c, dependent on the implanted ion [4-6]. In this work, the etch rate of the diamond has also been measured versus the *rf* power in the RIE system in order to evaluate any effect of prior ion implantation in relation to the etch mechanism.

EXPERIMENTAL DETAILS

The diamond films used in these experiments were grown by microwave plasma enhanced CVD on (111) oriented Si (76 mm). The polycrystalline films were ~15 µm in thickness as supplied by Sumitomo, Japan. Host samples of ~1 x 1 cm were mounted onto the base of the implantation target holder using silver paste in order to ensure a good thermal contact. The implantations were performed at a temperature of -196 °C to reduce the extent of dynamic annealing and hence maximise the net disorder. For implantation with C^+ ions, a series of sequential energies (60, 180, 330 and 525 keV) was used in order to create a near uniform profile of implantation-induced vacancies. Similarly, the Si^+ ions were implanted at energies of 200, 500 and 950 keV and the Sn^+ ions at 750 and 2,000 keV. Each of the ion species was implanted over the range of doses ($5 \times 10^{13} - 5 \times 10^{15}$ ions/cm^2). The implanted surfaces were then analysed with a Renishaw 1000 micro-Raman system equipped with an Ar^+ ion laser for excitation at either 488 or 514.5 nm. The laser radiation was focused onto the sample using a x 50 microscope objective which also acted to collect the scattered light from the surface. The conformal depth of sampling of the spectrometer was ~2 µm. Micro-Raman spectra (400-1800 cm^{-1}) were acquired from the as-grown and implanted surfaces. The subsequent reactive ion etching of the surfaces was performed in a commercial STS320 system which was *rf* powered at 13.56 MHz. The cathode was continuously cooled to 20 °C during etching. A gas mixture of CF_4 (28 sccm)/ O_2 (2 sccm) was selected because of the smooth surface and moderate etch rate produced by this plasma during the etching of diamond films [3]. The etch rate was measured as a function of *rf* power (200-400 W) at 50 mT. The depth of etching in the surfaces was measured by a Dektak stylus profilometer.

RESULTS AND DISCUSSION

Figs. 1-3 have compared typical Raman spectra for the as-deposited and the C^+, Si^+ and Sn^+ implanted diamond in the spectral range 400-1800 cm^{-1}. For the unimplanted samples, the Raman line in the vicinity of 1332 cm^{-1} was characteristic of sp^3 bonded diamond. A further broad Raman band at ~1510 cm^{-1} in the as-deposited samples has been assigned to the presence of diamond-like carbon. The presence of DLC in grain boundaries has been a feature of diamond films deposited by PECVD [7]. For the samples implanted with C^+ ions at a dose of 5×10^{13} ions/cm^2, the spectra in Fig. 1 have shown little change from the as-deposited samples. At higher doses ($\geq 5 \times 10^{14}$ ions/cm^2) of C^+ ions, the spectra have shown a progressive decrease in intensity of the sp^3 diamond peak. However, the ratio of intensity of the sp^3 peak to the sp^2 peak remained approximately constant at ~11 over the range of implant doses. For the Si^+ implants, the spectra in Fig. 2 have also shown a decrease in height of the sp^3 diamond peak with dose. At the highest dose of 5×10^{15} ions/cm^2, there was no evidence of an sp^3 peak. At this dose, only a broad Raman band was present in the spectrum at ~1550 cm^{-1}. A similar broad peak at 1550 cm^{-1} has previously been identified following the high dose implantation of diamond with Ar^+ (2×10^{16} ions/cm^2) [8] and B^+ (1×10^{16} ions/cm^2) [9]. The peak was attributed to the formation of non-diamond carbon as either amorphised [8] or graphitic carbon [9]. For the diamond films implanted with Sn^+ ions, the spectra in Fig. 3 have also shown a reduction in the sp^3 peak height with increasing dose. The sp^3 peak had almost disappeared after implantation at 5×10^{14} Sn^+/cm^2 and was completely absent following a dose of 5×10^{15} Sn^+/cm^2. At doses of $\geq 5 \times 10^{14}$ Sn^+/cm^2, a broad band was also evident in the spectra at ~1550 cm^{-1}.

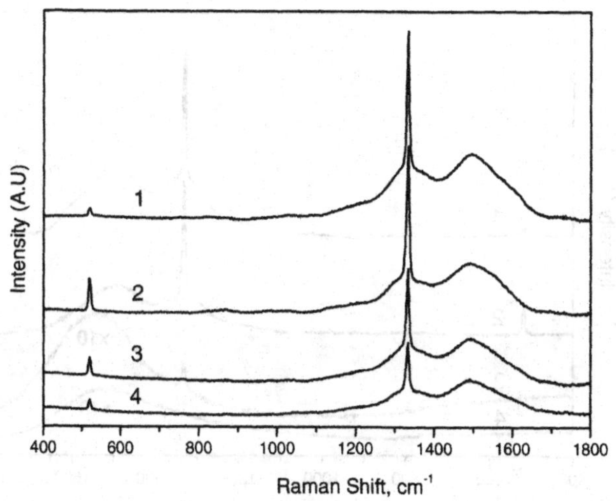

Figure 1. Raman spectra of unimplanted diamond (curve 1) and after implantation at 5 x 10^{13} C^+/cm^2 (curve 2), 5 x 10^{14} C^+/cm^2 (curve 3) and 5 x 10^{15} C^+/cm^2 (curve 4).

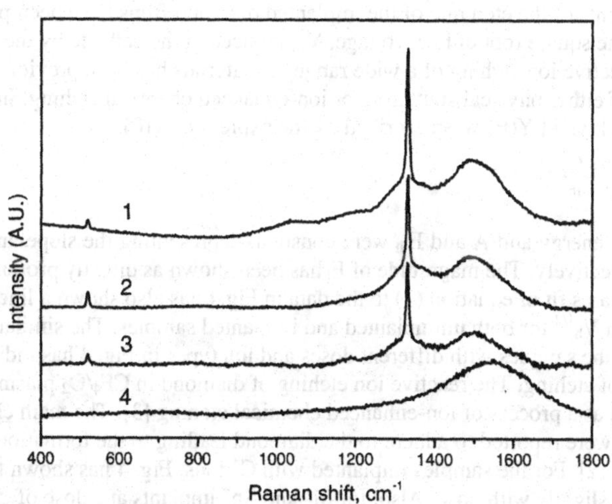

Figure 2. Raman spectra of unimplanted diamond (curve 1) and after implantation at 5 x 10^{13} Si^+/cm^2 (curve 2), 5 x 10^{14} Si^+/cm^2 (curve 3) and 5 x 10^{15} Si^+/cm^2 (curve 4).

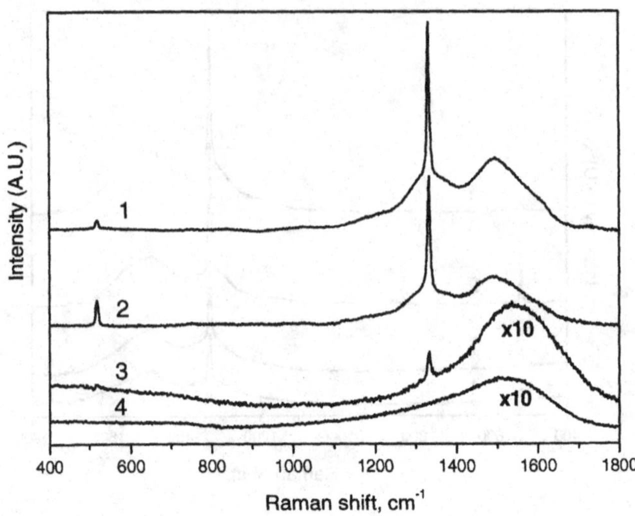

Figure 3. Raman spectra of unimplanted diamond (curve 1) and after implantation at 5 x 10^{13} Sn^+/cm^2 (curve 2), 5 x 10^{14} Sn^+/cm^2 (curve 3) and 5 x 10^{15} Sn^+/cm^2 (curve 4).

Measurements of the etch rate of the implanted diamond films have been plotted in Fig. 4 as a function of the square root of bias voltage, V_b, induced at the cathode by the application of rf power. The reactive ion etching of a wide range of materials has been previously shown to occur by a process of either physical sputtering or ion-enhanced chemical etching. In both types of process, the etch yield Y(E) was described by the expression [10]:

$$Y(E) = A(E_i^{1/2} - E_{th}^{1/2}) \quad (1)$$

where E_i = ion energy and A and E_{th} were constants representing the slope and threshold energy of etching respectively. The magnitude of E_i has been shown as directly proportional to V_b [11].

A least squares fit of equation (1) to the data in Fig.4 has also shown a linear dependence of the etch rate on $V_b^{1/2}$ for both unimplanted and implanted samples. The similarity in the slope of these plots for the samples with different doses and ion types in Fig. 4 has indicated the same basic process of etching. The reactive ion etching of diamond in CF_4/O_2 plasmas has previously been identified as a process of ion-enhanced chemical etching [3]. The main chemical reactants of C and CF_n^+ were reported to adhere to the diamond leading to the formation of highly volatile etch products [12]. For the samples implanted with C^+ ions, Fig. 4 has shown that the etch rate only increased slightly with dose. Also, for Si^+ and Sn^+ implants at a dose of 5 x 10^{13} $ions/cm^2$, the etch rates were only slightly higher in the implanted than for the unimplanted diamond. However, at doses of $\geq$5 x 10^{14} $ions/cm^2$, the etch rate increased significantly with either Si^+ or Sn^+ implantation. At the highest dose of 5 x 10^{15} $ions/cm^2$, the etch rate at 400 W increased in the order of C^+ (32 nm/min), Si^+ (48 nm/min) and Sn^+ (53 nm/min). As a reference, the etch rate of unimplanted diamond at 400 W was 25 nm/min.

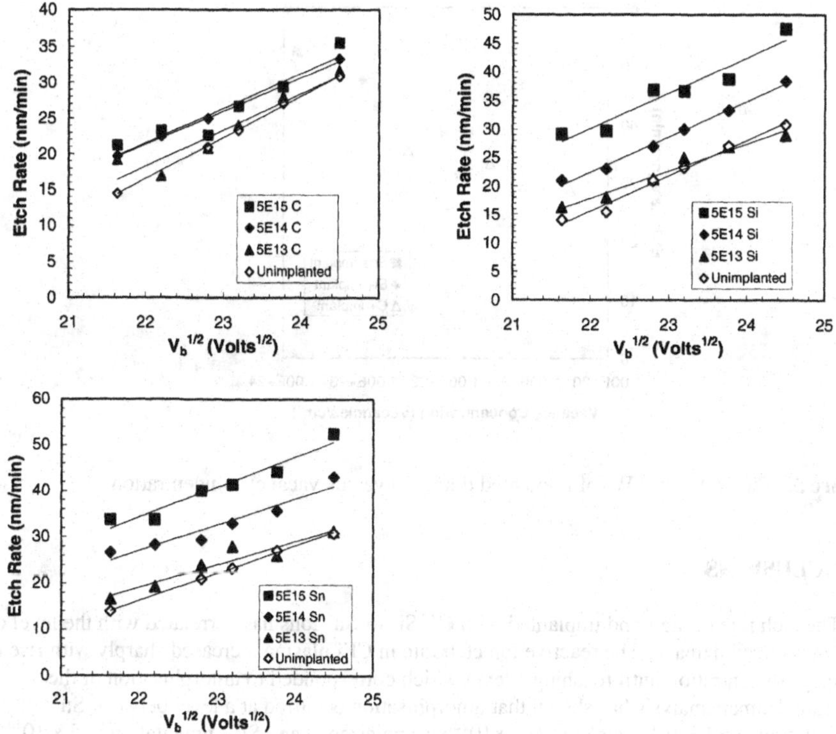

Figure 4. Etch rate versus $V_b^{1/2}$ for diamond implanted with a) C^+, b) Si^+ or c) Sn^+ ions at doses of 5×10^{13}, 5×10^{14} and 5×10^{15} ions/cm^2.

In Fig.5, the etch rate (at 400 W power) for the various implants of diamond have been plotted versus a calculation of the vacancy concentration. This parameter was determined for each combination of dose and implant species using the TRIM (TRansport of Ions in Matter) simulation code. The etch rate in Fig. 5 has shown a steep increase with the gross, athermal vacancy concentration until reaching a plateau at ~1×10^{23} vacancies/cm^3. Raman spectra have shown that the threshold for amorphisation was located between Sn^+ implantation at a dose of 5×10^{14} ions/cm^2 (6.8×10^{22} vacancies/cm^3) and Si^+ implantation at 5×10^{15} ions/cm^2 (1.1×10^{23} vacancies/cm^3). Hence, the onset of amorphisation corresponded to a level of vacancy production which was located between 6.8×10^{22} vacancies/cm^3 and 1.1×10^{23} vacancies/cm^3. The levelling-off in the etch rate in Fig. 5 at ~1×10^{23} vacancies/cm^3 has correlated with the onset of amorphisation. A dependence of etch rate on vacancy concentration was consistent with the results of Prins [13]. He established that the damage induced in diamond by implantation with a wide variety of ions including massive ions such as As^+ and In^+ consisted primarily of vacancies and interstitial atoms. These defects were formed within the individual collision cascades [13].

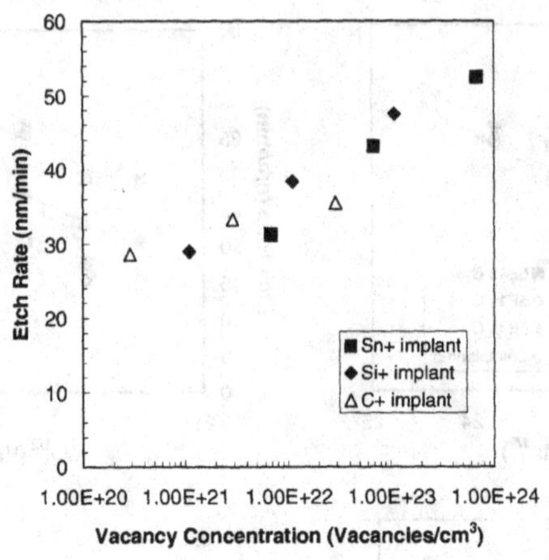

Figure 5. Etch rate (400 W) of implanted diamond versus vacancy concentration.

CONCLUSIONS

The etch rate of diamond implanted with C^+, Si^+ or Sn^+ ions has correlated with the level of the ion-induced damage. The reactive ion etch rate in CF_4 plasma increased sharply with rise in vacancy concentration until reaching a level which corresponded to amorphisation of the structure. Raman analysis has shown that amorphisation occurred at a level between Sn^+ implantation at $\geq 5 \times 10^{14}$ ions/cm^2 (6.8×10^{22} vacancies/cm^3) and Si^+ implantation at 5×10^{15} ions/cm^2 (1.1×10^{23} vacancies/cm^3). The dependence of etch rate on implant species and dose was controlled through the effect of these parameters on vacancy concentration.

REFERENCES

1. A.P. Malshe, B.S. Park, W.D. Brown and H.A. Naseem, *Diamond Rel. Mater.* **8,** 1198 (1999).
2. C. Vivensang, L. Ferlazzo-Manin, M.V. Ravet, G. Turban, F. Rosseaux and A. Gicquel, *Diamond Rel. Mater.*, **5**, 840 (1996).
3. P.W. Leech, G.K. Reeves and A.S. Holland, *Journal Materials Science*, **36**, 3453 (2001).
4. S. Prawer, A, Hoffman and R. Kalish, *Appl. Phys. Lett.* **57** (21) (1990) 2187.
5. S. Prawer, R. Kalish, *Phys. Rev.*, **B51**, (1995), 15711.
6. N. Jiang, M. Deguchi, C.L. Wang, J.H. Won, H.M. Jeon, Y. Mori, A. Hatta, M. Kitabatake, T. Ito, T. Hirao, T. Sasaki and A. Hiraki, *Appl. Surf. Sci.* **113/114** (1997) 254.
7. Z. Sun, J.R. Shi, B.K. Tay and S.P. Lau, *Diamond Rel. Mater.* **9** (2000) 1979.
8. S. Sato and M. Iwaki, *Nuclear Instruments Methods in Physics Res.*, **B32**, 145 (1988).
9. Y. Avigal, V. Richter, B. Fizgeer, C. Saguy and R. Kalish, *Diamond Rel. Mater.* **13** (2004) 1674.
10. C. Steinbruchel, *Appl. Phys. Lett.* **55**(19) (1989) 1960.
11. C. Steinbruchel, *MRS Symp.* Proc. Laser and particle beam chemical processes on surfaces, Pittsburgh PA, **129**, 477 (1989).
12. K. Kobayashi, N. Mutsukura and Y. Machi, *Thin Solid Films*, **200**, 139 (1991).
13. J.F. Prins, *Mater. Sci.Eng.*, B11, 219 (1992).

Mater. Res. Soc. Symp. Proc. Vol. 843 © 2005 Materials Research Society T3.15

Photochemical Adhesion of Fused Silica Glass Lens by Silicone Oil

Takayuki Funatsu*, Masanori Kobayashi *, Yoshiaki Okamoto** and Masataka Murahara *
*Department of Electrical Engineering, Tokai University,
1117 Kitakaname, Hiratsuka, Kanagawa, 259-1292, JAPAN
**Okamoto Optic Co.
8-34 Haramachi, Isogo-ku, Yokohama, Kanagawa, 235-0008 JAPAN.

ABSTRACT

The optical system that is pervious to ultraviolet light of 200nm and under in the wavelength has been developed by putting one silica glass to another with the silicone oil photo-oxidized in oxygen atmosphere.

Quartz has siloxane bonds, while silicone oil (dimethyl siloxane) is composed of siloxane bonds of the main chain and methyl groups of the side chain. Therefore, the organic silicone oil has been photo-oxidized by irradiating UV rays in oxygen atmosphere to change into inorganic glass. That is, the silicone oil was poured into the thin gap between the two pieces of silica glass in oxygen atmosphere and was irradiated with the Xe_2 excimer lamp while heating at temperature above 150°C. Consequently, the siloxane of the silicone oil was linked with the O atoms that had been absorbed on the glass surface to form SiO_2.

The UV and IR spectrum analysis was conducted on the silicone oil before and after lamp irradiation. The results revealed that as the time of lamp irradiation increased, the absorption peak of the CH_3 group in the region of 2960 cm^{-1} decreased but the transmittance of the light in the 190nm wavelength conversely became high. The UV transmittance of the silicone oil was 50% before the lamp irradiation; which improved to 87% after the irradiation for 60 minutes. Furthermore, the tensile strength of the bonded sample was measured. It confirmed that the adhesive strength of the silicone oil was enhanced from 0 kgf/cm^2 of before-irradiation to 180 kgf/cm^2 of after-irradiation.

INTRODUCTION

In general, balsam, epoxy, unsaturated polyester resins, silicone adhesives and UV hardening agent are used as adhesives for lenses and optical materials. The epoxy resin is used in a wide field because it is excellent in adhesive property and mechanical properties. Recently, it has been researched to improve heat resistance, adhesive strength, and the moistureproof. Hasegawa et al. mixed a generic epoxy and a siloxane modified-epoxy, which contained the silicone in main chain, the adhesive strength was improved to occur the micro phase separation silicone and epoxy [1-2]. However, after hardening, the modified silicone got opaque. In order to solve the problems, Ochi et al. reported that improved adhesive strength and transparency became high by using silicone- poly(methylmethacrylate) PMMA graft for the a modifying agent [3]. In these methods, however UV rays under 250nm are not transmitted because of being high polymer. An optical contact method is the only way to transmit UV rays under the 250nm. It, however, requires a high precision roughness to use the intermolecular force that is called air contact. Moreover, the optical contact method is so weak in slight vibration that its field of use is limited. Therefore, the new, strong adhesion method was required for the optical materials to transmit V-UV rays under 250nm. Thus, we developed new bonding method of the fused silica glasses together with photo-excited silicone oil by Xe_2 excimer lamp irradiation.

In our previous study, low refractive index and transparent SiO_2 thin film was laminated on the silicon or glass substrate with photochemical reaction of V-UV light [4]. Furthermore, to improve the transmittance of PMMA in the visible region, the PMMA surface was modified into hydrophobic and developed the fibrin free intraocular lens [5, 6]. And we have demonstrated the bonding of fluorocarbon and a fused silica glass with silicone oil as an adhesive by using an excimer laser [7]. Based on these results, in this study, we bonded two fused silica glasses with photo-excited silicone oil.

PRINCIPLE OF FUSED SILICA GLASS BONDING

Figure 1 shows the principle of the photochemical adhesion. Silicone oil is poured between the two silica glasses, and Xe_2 lamplight is vertically irradiated. The Si-C bond of silicone oil is photo- dissociated because the bonding energy of Si-C (105kcal/mol) is lower than the photon energy of Xe_2 lamp (165kcal/mol). And at the boundary between the silicone oil and the silica glass, the oxygen adsorbed on the glass surface are photo-excited by the UV to produce O^{1d}, as shown in Equation (1). This active oxygen reacts with the silicone oil to be modified in (SiO_2) n. The methyl group, which was photo-dissociated from silicone oil, also reacts with the active oxygen to form CO_2 or H_2O. With the reactions described above, the silica glasses are bonded; the organic silicone oil turns to be an inorganic glass in the presence of excited oxygen.

$$O_2 + h\nu \rightarrow O^{1d} \quad (\lambda < 200[\mathrm{nm}]) \tag{1}$$

$$[SiO(CH_3)_2]_n + O_2 + h\nu \rightarrow (SiO_2)_n + CO_2 + H_2O \tag{2}$$

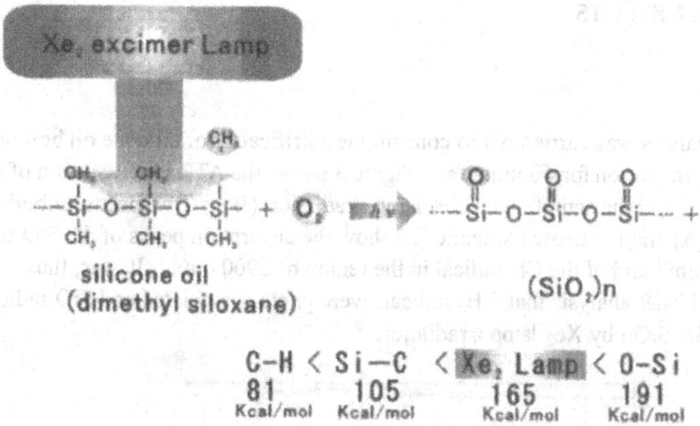

Xe_2 excimer Lamp

$$\text{—Si(CH}_3)_2\text{—O—Si(CH}_3)_2\text{—O—Si(CH}_3)_2\text{—} + O_2 \xrightarrow{h\nu} \text{—Si(=O)—O—Si(=O)—O—Si—} + CO_2 + H_2O$$

silicone oil (dimethyl siloxane) $\qquad (SiO_2)n$

$$\text{C–H} < \text{Si–C} < Xe_2\ \text{Lamp} < \text{O–Si}$$

81 Kcal/mol, 105 Kcal/mol, 165 Kcal/mol, 191 Kcal/mol

Figure 1. Principle of photochemical adhesion

EXPERIMENTAL METHOD

Silicone oil, which was heated at 160°C by the infrared lamp in oxygen atmosphere, is poured between the fused silica glasses. In this condition, Xe_2 lamplight is irradiated on the sample. The silicone oil, turned to the SiO_2 with the Xe_2 excimer lamp irradiation, as shown in Figure.2

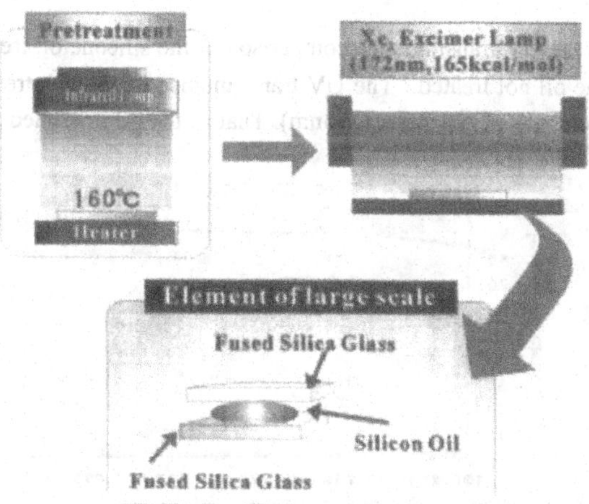

Figure 2. Experimental setup.

EXPERIMENTAL RESULTS

FT-IR Analysis

ATR FT-IR analysis was carried out to confirm the vitrification of silicone oil before and after the Xe_2 lamp irradiation for 60 minutes. Figure 3 shows the ATR FT-IR spectra of the untreated sample (A) and the sample with Xe_2 lamp irradiation (B). The spectra of both the untreated silicone (A) and the treated silicone (B) show the absorption peaks of the SiO radical in the region of 1030 cm^{-1} and of the CH radical in the region of 2900 cm^{-1}. It was, thus, confirmed by ATR FT-IR analysis that CH_3 radicals were photo-dissociated and SiO radical increased to form the SiOn by Xe_2 lamp irradiation.

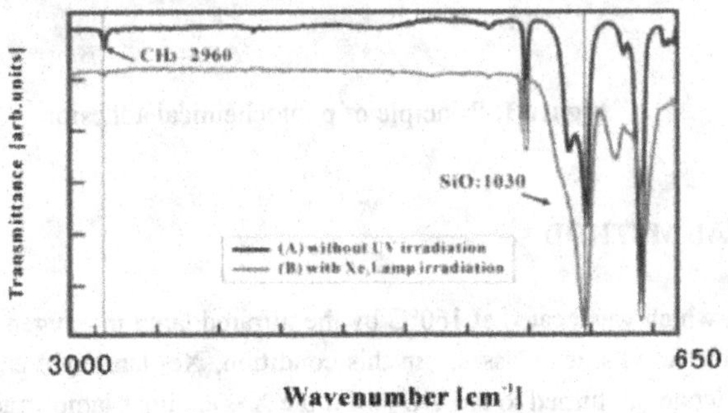

Figure 3. FT-IR Analysis

UV Transmittance Analysis

Figure 4 shows the UV transmittance comparison of the silicone oil treated with the Xe_2 lamp and the silicone oil not treated. The UV transmittance of the non-treatment silicone oil was 50% at the wavelength of ArF laser (193nm). That of the oil irradiated with the Xe_2 lamp, it further improved to 87%.

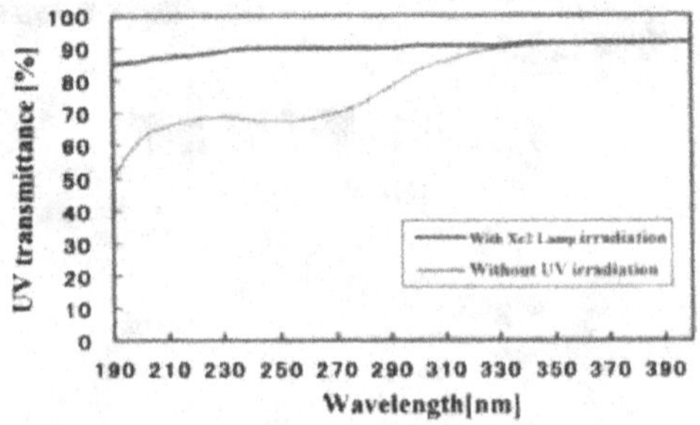

Figure 4. UV transmittance with and without Xe_2 lamp irradiation

3) Measurement of -CH Absorption Strength

Figure 5 shows the correlation between UV transmittance at the 193nm or CH_3 absorbance of FT-IR and the Xe_2 lamp irradiation time. The -CH functional groups were photo dissociated and decreased according to the time of Xe_2 lamp irradiation. As a result, the silicone oil vitrified and UV transmittance became possible.

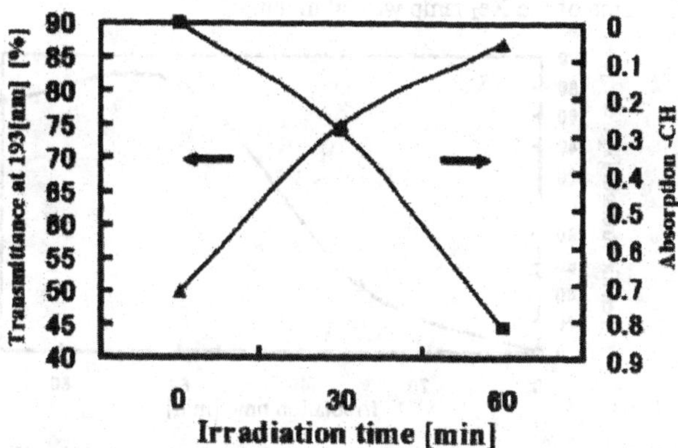

Figure 5. Correlation between lamp irradiation time and UV transmittance & absorption of CH group

Analysis of Sample Surface

Figure 6 displays the photograph of the microscope. The layer (a) is the fused silica glass, the layer (b) is the vitrified silicone oil and the layer (c) is the fused silica glass. The vitrified silicone oil thickness was measured 2 [μm].

Figure 6. A cross section of bonded sample by microscope photograph

Adhesive Strength Measurement

The tensile shear strength were measured the pasted-up sample with the Xe_2 lamp irradiation. Figure 7 shows the correlation between the tensile she ar strength and the irradiation time. The tensile shear strength was not obtained sample without excimer light irradiation. However, the tensile shear strength becomes large with the increase in irradiation time, The adhesion intensity of the pasted-up sample with the Xe_2 lamp irradiation was 180kgf/cm^2 that the optimum irradiation time of the Xe_2 lamp was 60 minutes.

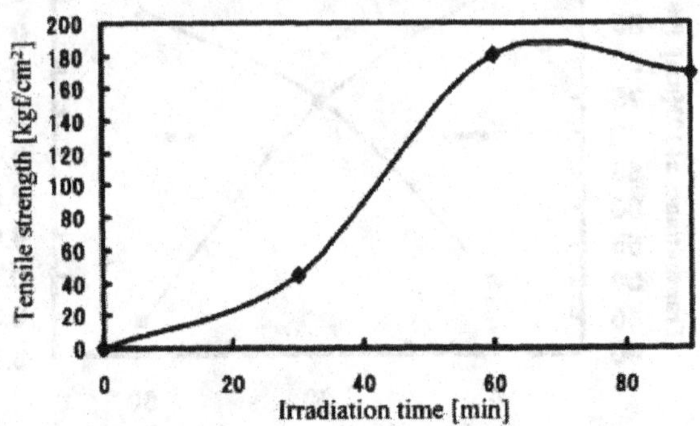

Figure 7. The correlation between the tensile shear strength and the Xe_2 irradiation time.

CONCLUSION

We have demonstrated the new bonding method for fused silica glass by using the Xe_2 excimer lamp. The tensile shear strength of the treated sample with the Xe_2 lamp irradiation was 180kgf/cm^2 and the UV transmittance of the treatment silicone oil was from 50% to 87% at the wavelength of ArF laser (193nm). These results strongly suggest the possibility to develop new bonding method that not only has high adhesive strength but also UV transmittance.

ACKNOWLEDGEMENT

This work is supported in part by grant #15360034 from the Ministry of Education and science Japan.

REFERENCES

1. K.Hasegawa, K,Ootsuka, A.Matsumoto, H.Kimura, A.Fukuda, M.Fujikawa, H.Fujiwara, *The Adhesion Society of Japan*, 36,355(2000)
2. K.Hasegawa, *The Adhesion Society of Japan*, 47(3)16(112)(2000)
3. M.Ochi,,K.Takemiya,O.kiyohara,T.Nakanishi,*Polymer*,41,195(2000)
4. H.Iizuka and M.Murahara, *Mater.Res.Soc.Symp.Proc.*98-23 446-451(1999)
5. Y.Sato, K.Tanizawa, F.Manns, J.M.Palel and M.Murahara,*Proc.SPIE*4951 92-102(2003)
6. K.Tanizawa and M.Murahara, *Mater.Res.Soc.Symp.Proc.*735 117-122(2003)
7. K.Asano and M.Murahara, *Mater.Res.Soc.Symp.Proc.* 796 133-138(2003)

Mater. Res. Soc. Symp. Proc. Vol. 843 © 2005 Materials Research Society T3.17

Comparison of -OH and -NH_2 Functional Group Substitution on PTFE Surface with V-UV Photon Irradiation for Protein Adsorption

Yuji Sato, Naoki Kobayashi and Masataka Murahara
Department of Electrical and Electronic Engineering, Tokai University
1117 Kitakaname Hiratsuka Kanagawa, 259-1292, JAPAN

ABSTRACT

Poly-tetrafluoroethylene [PTFE] presents few rejections in a living body but has low tissue affinity. Then, the soft tissue implant material that has not only high biocompatibility but also superb bondability has been developed by photo-chemically substituting the hydrophilic of –OH or –NH_2 groups on the PTFE surface with V-UV photon irradiation. The protein adsorption of the sample before and after treatment was also evaluated by scanning electron microscope [SEM] and attenuated total reflection Fourier-transform infrared [ATR FT-IR], using bovine serum albumin [ALB] and fibrin [FIB] solution as a protein index in biocompatibility test. From the results, it has been confirmed that the protein adsorption increased with the increase in the hydrophilic group's substitution density. The -OH incorporated sample adsorbed the ALB and FIB more than the -NH_2 incorporated sample; the amount of the ALB and FIB sticking became 2.3 times larger than that of the non-treatment sample.

INTRODUCTION

In general, when a plastic material is implanted in a human body, proteins are adsorbed on the plastic surface; on which a fibroblast or a tissue cell adheres to form a fibrous tissue. It is, therefore, essential for a biomaterial to be biocompatible, chemically resistant and have resistance to hostile environments. However, many plastics, which have been clinically employed as a soft tissue implant material, have low adhesiveness and affinity to tissue because of its low surface energy.

In order to improve its adhesive property, Okada and Ikada reported that the carboxyl groups in poly (acrylic acid) chains were grafted onto the silicone surface, and the collagen was able to immobilize there [1]. M.C.L. Martins et al. produced a poly(hydroxyethylmethacrylate) (PHEMA) and a poly(hydroxyethylacrylate)(PHEA) film by the graft polymerization onto the poly(etherurethane)(PU) films to absorb the ALB and fibrinogen [2]. Satriano and Marletta incorporated oxygen groups on a PHMS surface by O_2-plasma and Ar^+-ion beam irradiations and cultivated fibroblast on the modified surface [3]. Lee et al. formed the micropatterning on the oxygen plasma treated polystyrene surface by using KrF excimer laser ablation method and cultivated the NG 108-15 neuroblastoma *x* glioma hybrid cells [4]. They reported qualitatively that the protein adsorption depends on the functional groups and shape on the material surface.

There are, however, no previous reports to investigate quantitatively relationship between protein adsorption and the functional groups or its density of material surface. Thus, we photo-chemically incorporated the hydrophilic groups on the PTFE surface, which has the most water repellency and causes only a few rejections in a living body and clarified the correlation between hydrophilic groups substitution and protein adsorption on the same material.

In our previous studies, we incorporated desired functional groups or Cu atoms onto the polymer surface using an excimer laser to generate a hydrophilic or hydrophobic [5-12] and found that especially for PTFE, B, Al and H atoms were effective in defluorination [8-10]. Having studied the compatibility between the defluorination atoms and a living body, it was found that when using B and Al, the modified PTFE caused rejections; however, no rejection occurred on the hydrophilic PTFE modified with H, i.e., water (H_2O) [12]. This was confirmed by post-operated days [POD] test on the sample that had been implanted in the cornea of a rabbit.

Based on these results, in this study, we have clarified the correlation between the OH groups or NH_2 functional groups substitution density and the protein adsorption when the PTFE surface was photo-chemically modified to be hydrophilic using ammonia gas or water.

PRINCIPLE OF PHOTOCHEMICAL REACTION

The PTFE is composed of (CF_2- CF_2) n. In order to modify the PTFE to have hydrophilic properties, ammonia gas or H_2O are used as reaction agents. When the PTFE is irradiated with Xe_2 excimer lamp in the presence of reaction agents, the OH or NH_2 radicals are produced by photo-dissociation of H_2O or ammonia gas, as shown in Equation (1,3). At the same time, the Xe_2 lamp photon energy (165 kcal/mol) is higher than the C-F bond energy in PTFE (128 kcal/mol), the C-F bond can also be photo-dissociated by the UV photon [8,9]. As indicated in Equation (2,4), the H radical pulls out F atom from the photo-excited C-F bond to produce HF. The HF has absorption below 160 nm and is not absorbed at the 172 nm wavelength of Xe_2 lamp[13]. It is, therefore, not photo-dissociated by Xe_2 lamp. And the OH or NH_2 radical immediately combines with C- to form C-OH or C-NH_2, which develops a hydrophilic property.

$$H_2O + h\nu \rightarrow H^* + OH^* \quad (1)$$
$$(CF_2)n + H_2O + h\nu \rightarrow (CF\text{-}OH)n + nHF \quad (2)$$
$$NH_3 + h\nu \rightarrow H^* + NH_2^* \quad (3)$$
$$(CF_2)n + NH_3 + h\nu \rightarrow (CF\text{-}NH_2)n + nHF \quad (4)$$

EXPERIMENTAL SETUP

Figure 1a illustrates the –OH substitution procedure. The water was dropped on the PTFE surface and a fused silica glass was placed on it to form the H_2O thin layer by capillary phenomenon. Next, the PTFE sample was place in the reaction chamber where NH_3 gas was set, as shown in Figure 1b. In these conditions, the Xe_2 excimer lamp was vertically irradiated.

ATR FT-IR and contact angle measurement with water were carried out to investigate the substitution of the functional groups and density on the PTFE surface before and after the V-UV photon irradiation. Furthermore, the protein adsorption of the sample before and after treatment was also evaluated by scanning electron microscope [SEM] and FT-IR, using [ALB] and [FIB] solution as a protein index in biocompatibility test.

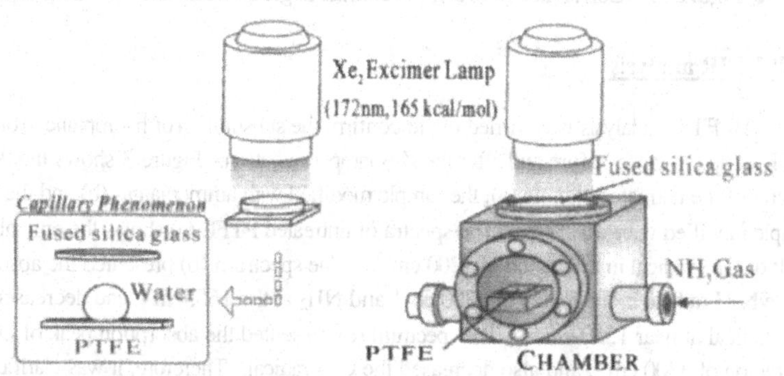

(a) –OH substitution procedure. (b) $-NH_2$ group substitution procedure.

Figure 1. Schematic diagram of experimental set-up

RESULTS AND DISCUSSION

Contact angle measurement

In order to evaluate the wettability of the modified sample, the contact angle with water was measured. Figure.2 shows the correlation between the contact angle and the lamp irradiation time. The contact angle on the untreated PTFE was 110 degrees. In the case of the sample modified with ammonia gas, the contact angle with water decreased, as the UV irradiation time increased. At the 20 minute Xe_2 lamp irradiation, the contact angle became 38 degrees. In case of the other sample modified with H_2O, the contact angle also decreased, as the UV irradiation time increased. At the 10 minute Xe_2 lamp irradiation, the contact angle became 36 degrees being smaller than the sample modified with ammonia gas.

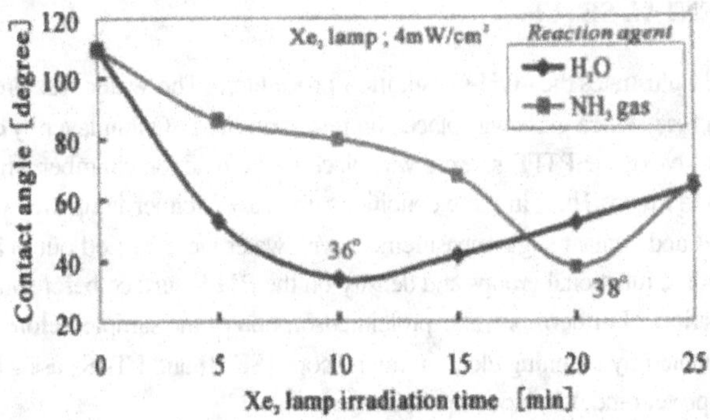

Figure 2. Correlation between the contact angle of water and Xe_2 lamp irradiation time.

ATR FT-IR analysis

ATR FT-IR analysis was carried out to confirm the substitution of hydrophilic groups on the PTFE sample surface before and after the Xe_2 lamp irradiation. Figure 3 shows the ATR-FT-IR spectra of the untreated sample (a), the sample modified with ammonia gas (b) and the other sample modified with water (c). The spectra of untreated PTFE (a) shows the only absorption peak of CF_2 radical in the region of 1200 cm^{-1}. The spectrum (b) presented the absorption peak of the N-H radical in the region of 3300 cm^{-1} and NH_2 at the 1655cm^{-1}, and decreased that of the CF_2 radical at near 1250 cm^{-1}. The spectrum (c) presented the absorption peak of OH radical in the region of 3300 cm^{-1}, and also decreased the CF_2 radical. Therefore, it was clarified by ATR FT-IR analysis that the $-NH_2$ groups or -OH groups were incorporated on the PTFE surface.

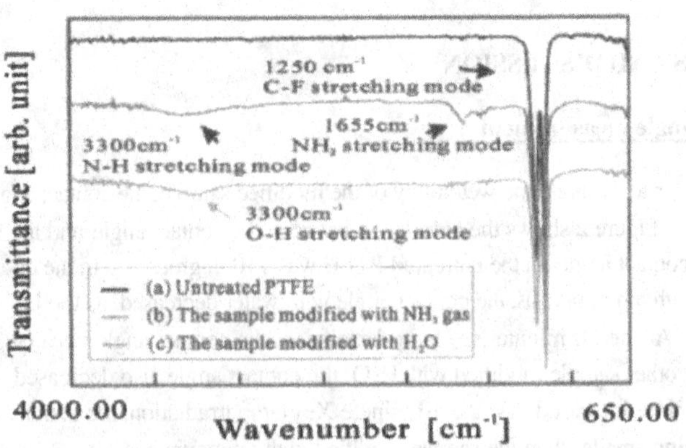

Figure 3. ATR FT-IR spectra of PTFE surface before and after treatment

Protein adsorption test

Figure 4shows the correlation between the absorption coefficient of the ALB and the water contact angle on modifying the PTFE surface from hydrophobic to hydrophilic by substituting hydrophilic groups with Xe_2 lamp irradiation. In ALB when the contact angle on the untreated PTFE is 110 degrees, the absorption coefficient of the amide band is 0.004, as shown in Figure.4 (a). As the PTFE surface modification to become hydrophilic progressed, the absorption coefficient of the amide band increased. In particular, the OH incorporated sample, which became 36 degrees of the contact angle, became 0.009 in ALB, increasing 2.25 times than the non-treatment sample. In FIB, as the PTFE surface modification to become hydrophilic progressed, the absorption coefficient of the amide band also increased, as shown in Figure 4 (b). As these results, it is found that the quantity of the protein adsorption depended the hydrophilic group density. Furthermore, The -OH incorporated sample adsorbed the ALB and FIB more than the $-NH_2$ incorporated sample.

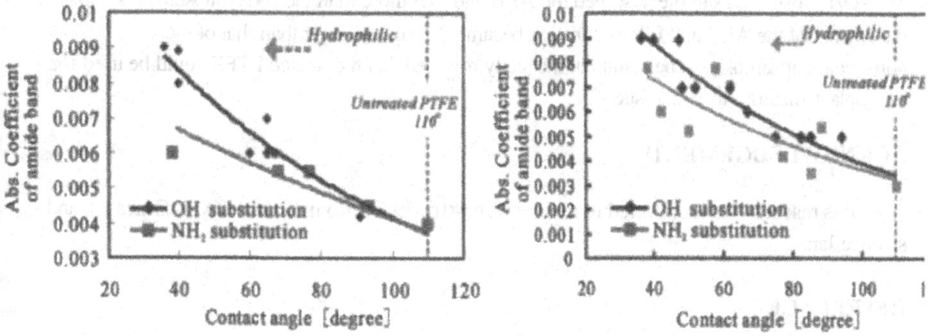

(a) The adsorption coefficient of ALB dependence upon the contact angle.

(b) The adsorption coefficient of FIB dependence on the contact angle.

Figure 4. Correlation between absorption coefficient of amide band and contact angle; the sample was soaked in a 0.1% fibrin or albumin water solution for 48 hours at 30 , after rinsing with water and vacuum drying, the sample was investigated by ATR FT-IR

Figure 5 displays SEM micrographs of the PTFE surface before and after modification. Only slight amounts of FIB were adsorbed on the untreated, hydrophobic PTFE surface, and FIB was adsorbed considerably on the treated, hydrophilic PTFE surface. This agrees with the results shown in Figure 4

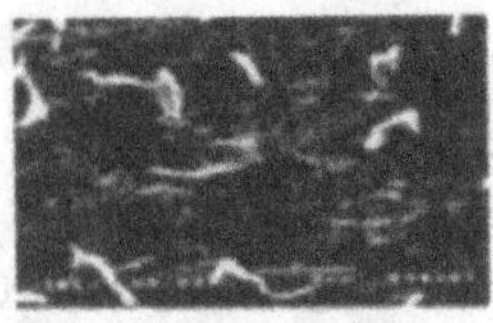

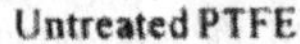

Untreated PTFE

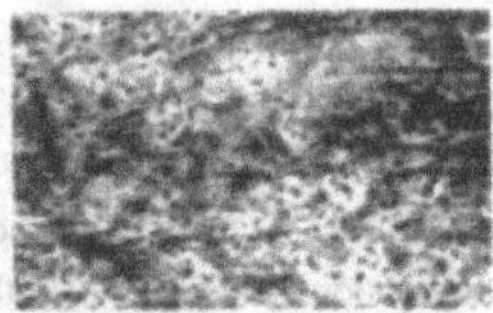

The sample modified with ammonia gas (contact angle is 38°)

The sample modified with H_2O (contact angle is 36°)

Figure 5. SEM Photographs of FIB adsorption on PTFE surface before and after modification

CONCLUSIONS

The -NH_2 or -OH groups were photo-chemically incorporated on the PTFE surface by the Xe_2 excimer lamp irradiation to generate the hydrophilic property. As the result, it was found that as the hydrophilic groups density became higher, the ALB and FIB adsorption increased. The -OH substituted sample adsorbed the ALB and FIB more than the -NH_2 substituted sample; the amount of the ALB and FIB adsorption became 2.25 times larger than that of the non-treatment sample. The results of the study revealed that the treated PTFE could be used for an implant material to soft tissue.

ACKNOWLEDGEMENTS

This research was supported in part by grant #16659235 from the Ministry of Education and science Japan.

REFERENCE

1. T. Okada and Y. Ikada, *J. Biomater. Sci. Polymer Edn*, **7 No.2**, 171-180 (1999)
2. M.C.L. Martins, D. Wang, J. Ji, L. Feng, M.A. Barbosa, *Biomaterials* **24,** 2067-2076 (2003)
3. C. Satriano and G. Marletta, *J. Mater. Sci., Mater. in Medicine* **14**, 663-670 (2003)
4. J. S. Lee, K. Sugioka and K. Toyoda, *Appl. Phys. Lett.* **65 No.4**, 400-402 (1994)
5. Y. Sato, J. M. Parel and M. Murahara, *Mater. Res. Soc. Symp. Proc.* **752**, 83-88 (2003)
6. Y. Sato, K. Tanizawa, F. Manns, J. M. Parel and M. Murahara, *Proc. SPIE* **4951**, 92-102 (2003)
7. H. Omuro, K. Hamada, T. Nakajima, E. Sinpuku, M. Nakagawa, H. Fukuda and M.Murahara, *Mater. Res. Soc. Symp. Proc.* **711,** 85-90 (2002)
8. M. Okoshi, M. Murahara and K. Toyoda, *J. Mater. Res.* **7**, 1912-1916 (1992)
9. M. Murahara and M. Okoshi, *J. Adhesion Sci. Technol.* **9**, 1593-1599 (1995)
10. M. Murahara and K. Toyoda, *J. Adhesion Sci. Technol.* **9**, 1601-1609 (1995)
11. M. Okoshi and M.Murahara, *App. Phys. Lett.* **72**, 2616-2618 (1998)
12. Y. Sato, J. M. Parel and M. Murahara, *Mater. Res. Soc. Symp. Proc.* **711**, 277-282 (2002)
13. H. Okabe, *Photochemistry of Small Molecules*, 62-163, a Wiley Interscience, N.Y. (1978)

Mater. Res. Soc. Symp. Proc. Vol. 843 © 2005 Materials Research Society T3.18

PHOTOCHEMICAL DEPOSITION OF TRANSPARENT LOW REFRACTIVE INDEX SiO_2 TOPCOAT FOR LASER HEAD AT ROOM TEMPERATURE

Y.Tezuka and M.Murahara
Department of Electrical Engineering, Tokai University
1117 Kitakaname, Hiratsuka, Kanagawa 259-1292, JAPAN

ABSTRACT

A transparent, low refractive index SiO_2 film was photo-chemically laminated on a glass slab laser head by the Xe_2^* excimer lamp in the atmosphere of NF_3 and O_2 mixed gas at room temperature; which made it possible to inhibit the decrease in the laser output power caused by the evanescent wave leakage.

The transparent SiO_2 film of 260nm thickness was laminated on the non-heated substrate by the lamp irradiation for one hour. The hardness of the film before annealing was 3 by Mohs' Scale of Hardness, and its hardness improved to 5 after annealing at 250 degrees centigrade for one hour. The refractive index of the film was 1.42, being lower than the index of silica glass that is 1.46. Furthermore, the transmittance in the visible region increased by 2% with its antireflection coating.

INTRODUCTION

The evanescent waves arise on the boundary between cooling water and laser head when the lights reflect perfectly from the laser head, of which leakage causes the decrease in the laser output power. To inhibit the decrease, we have tried to laminate the thin protective coat of 2-micron meter with low refractive index on the contact surfaces of cooling water and glass slab laser head.

For a protective coat formation, there are a vacuum deposition method as a dry process and a spin coating method as a wet process. The vacuum deposition method requires the temperature of 500 degrees centigrade and above; therefore, the coat becomes hard but the thermal denaturation of the substrate cannot be avoided. The spin coating method cannot attain a good adhesion to the substrate due to its low density of the topcoat.

C. Calmes of Univ. Paris has fabricated the Si-Ge-C film by ultra high vacuum chemical vapor deposition [1]. This method, however, requires high temperature. In addition, T. Kohoutek of Univ. of Pardubice has fabricated amorphous Ag–As–S films by standard spin-coating techniques from their organic solution [2]. However, Kohoutek's study doesn't refer to adhesion to the substrate.

For this reason, a new method to form a hard protective coat at room temperature is required. We have successfully fabricated a transparent and low refractive index SiO_2 film at room temperature by using Xe_2^* excimer lamp [3-8]. In this paper, we reported the improvement of film hardness with annealing.

A GROWTH PROCESS OF THE SiO_2 FILM

Figure 1 shows a growth process of the SiO_2 film. The NF_3 and O_2 mixed gas was photo-dissociated by the Xe_2^* excimer lamp to generate NO_2 and F radicals because the photon energy of Xe_2^* excimer lamp was higher than the binding energy of N-F and O=O. The F radicals etched the Si wafer and produce SiF_4, which adheres on the sample substrate. The SiF_4 adhered reacted with the NO_2 photo-dissociated, laminating a SiO_2 monomolecular layer on the sample substrate repeatedly. The SiF_4 was highly absorbent to the sample surface and easy to laminate a monomolecular layer. But, the SiF_4 and moisture interacted rapidly and vigorously, which made the SiF_4 control difficult. Our method, however, made good use of photochemical reaction in a chamber without interfering with the property of SiF_4.

However, the SiO_2 film also covered the surface of the Si wafer. The SiF_4 could not be supplied in this condition. Therefore, the piece of Si wafer was heated at 60 degrees centigrade in order not to laminate the SiO_2 film on the surface. The density of the SiF_4 on the Si wafer rarefied. Reversely, the glass substrate was cooled down in order to increase deposition rate of the SiO_2 film on the glass substrate. The adsorption density of the SiF_4 on the glass substrate improved.

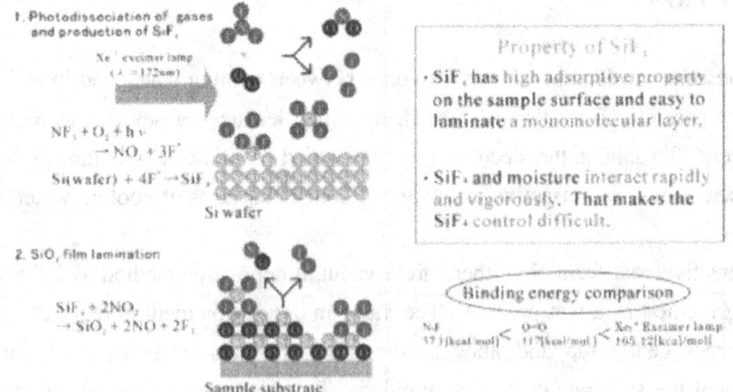

Figure 1. A growth process of the SiO_2 film

EXPERIMENTAL SET-UP

Figure 2 shows the Experimental Set-up. The glass substrate and a piece of Si wafer were placed in the reaction chamber, which was filled with the NF_3 and O_2 gases in the ratio of 10:1 and the total gas pressure was set at 330 Torr. The glass substrate was cooled down by the peltier device, and the Si wafer is heated at 60 degrees centigrade by the heater. Then, the Xe_2^* excimer lamp was irradiated for 60 minutes.

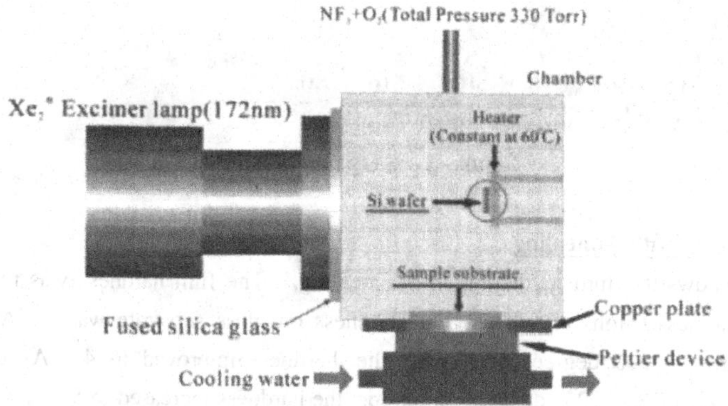

Figure 2. Experimental set-up

EXPERIMENTAL RESULTS

SiO_2 Film Thickness

Figure 3 shows the relationship between the film thickness and the lamp irradiation time of the samples with and without treatments. The film thickness was measured by the 3-dimension roughness meter ZYGO. With the substrate cooled down and the Si wafer heated, the film thickness increased by 3 times compared with that of the one without cooling and heating. The SiF_4 density on the Si wafer rarefied with the Si wafer heated. Therefore, the SiF_4 production was continued during the lamp irradiation. On the other hand, the SiF_4 density on the glass substrate became dense by cooling the substrate. Consequently, the adsorption density of the SiF_4 on the glass substrate increased.

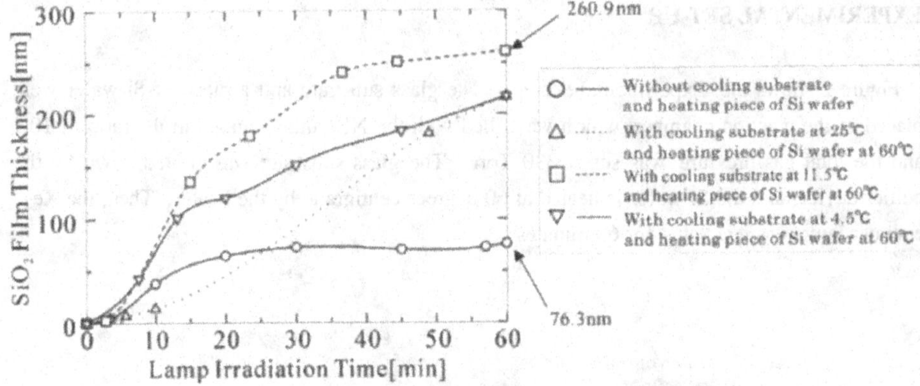

Figure 3. SiO_2 film thickness

Film Hardening with Annealing

Figure 4 shows the film hardening with annealing. The film hardness was measured with [Scale of Hardness (Mohs's)]. This film Hardness on glass substrate was 3. And after one hour annealing at 200 degrees centigrade, the hardness improved to 4. As the annealing temperature was raised at 250 degrees centigrade, the hardness increased to 5.

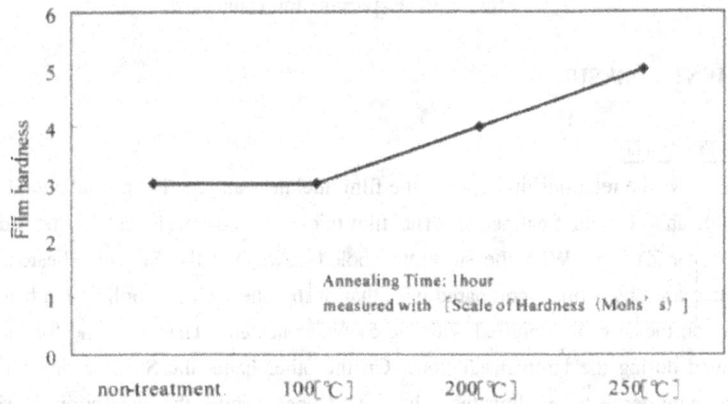

Figure 4. Film hardening with annealing

Annealing Dependence of Refractive Index and F Atom Density of SiO_2 Film

Figure 5 shows the annealing temperature dependence of the refractive index of the SiO_2 film and atom density in the SiO_2 film. The refractive index of the SiO_2 film was measured by the ellipsometry. The refractive index of the SiO_2 film was 1.36 without annealing. With annealing at 200 degrees centigrade for one hour, it increased to 1.42. The refractive index of the film was lower than the index of silica glass that was 1.46. Then, the F atoms density in the SiO_2 film was evaluated by the ESCA; which decreased from 1600 to 550.

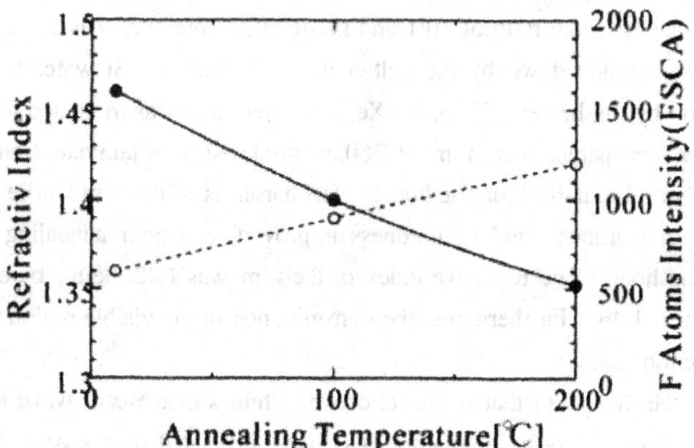

Figure 5. Comparison with refractive index

Transmittance of Glass Substrate with and without Laminated SiO_2 Film

Figure 6 shows the transmittance of glass substrate with and without laminated SiO_2 film. The relative transmittance of the glass substrate with the film was increased by 3% when the wavelength becomes longer than 400nm. It was clear that the film of low refractive index avoids the reflection.

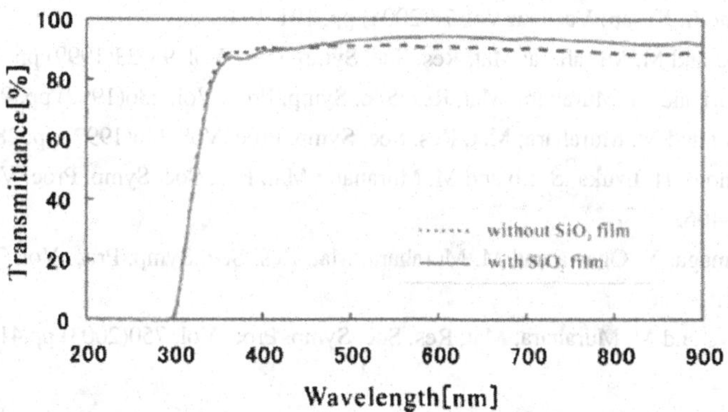

Figure 6. Transmittance of glass substrate with and without laminated SiO_2 film

CONCLUSION

A glass substrate and a piece of Si wafer were placed in the reaction chamber which was filled with NF_3 and O_2 gases in the ratio of 10:1 and the total gas pressure was set at 330 Torr. The glass substrate was cooled down by the peltier device. And, the Si wafer was heated to 60 degrees centigrade by the heater. Then, the Xe_2^* excimer lamp was irradiated for 60 minutes.

As a result, the transparent SiO_2 film of 260nm thickness was laminated on the non-heated substrate by the lamp irradiation for one hour. The hardness of the film before annealing was 3 by Mohs' Scale of Hardness, and its hardness improved to 5 after annealing at 250 degrees centigrade for one hour. The refractive index of the film was 1.42, being lower than the index of silica glass that is 1.46. Furthermore, the transmittance in the visible region increased by 2% with its antireflection coating.

In conclusion, the SiO_2 film that is antireflective, inhibits evanescent wave leakage, protects from cooling water and is not heat-denatured has been laminated on the optical material. This new technology contributes to improvement in the output power of glass slab laser.

REFERENCES

1. C. Calmes, D. Bouchier, D. Debarre, V. Le Thanh and C. Clerc; Thin Solid Films. Vol. 428(2003) pp. 150-155
2. T. Kohoutek, T. Wagner, J. Orava, M. Frumar, V. Perina, A. Mackova, V. Hnatowitz, M. Vlcek and S. Kasap; Vacuum Vol. 76(2004) pp. 191-194
3. H. Iizuka and M. Murahara; Mat, Res. Soc. Symp. Proc. Vol. 98-23(1999) pp. 446-451.
4. S. Aomori and M. Murahara; Mat, Res. Soc. Symp. Proc. Vol. 236(1992) pp. 9-14.
5. T. Suzuki and M. Murahara; Mat, Res. Soc. Symp. Proc. Vol. 446(1997) pp. 285-290.
6. T. Okamoto, H. Iizuka, S. Ito and M. Murahara; Mat, Res. Soc. Symp. Proc. Vol. 495(1998) pp. 401-406.
7. H. Tokunaga, Y. Ogawa and M. Murahara; Mat, Res. Soc. Symp. Proc. Vol. 700(2001) pp. 185-191.
8. Y. Ogawa and M. Murahara; Mat, Res. Soc. Symp. Proc. Vol. 750(2003) pp.419-424

Mater. Res. Soc. Symp. Proc. Vol. 843 © 2005 Materials Research Society T3.19

Copper Pattern Formation on Fluorocarbon Film by Single Shot of ArF Laser

T. Mochizuki and M. Murahara
Department of electrical engineering, Tokai University
1117 Kitakaname, Hiratsuka, Kanagawa 259-1292, Japan

ABSTRACT

Copper nuclei have been photo-chemically patterned on the fluorocarbon film surface in copper sulfate atmosphere via the reticle with just one shot of the ArF laser. Previously, we have reported that it required more than 3,000 shots of the ArF laser to substitute the Cu atoms on the fluorocarbon surface.

Fluorocarbon is regarded as promising as a printed wiring board material in the high-frequency band. However, the fluorocarbon is chemically stable making it difficult to bond to copper foil. Generally, a copper foil is formed on the fluorocarbon surface by electroless plating with a catalyst core. In this method, however, the surface is made rough in pretreatment, which impairs its characteristics. Using the catalysts also causes differences in the dielectric constant, generating a high frequency noise. Thus, we demonstrated the direct formation of copper nuclei on the fluorocarbon surface photo-chemically by using the Xe_2 excimer lamp and the ArF excimer laser.

The sample surface was first irradiated by Xe_2 excimer lamplight to make it hydrophilic. The surface was then irradiated with circuit patterned ArF laser light in the presence of copper-sulfate ($CuSO_4$) aqueous solution to substitute the Cu atom. The modified sample was immersed in the electroless plating solution at 60 degrees Celsius for 15 minutes, and the copper foil was locally formed on the areas exposed to light.

INTRODUCTION

Fluorocarbons such as polytetrafluoroethylene (PTFE), have excellent properties including heat resistance, chemical resistance, and water or oil repellency. Besides high electrical insulation, a low dielectric constant of 2.1 is required for a printed circuit substrate for a high frequency bond. Thus, the fabrication of metal films such as a copper on the fluorocarbon surface is needed.

On electric circuit printed boards, the copper foils have been generally produced by electroless plating after making the sample rough and using palladium (Pd) or platinum (Pt) catalyst cores for adhesion. Fluorocarbons are mainly composed of C-F bonds. However, the chemical stability of the fluorocarbons makes it difficult to create a chemical bond on the surface Thus a surface modification of fluorocarbons is required to improve the formation of copper thin foil.

Many studies have been done in the polymer surface modification with plasma or laser-light for electroless plating or electroplating treatment. Weber and co-workers tried to deposit of the titanium nitride (TiN) on the PTFE surface with electron cyclotron resonance (ECR) plasma in low temperature [1]. Zhao and co-workers tried to modify the polymide surface with KrF laser abrasion. The modified surface was deposited on the Pd catalyst by immersing in Pd solution [2]. Wang demonstrated that Q-switched Nd:YAG laser induced deposition of Cu atoms on the polymide surface from copper electrolyte solution [3]. In these methods, the polymer surface was made rough. When the printed board is used at high frequency, the roughness of the polymer surface causes noise.

We have developed a new method of directly depositing copper on a polymer using a photochemical process [4-9]. In this method, we have demonstrated that the surface modification of polymide or PTFE by substituting the metal atoms (Cu atoms, Ni atoms) with ArF excimer laser [4]. We have also reported the replacement of the Cu atoms on PTFE [6-9]. However, they required extensive laser pulse irradiation for surface modification [6]. On the other hand, we have demonstrated the substitution of Cu atoms on the polyethylene-terephthalate (PET) surface with one shot of ArF excimer laser irradiation after Xe_2 excimer lamp pre-treatment [9].

In his paper, we report growth of Cu nuclei on PTFE surfaces with a single shot (10ns) of ArF excimer laser irradiation after having oxidized the sample surface with the Xe_2 excimer lamp in pre-treatment in the presence of ozone.

EXPERIMENTAL

Experimental Set-up for Pre-Treatment

Figure 1 shows the schematic diagram of the experimental set-up [10]. The PTFE surface was photo-oxidized in the presence of active oxygen by the irradiation of Xe_2 excimer lamp and the discharge.

Experimental Procedure

Figure 2 shows the experimental set-up. A fused silica glass was placed on the PTFE surface, and a sulfate water solution was poured into the gap between the glass and the sample to form a thin liquid layer. The PTFE surface was irradiated with the ArF excimer laser light through the $CuSO_4$ solution. The homogenized ArF excimer laser beam was irradiated vertically on to the surface via the electric circuit pattern to photo-excite the PTFE and the thin liquid layer. The experimental conditions were as follow: the ArF excimer laser fluence of 18~22 mJ/cm^2; the shot number of 1~4; and the $CuSO_4$ concentration of 0.3, 0.5 and 1.0 %.

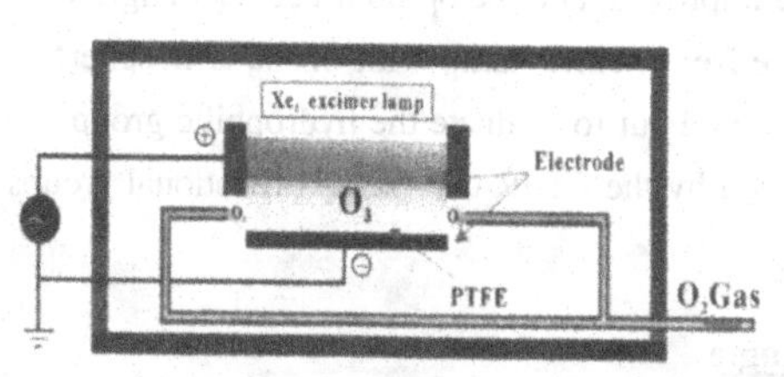

Figure 1. Pre-treatment set-up

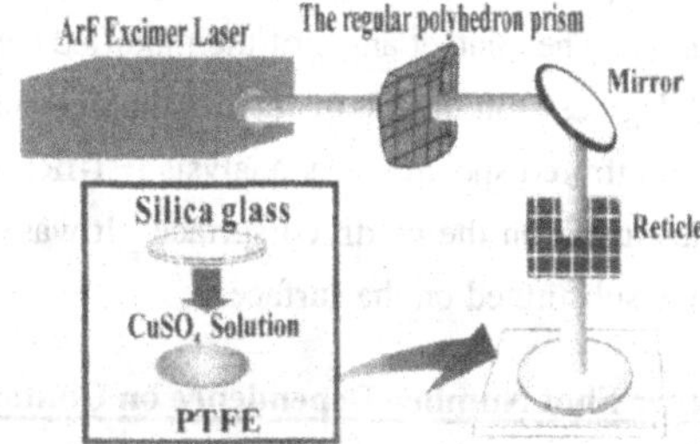

Figure 2. Experimental set-up

Substitution of Cu Nuclei on the Fluorocarbon Surface

First, the Xe_2 excimer lamp was irradiated on the PTFE surface in the presence of oxygen. The Xe_2 excimer lamp, which has a photon energy of 165 kcal/mol higher than the C-F binding energy of 128 kcal/mol, is employed for the F atom pulled out from C-F binding. The ozone is produced because of the oxygen absorption at the wavelength of 172 nm (the Xe_2 excimer lamp). The ozone was photo-dissociated to produce the active oxygen. By the active oxygen production, the OH functional groups were substituted on PTFE surface (Eqn. 1). As a result, the highly hydrophilic and highly reactive PTFE surface was modified.

Next, an ArF excimer laser was irradiated on the modified surface and the thin $CuSO_4$ solution layer. An ArF excimer laser, which has the photon energy of 147 kcal/mol higher than the C-F binding energy of 128 kcal/mol, was employed for modification. The modified PTFE surface was photo-excited and decomposed to C-O-H atoms, and the $CuSO_4$ solution was photo-dissociated to produce H, Cu and O atoms (Eqn. 2). The photo-dissociated H atom pulled out the H atom from the surface to generate H_2 gas. By the photochemical reaction, the copper atom combined with the dangling bond of C atom using the oxygen atom as the media As a result, the C–O–Cu bond was substituted on the fluorocarbon surface.

$$\left.\begin{aligned} O_2 + h\nu\,(172\text{nm}) &\rightarrow O_3 \\ O_3 + h\nu\,(172\text{nm}) &\rightarrow O(^1D) + O_2(^1\Delta) \\ O(^1D) + H_2O &\rightarrow 2OH \end{aligned}\right\} \quad (1)$$

$$\left.\begin{aligned} &CuSO_4 + H_2O + h\nu\ (193\text{nm}) \\ &\quad \rightarrow Cu^* + O^* + H^* \end{aligned}\right\} \quad (2)$$

RESULTS

Contact Angle Dependence on Irradiation Time

Figure 3 shows the contact angle dependence on the irradiation time of the Xe_2 excimer lamp. The contact angle of the modified surface improved, and the optimum contact angle was 68 degrees with the 20-minute irradiation using the Xe_2 excimer lamp with ozone atmosphere. And infrared spectroscopy analysis (FT-IR) was carried out to evaluate the hydrophilic groups substituted on the modified surface. It was clarified by the FT-IR that the OH functional groups were substituted on the surface.

Laser Shot Number Dependence on Contact Angle

Figure 4 shows the dependence of the number of laser shots leading to the highest Cu nucleus density on the contact angle with the Xe_2 excimer lamp. The number of laser shots decreased with the contact angle of the improved surface. The optimum number of laser shots was 4 with a remarkably improved contact angle of 68 degrees.

Laser Fluence Dependence on Cu Nucleus Density

Figure 5 shows the laser fluence dependence on the Cu nucleus density with the 1.0 % concentration of $CuSO_4$ solution. With the shot number of 4, the highest density was achieved at the laser fluence of 18 mJ/cm^2. However, the density decreased when the laser fluence exceeded 18 mJ/cm^2.

$CuSO_4$ Solution Concentration Dependence on Cu Nucleus Density

Figure 6 shows the $CuSO_4$ solution concentration dependence upon the Cu nucleus density. The highest Cu nucleus density was achieved with the 1.0 % concentration of $CuSO_4$ solution. When the concentration of $CuSO_4$ solution went below 0.5 %, the Cu nucleus was not sufficiently formed.

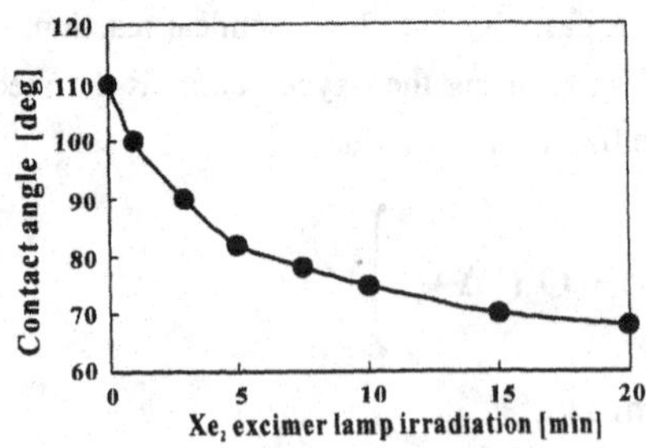

Figure 3. The contact angle dependence upon the irradiation time of Xe_2 excimer lamp

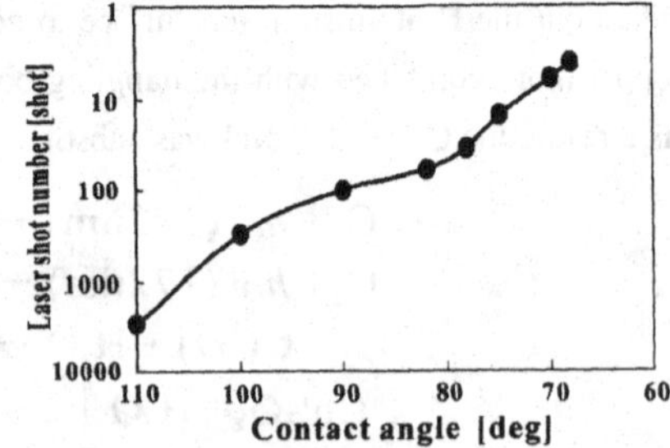

Figure 4. The laser shot number dependence upon the contact angle

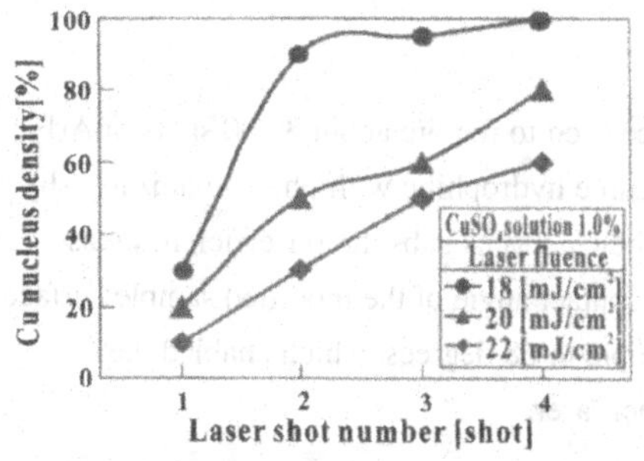

Figure 5. The laser fluenece dependence upon the Cu nucleus density

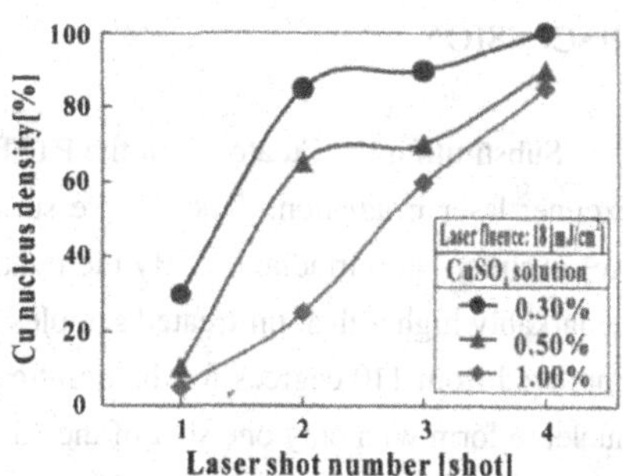

Figure 6. The $CuSO_4$ solution concentration upon the Cu nucleus density

Modified Sample Surface After Electroless Plating Treatment

Figure 7 shows the electroless plating treatment and the extension photographs. The Cu atoms were substituted on the PTFE surface by the irradiation of the electric circuit patterned ArF excimer laser. During this treatment, no change in the sample surface could be detected. Then, the modified sample was immersed in the electroless plating solution at 60 degrees Celsius for 15 minutes. The copper thin film was grown only on the Cu nuclei by this electroless plating treatment.

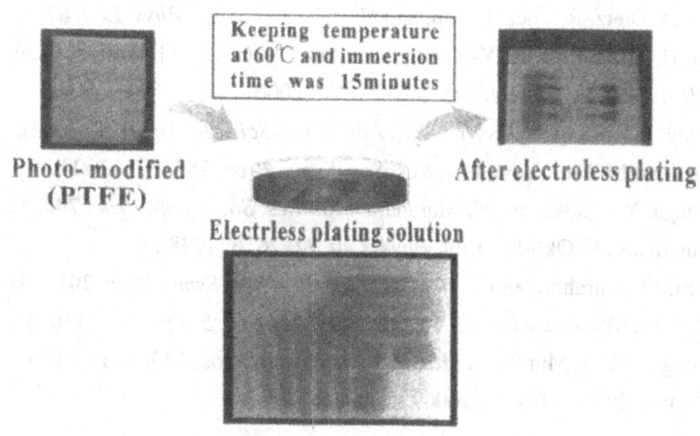

Figure 7. Electroless plating treatment

DISCUSSION

Substituting the Cu atoms on the PTFE surface used to require about 3,000 shots of ArF excimer laser irradiation. The sample surface became hydrophilic with photo-oxidization by Xe_2 excimer lamp irradiation. By the treatment, the Cu atoms substitution efficiency was remarkably higher than un-treated samples. The contact angle of the modified sample surface improved from 110 degrees for the non-treated sample to 68 degrees, which enabled the Cu nuclei to form with only one shot of the ArF excimer laser.

CONCLUSION

The PTFE surface was photo-oxidized with the Xe_2 excimer lamp in pre-treatment. The optimum contact angle was 68 degrees, with the 20-minute irradiation. The sample was vertically irradiated with the circuit patterned ArF excimer laser light with just one shot. The modified sample was then immersed in the electroless plating solution at 60 degrees Celsius for 15 minutes. As a result, the copper thin film was grown only on the Cu atom substitution sites. The intensity of Cu nucleus became the most effective at the 18 mJ/cm^2 of ArF excimer laser fluence, using 4-shots and a concentration of 1.0 %.

REFERENCES

1. A. Weber, A. Dietz, R. Pockelmann, and C. P. Klages, *Appl. Phys. Lett.* **67**, 2311 (1995).
2. G. Zhano, H. M. Phillips, H. Y. Zheng, S. C. tam, W. Q. Liu, G. Wen, Z. Gomg, and Y. L. Lam, *SPIE Int. Soc. Opt. Eng. Proc.* **3933**, 505 (2000).
3. W.C. Wang, E.T. Kang, and K.G. Neoh, *Appl. Surf. Sci.* **200,** 165-171 (2002).
4. M. Tomita and M. Murahara, *Mat. Res. Soc. Symp. Proc.* **585**, 197 (2000).
5. H. Tokunaga, Y. Ogawa and M. Murahara, *Mat. Res. Soc. Symp. Proc.* **700**, 185 (2001).
6. M. Murahara and M.Okoshi, *Appl. Phys. Lett.* **72**, 2616 (1998).
7. M. Okoshi, M. Murahara and K. Toyoda, *Mat. Res. Soc. Symp. Proc.* **201**, 451 (1991).
8. M. Okoshi, M. Murahara and K. Toyoda, *Mat. Res. Soc. Symp. Proc.* **158**, 33 (1990).
9. H. Tokunaga and M. Murahara, *Mat. Res. Soc. Symp. Proc.* **734**, 443 (2002).
10. M. Murahara, JP. Patent No 3316069 (7 June 2002).

Mater. Res. Soc. Symp. Proc. Vol. 843 © 2005 Materials Research Society T3.20

Photochemical Moisture Proof Coating on Nonlinear Optical Crystal by ArF Excimer Laser

M. Kojima and M. Murahara:
Department of Electrical Engineering, Tokai University, 1117 Kitakaname, Hiratsuka, Kanagawa 259-1292, Japan

ABSTRACT

Silicone oil was photo-chemically oxidized to change into SiO_2 on the crystal by using an ArF excimer laser; the protective moistureproof film has been developed for a nonlinear optical crystal that is deliquescent.

The nonlinear optical crystals such as $CsLiB_6O_{10}$ (CLBO) and KH_2PO_4 (KDP) are deliquescent, which causes their surfaces to be cloudy by absorbing moisture in the air. We, therefore, demonstrated the growth of the SiO_2 film directly on the crystal so as to be moistureproof.

Firstly, dimethylsiloxane silicone oil (-O-Si(CH_3)$_2$-O-)n was poured on the substrate and coated by a spinner for making the silicone oil thin layer. Then, the ArF excimer laser was vertically irradiated on the sample in oxygen atmosphere. The O atom on the substrate surface was photo-excited by the laser to generate a high active O atom. At the same time, the Si-CH_3 bond of the silicone oil was photo-dissociated and the dangling bond of Si was linked with the active O atom to form a SiO_2 film on the crystal surface.

In short, the film formed by the new technology can be used as a protective coating, which has the moisture resistance and the UV permeability, for a nonlinear optical crystal.

INTRODUCTION

The SiO_2 films have been widely used as the transparent protective film, the protective chemicalproof film and the insulated film, which is used for semiconductors or electric parts. The thermal oxide, chemical vapor deposition (CVD), vacuum vapor deposition and sputtering methods are generally used for making the optical thin films.

M.Tsuji of Kyushu Univ. has fabricated the SiO_2 film by LCVD (laser-induced chemical vapor deposition) method in the presence of mixtures of SiH_4 and N_2O at substrate temperature of 30-500 degrees centigrade [1]. G.Reisse Engineering Mittweida Univ. has also fabricated the SiO_2 film by LCVD method in the presence of mixtures of SiH_4 and N_2O [2].

E.Fogarassy has fabricated the SiO_2 film by laser ablation from silicon and silicon monoxide targets, performed under vacuum and in oxygen atmosphere at substrate temperature of 20-450 degrees centigrade [3].

K.Sano of Suzuki Motor Co. has fabricated the SiO_2 film by TEOS/O_2 ECR plasma at substrate temperature of 25-350 degrees centigrade [4]. However, these methods need degree of vacuum to use the reaction gases, which are used to form the film. In addition, the substrate is heated at certain amount of temperature. Therefore, the difference between the substrate and the film in thermal expansion rate causes cracking.

On the other hand, our method does not need the degree of vacuum and the heat. We have developed the new film formation method to inhibit cracking by irradiating the ArF laser on the silicone oil to be photo-oxidized in the air [5-7].

Because the nonlinear optical crystals are deliquescent, the crystal, in general, is placed in a cell, which is heated to control the temperature and protect from moisture [8]. However this method always needs to control the temperature in a cell. We, therefore, demonstrated the growth of the SiO_2 film directly on the crystal so as to be moistureproof.

Then, in this paper, we report the SiO_2 film formation method with photochemical oxidization of silicone oil on the crystal, and the assay of the formed film.

PRINCIPLE OF GROWTH OF SiO_2 FILM

As in shown Figure 1, an ArF laser beam or Xe_2 lamplight is vertically irradiated on the Silicone oil that is coated on the substrate by spinning in an oxygen atmosphere. The C-H and the Si-C bonds of the silicone oil are photo-dissociated because the photon energy of UV photon is higher than the bonding energies of the C-H (81kcal/mol) and Si-C (105kcal/mol). At the same time, the O atom in the atmosphere is photo-excited by the UV photon to generate a high active O atom. This active oxygen reacts with the silicone oil to produce SiO_2 on the substrate. The separated methyl group, which is disposed from the series, also reacts with the active oxygen to produce CO_2 and H_2O. With the reactions described above, the organic silicone oil turns to be ceramic glass in the presence of excited oxygen.

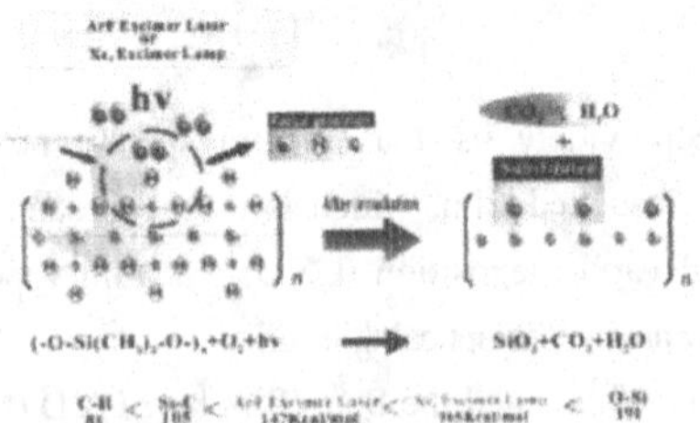

Fig 1. Principle of Growth of SiO_2 Film

EXPERIMENTAL SET-UP

The schema of the experimental set up by an ArF excimer laser is shown in (a) of Figure 2. Firstly, the silicone oil was put on the substrate. The substrate was spun to have the silicone oil spread thin and even. Then, an ArF excimer laser was vertically irradiated on the sample in an oxygen atmosphere.

The schema of the experimental set up by the Xe_2 excimer lamp is shown in (b) of Figure 2. In the same way as an ArF excimer laser irradiation, the Xe_2 excimer lamplight was irradiated on the sample in an oxygen atmosphere.

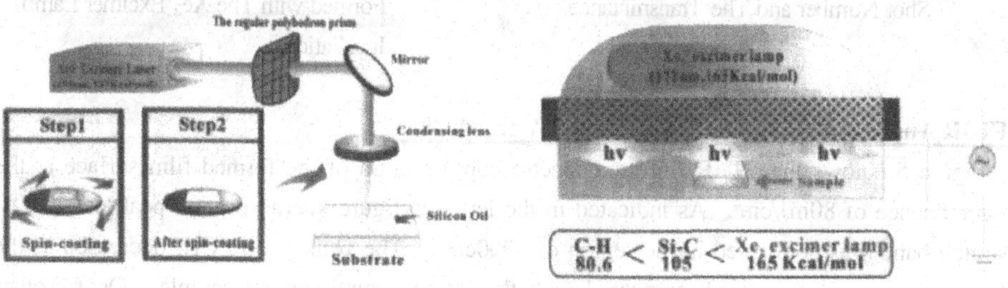

(a) With ArF Laser Irradiation (b) With Xe_2 Lamp Irradiation

Fig 2. Experimental Set-Up

RESULTS AND DISCUSSION

Transmittance of SiO_2 Film

When the wavelength (1064nm) of the YAG laser is converted with the nonlinear optical crystal, it is converted to the wavelength of 532nm, 355nm, 266nm and so on. Then we measured transmittance of the formed film at the wavelength of 266nm. Figure 3. shows the relationship between the laser shot number and the transmittance at 260nm. As indicated in the figure, the transmittance became high as the laser shot number increased at the laser fluence of $30mJ/cm^2$ and $80mJ/cm^2$. The transmittance became the highest when the laser fluence was $80mJ/cm^2$ and the laser shot number, 20,000.

On the other hand, Figure 4. shows the transmittance of the fabricated film with the Xe_2 excimer lamp irradiation. As indicated in the figure, the UV transmittance became high when the irradiation time of Xe_2 excimer lamp increased, compared with the non-treatment sample's. It reached the highest when the irradiation time was 90 minutes.

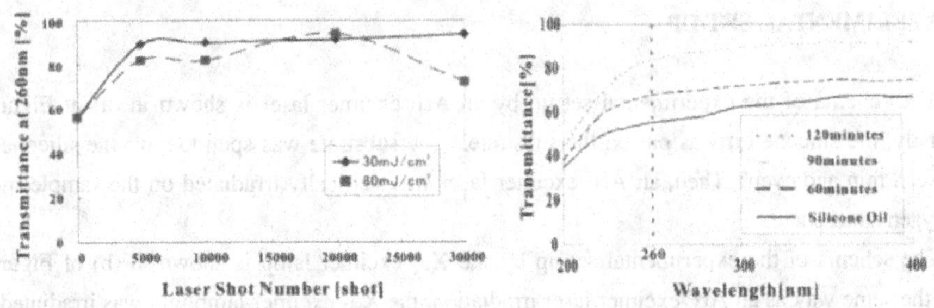

Fig 3. The Relationship between The Laser Shot Number and The Transmittance

Fig 4. The Transmittance of The Film Formed with The Xe_2 Excimer Lamp Irradiation

FT-IR Analysis of SiO_2 Film Formed with UV Irradiation

Figure 5. shows the FT-IR (infrared spectroscopy) spectra of the formed film surface at the laser fluence of 80mJ/cm^2. As indicated in the left side figure, the absorption peak of the CH_3 stretch band was confirmed in the region of 2900cm^{-1}. The peak of the CH_3 decreased as the laser shot number increased, compared with that of the non-treatment sample. On the other hand, the absorption peak of the Si-O stretch band was confirmed in the region of 1030cm^{-1}, as shown in the right side figure. The peak of the Si-O increased as the laser shot number increased, compared with that of the non-treatment sample. The results reveal that the silicone oil vitrifies as the laser shot number increases.

The FT-IR analysis of SiO_2 film formed with Xe_2 excimer lamp irradiation indicated that the silicone oil vitrified when the lamp irradiation time increased as it as in the ArF excimer laser irradiation.

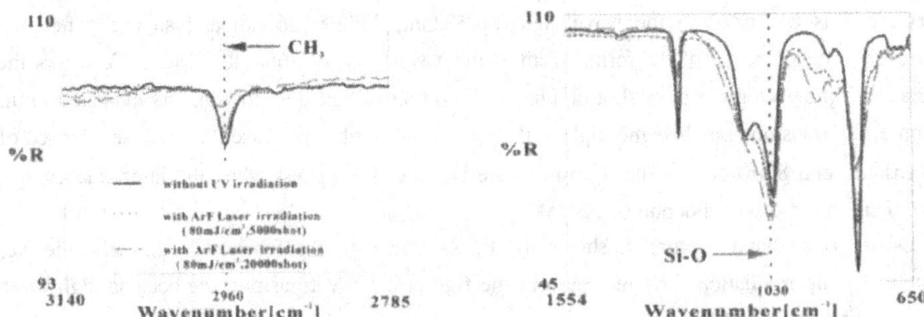

Fig 5. FT-IR Analysis of SiO_2 Film Formed with UV Irradiation

Relationship between Laser Shot Number and UV Transmittance/Absorption of Methyl Group

The -CH absorption strength of the silicone oil surface was measured by FT-IR and compared with its UV transmittance. The results are shown in Figure 6. As the laser shot number increased, the -CH functional groups were photo-dissociated to decrease the absorption strength. On the contrary, the UV transmittance increased. As a result, the silicone oil was vitrified to make UV transmittance possible.

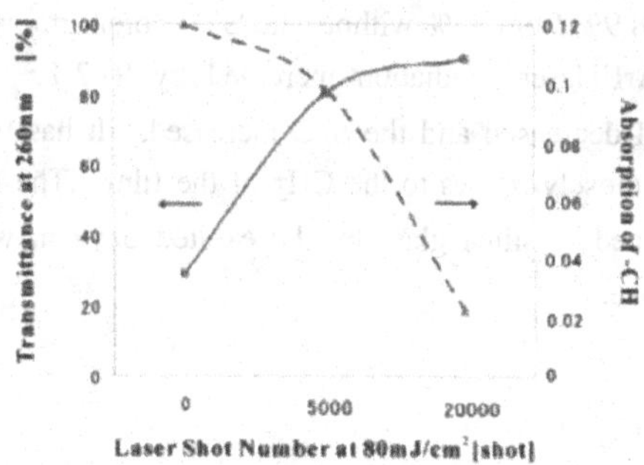

Fig 6. Relationship between Laser Shot Number and UV Transmittance /Absorption of Methyl Group

UV Transmittance With and Without Excimer Light Irradiation

Figure 7. exhibits the comparison of the UV transmittance of the silicone oil treated with the ArF excimer laser or Xe_2 excimer lamp and the silicone oil not treated. The UV transmittance of the non-treatment silicone oil was 56% at 260nm. That of the silicone oil irradiated with the Xe_2 lamp increased to 88.9%, and of the one irradiated with the ArF laser, it further improved to 94.2%.

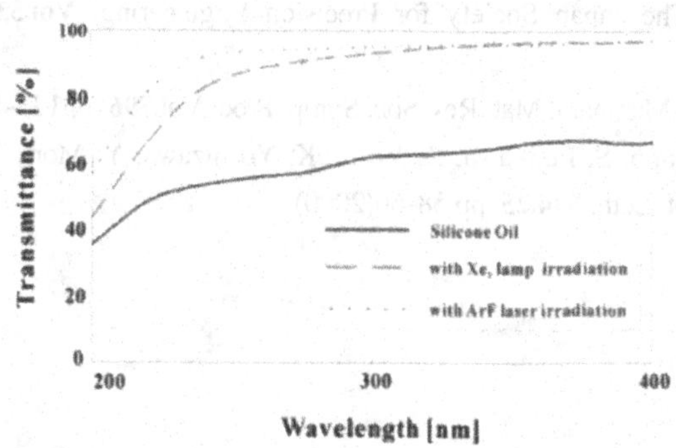

Fig 7. UV Transmittance With and Without Excimer Light Irradiation

CONCLUSION

The transparent SiO_2 film and permeable to ultraviolet rays has been fabricated photo-chemically with the organic silicone oil by the ArF excimer laser or Xe_2 excimer lamp in an oxygen atmosphere. The UV transmittance became high depending with the laser shot number or the lamp irradiation time. The transmittance of vitrified silicone oil was measured at the wavelength of 260nm. The transmittance of silicone oil modified with the Xe_2 lamp irradiation increased by 88.9% from 56% without the UV photon irradiation, and that of silicone oil modified with the ArF laser irradiation increased by 94.2%. At that time, the CH_3 consisting the silicone oil decreased and the Si-O increased. It has been proved that the UV transmittance of the film closely relates to the CH_3 of the film. The results indicated that the organic silicone oil changed to silica glass by the excited oxygen, which was considered to improve the UV transmittance.

REFERENCE

1. M. Tsuji, M. Sakumoto, N. Itoh, H. Obase and Y. Nishimura, Applied Surface Science, **Vol.51**, pp.171-176(1991)
2. G. Reisse, F. Gaensicke, R. Ebert and U. Illmann, Applied Surface Science, **Vol.54**, pp.84-88(1992)
3. E. Fogarassy, A. Slaoui, C.Fuchs and J.P. Stoquert, Applied Surface Science, **Vol.54**, pp.180-186(1992)
4. K. Sano, S. Hayashi, S. Wickramanayaka, Y. Hatanaka, Thin Solid Films, **Vol.281-282**, pp.397-400(1996)
5. M. Murahara, Journal of the JSTP, **Vol.27**, No.307, pp.934-942(1986) in Japanese
6. M. Murahara, The Japan Society for Precision Engineering, **Vol.53**, pp.34-38(1987) in Japanese
7. K. Asano and M. Murahara, Mat. Res. Soc. Symp. Proc. **Vol.796**, pp133-138(2004)
8. T. Kojima, S. Konno, S. Fujikawa, K. Yasui, K. Yoshizawa, Y. Mori, T. Sasaki, M. Tanaka and Y.Okada, Opt. Lett., **Vol.25**, pp.58-60(2000)

Mater. Res. Soc. Symp. Proc. Vol. 843 © 2005 Materials Research Society T3.36

High-speed Processing with Reactive Cluster Ion Beams

Toshio Seki[1,2] and Jiro Matsuo[1]
[1]Quantum Science and Engineering Center, Kyoto Univ., Gokasyo, Uji, Kyoto 611-0011, Japan
[2]Osaka Science and Technology Center, Utsubo Honmachi Nishi-ku, Osaka 550-0004, Japan

ABSTRACT

Cluster ion beam processes can produce high rate sputtering with low damage in comparison with monomer ion beam processes. Especially, it is expected that extreme high rate sputtering can be obtained using reactive cluster ion beams. Reactive cluster ion beams, such as SF_6, CF_4, CHF_3, and CH_2F_2, were generated and their cluster size distributions were measured using Time-of-Flight (TOF) method. Si substrates were irradiated with the reactive cluster ions at the acceleration energy of 5-65 keV. Each sputtering yield was increased with acceleration energy and was about 1000 times higher than that of Ar monomer ions. The sputtering yield of SF_6 cluster ions was about 4600 atoms/ion at 65 keV. With this beam, 12 inches wafers can be etched 0.5 μm per minute at 1 mA of beam current. The TOF measurement showed that the size of SF_6 cluster was about 550 molecules and the number of fluorine atoms in a SF_6 cluster was about 3300. If the sputtered product was SiF, the yield has to be less than 3300 atoms/ion. These results indicate that the reactive cluster ions etch targets not only chemically, but also physically. This high-speed processing with reactive cluster ion beam can be applied to fabricate nano-devices.

INTRODUCTION

Cluster ion beam process has high potential for material processing in nano-technology, such as photonic crystal and MEMS. In order to fabricate the devices, it is needed to etch targets with high-speed, low-damage, and ultra-smooth process. A cluster is an aggregate of a few to several thousands atoms. When many atoms constituting a cluster ion bombard a local area, high-density energy deposition and multiple-collision processes are realized. Because of the unique interaction between cluster ions and surface atoms, new surface modification processes, such as surface smoothing [1-3], shallow implantation [4,5] and high rate sputtering [6], have been demonstrated using gas cluster ions. The cluster ion beam processes can produce high rate sputtering with low damage in comparison with monomer ion beam processes. Moreover, cluster ion beam etching can smooth both bottom surface and sidewall. Especially, it is expected that the extreme high rate sputtering can be realized with using reactive cluster ion beams.

Figure 1 shows a schematic diagram of the cluster ion beam irradiation system. Adiabatic expansion of a high-pressure gas through a Laval nozzle is utilized for the formation of gas cluster beams [7]. The nozzle was made from glass. The diameter of the throat of nozzle is 0.1 mm. The neutral beam intensity can be measured with an ion gauge. The pressure measured by the ion gauge on the beam line in process chamber was regarded as the neutral beam intensity. The neutral clusters were ionized by the electron bombardment method. The size distributions of cluster ion beams can be measured using Time of Flight (TOF) method.

It has been reported that when fluorine clusters bombard a Si surface, volatile compounds of SiF_x are produced [8]. As for experiments, SF_6 cluster beams have been generated by mixing with He gas and higher sputtering yield was demonstrated [9,10]. In this paper, reactive fluorine cluster ion beams, such as SF_6, CF_4, CHF_3, and CH_2F_2, were generated and their cluster size

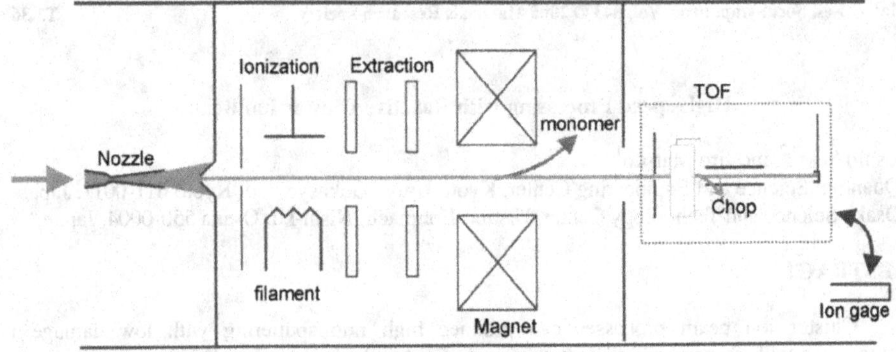

Figure 1. Schematic diagram of the cluster ion beam irradiation system.

distributions were measured using Time-of-Flight (TOF) method. Si substrates were irradiated with the reactive cluster ions at the acceleration energy of 5-65 keV and sputtering yields were measured.

RESULTS AND DISCUSSION

Figure 2 shows the mol fraction dependence of neutral cluster beam intensity. CF_4 cluster beam and CHF_3 cluster beam were generated from high-pressure CF_4 or CHF_3 gas mixed with He gas through a Laval nozzle. The mixed gas pressure was fixed to 4000 Torr or 6000 Torr, and

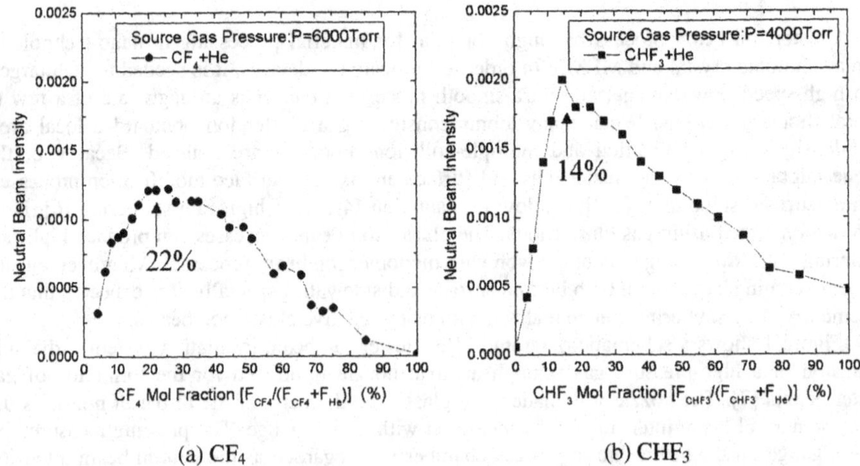

(a) CF_4 (b) CHF_3

Figure 2. Mol fraction dependence of neutral cluster beam intensity.

CF_4 or CHF_3 mol fraction was varied. CF_4 mol fraction (X_{CF4}) and CHF_3 mol fraction (X_{CHF3}) were as follows

$$X_{CF4} = F_{CF4}/(F_{CF4} + F_{He}),$$

$$X_{CHF3} = F_{CHF3}/(F_{CHF3} + F_{He}),$$

where F_{CF4} was flux of CF_4 gas, F_{CHF3} was flux of CHF_3 gas, and F_{He} was flux of He gas. The maximum beam intensity of CF_4 was obtained at a CF_4 mol fraction of 22 %, and the intensity was about 100 times higher than that of pure CF_4 gas. The maximum beam intensity of CHF_3 was obtained at a CHF_3 mol fraction of 14 %, and the intensity was about 4 times higher than that of pure CHF_3 gas. These results indicated that a high intensity CF_4 cluster beam and CHF_3 cluster beam can be generated by using a high-pressure gas mixed with He gas.

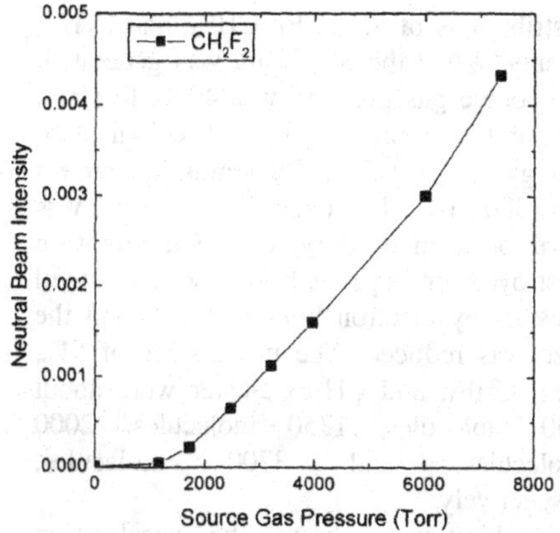

Figure 3. Source gas pressure dependence of neutral CH_2F_2 cluster beam intensity.

Figure 3 shows the source gas pressure dependence of neutral CH_2F_2 cluster beam intensity. CH_2F_2 cluster beam was generated from high-pressure pure CH_2F_2 gas through a Laval nozzle. The neutral beam intensity of CH_2F_2 increased with the source gas pressure. This result indicated that a high intensity CH_2F_2 cluster beam can be generated by using a high-pressure pure CH_2F_2 gas without He gas.

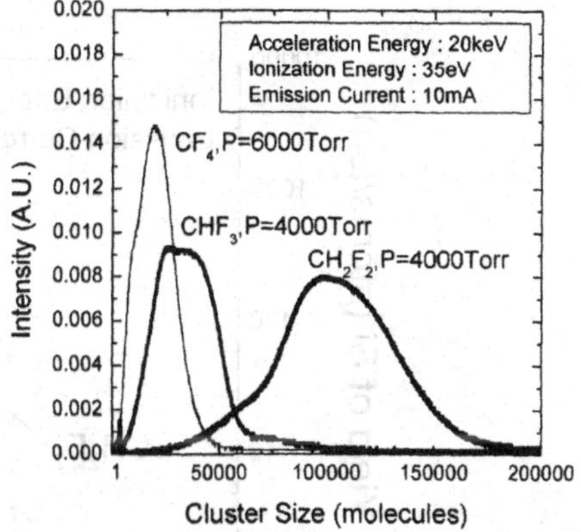

Figure 4. Cluster size distributions of CH_4, CHF_3, and CH_2F_2 beams under soft ionization condition.

In order to confirm that the neutral beams include clusters, the mass distributions of CF_4, CHF_3, and CH_2F_2 beams were measured with a TOF system. Figure 4 shows the cluster size distributions of CF_4, CHF_3, and CH_2F_2 beams. When the CF_4 beam was generated, the source gas pressure was 6000 Torr and CF_4 mol fraction was 22 %. When the CHF_3 beam was generated, the source gas pressure was 4000 Torr and CHF_3 mol fraction was 14 %. When the CH_2F_2 beam was generated, the source gas pressure was 4000 Torr without He gas. The ionization energy was 35 eV and the emission current was 10 mA. Because both ionization energy and emission current were low, the size reduction by ionization could be less pronounced. The size distributions would be close to the initial size distributions of neutral beams. The size distributions of the neutral beams were in the range up to 170,000 molecules. This result shows that the neutral beams include the large size of clusters.

Figure 5 shows the cluster size distributions of SF_6, CF_4, CHF_3, and CH_2F_2 beams. When the SF_6 beam was generated, the source gas pressure was 4000 Torr and SF_6 mol fraction was 8 %. The ionization energy was 500 eV and the emission current was 300 mA. This ionization condition was used for irradiations, typically. Clusters were destroyed or separated into several small clusters by electron bombardment, and the size was reduced. The mean sizes of SF_6, CF_4, CHF_3, and CH_2F_2 cluster were about 550 molecules, 1350 molecules, 2000 molecules, and 3700 molecules, respectively.

Figure 6 shows the acceleration energy dependence of the sputtering yield of Si with SF_6 cluster, CF_4 cluster, CHF_3 cluster, CH_2F_2 cluster, Ar cluster and Ar monomer. Sputtering yield with Ar

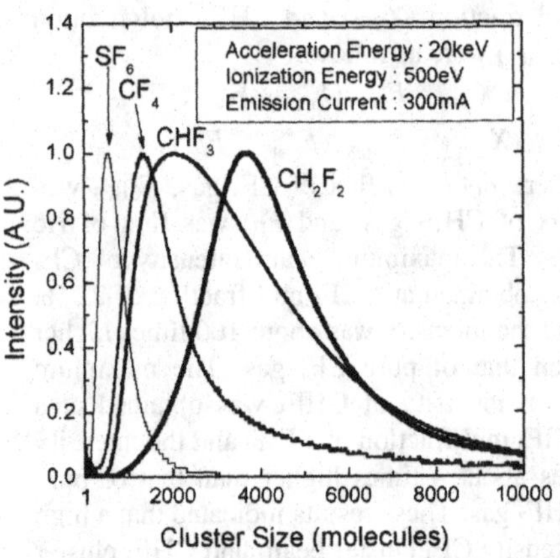

Figure 5. Cluster size distributions of the SF_6, CF_4, CHF_3, and CH_2F_2 beams used for sputtering measurement.

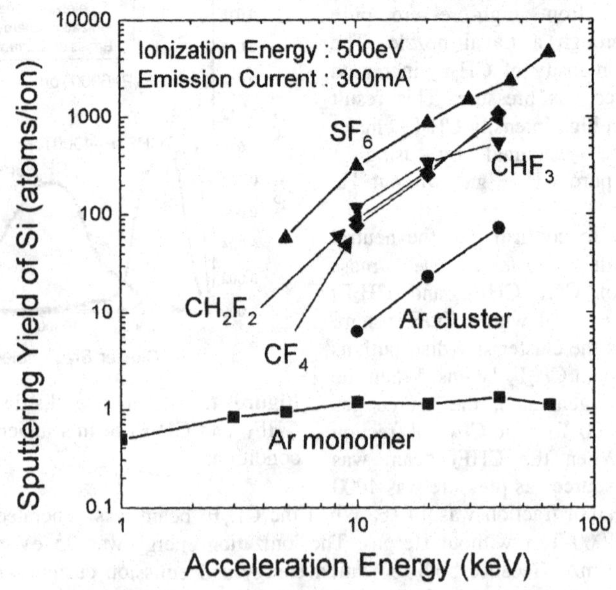

Figure 6. Sputtering yield of Si with SF_6 cluster, CF_4 cluster, CHF_3 cluster, CH_2F_2 cluster, Ar cluster and Ar monomer.

monomer was calculated using TRIM [11]. The sputtering yields with SF_6, CF_4, CHF_3, and CH_2F_2 cluster ions increased with acceleration energy. Especially, the sputtering yield with SF_6 cluster was about 4600 atoms/ion at 65 keV. The sputtering yield was about 4000 times higher than that of Ar monomer ions and also higher than that of Ar cluster ions. It was found that reactive sputtering occurred with SF_6, CF_4, CHF_3, and CH_2F_2 cluster ion irradiations. These results indicate that high-speed etching can be realized with reactive cluster ion irradiation at high energy.

When carbon atoms are included in projectiles, the carbon atoms accumulate on a surface and reactive sputtering is hindered. If there are hydrogen atoms in a molecule including carbon atom, the carbon atoms can be sputtered as CH_x and the accumulation of carbon atoms on a surface can be suppressed. As a result, the sputtering yield increases. However, the sputtering yields with CF_4, CHF_3, and CH_2F_2 cluster ions were almost same yields regardless of the number of hydrogen atoms in a molecule. This result indicates that carbon atoms can be sputtered physically, if the surface is bombarded by cluster ions. Because the size of SF_6 cluster was about 550 molecules, the number of fluorine atoms in a SF_6 cluster was about 3300. If the sputtered product was SiF, the yield has to be less than 3300 atoms/ion. However, the yield reached 4600 atoms/ion at high acceleration energy. These results indicate that the reactive cluster ions etch targets not only chemically, but also physically. It is expected that the yields with reactive cluster ions increase at higher than 65 keV of acceleration energy without restriction by the number of fluorine atoms in a cluster. The reactive cluster ion irradiation at high energy can be used for high-speed fabrication processes.

CONCLUSIONS

Reactive cluster ion beams, such as SF_6, CF_4, CHF_3, and CH_2F_2, were generated and their cluster size distributions were measured using Time-of-Flight (TOF) method. The sputtering yields of Si with the reactive cluster ion beams increased with acceleration energy. Especially, the sputtering yield with SF_6 cluster was about 4600 atoms/ion at 65 keV. The reactive cluster ions etch targets not only chemically, but also physically. These results indicate that reactive cluster ion irradiation at high energy can be used for high-speed fabrication processes.

ACKNOWLEDGEMENTS

This work is supported by the New Energy and Industrial Technology Development Organization (NEDO) of Japan.

REFERENCES

[1] H.Kitani, N.Toyoda, J.Matsuo and I.Yamada, Nucl. Instr. and Meth. **B121** (1997) 489.

[2] N.Toyoda, N.Hagiwara, J.Matsuo and I.Yamada, Nucl. Instr. and Meth. **B148** (1999) 639.

[3] A.Nishiyama, M.Adachi, N.Toyoda, N.Hagiwara, J.Matsuo and I.Yamada, AIP conference proceedings (15-th International Conference on Application of Accelerators in Research and Industry) **475** (1998) 421.

[4] D.Takeuchi, J.Matsuo, A.Kitai and I.Yamada, Mat. Sci. and Eng. **A217/218** (1996) 74

[5] N.Shimada, T.Aoki, J.Matsuo, I.Yamada, K.Goto and T.Sugui, J. Mat. Chem. and Phys. **54** (1998) 80.

[6] I.Yamada, J.Matsuo, N.Toyoda, T.Aoki, E.Jones and Z.Insepov, Mat. Sci. and Eng. **A253** (2000) 249.

[7] T.Seki, J.Matsuo, G.H.Takaoka and I.Yamada, Nucl. Instr. and Meth. **B206** (2003) 902.

[8] T.Aoki, J.Matsuo and I.Yamada, Nucl. Instr. and Meth. **B180** (2001) 164.

[9] N.Toyoda, H.Kitani, J.Matsuo and I.Yamada, Nucl. Instr. and Meth. **B121** (1997) 484.

[10] T.Seki and J.Matsuo, Mat. Res. Soc. Symp. Proc. **792** (2004) 593.

[11] J.P.Biersack and L.G.Haggmark, Nucl. Instr. and Meth., **174** (1980) 257.

Surface Indentation and Phase Transformations

Mater. Res. Soc. Symp. Proc. Vol. 843 © 2005 Materials Research Society T6.3/R10.3

Nanoindentation of Ion Implanted and Deposited Amorphous Silicon

J. S. Williams, B. Haberl, J. E. Bradby
Department of Electronic Materials Engineering, Research School of Physical Sciences and Engineering, The Australian National University, Canberra, ACT 0200, Australia

ABSTRACT

The deformation behavior of both ion-implanted and deposited amorphous Si (a-Si) films has been studied using spherical nanoindentation, followed by analysis using Raman spectroscopy and cross-sectional transmission electron microscopy (XTEM). Indentation was carried out on both unannealed a-Si films (the so-called unrelaxed state) and in ion implanted films that were annealed to 450°C to fully relax the amorphous film. The dominant mode of deformation in unrelaxed films was via plastic flow of the amorphous phase rather than phase transformation, with measured hardness being typically 75-85% of that of crystalline Si. In contrast, deformation via phase transformation was clearly observed in the relaxed state of ion implanted a-Si, with the load–unload curves displaying characteristic discontinuities and Raman and XTEM indicating the presence of high-pressure crystalline phases Si-III and Si-XII following pressure release. In such cases the measured hardness was within 5% of that of the crystalline phase.

INTRODUCTION

It is well known that diamond cubic Si-I undergoes a series of pressure-induced phase transformations during mechanical loading using both diamond anvil and indentation apparatus. It has been shown that Si-I first transforms to a metallic β-Sn (Si-II) phase during loading at a pressure of ~11 GPa [1-5]. During pressure release, the material undergoes further transformation to crystalline phases (Si-III and Si-XII) and, during fast unloading in indentation experiments, can transform to an a-Si phase [3].

For a-Si, early indentation measurements on ion-implanted amorphous films showed that the mechanical deformation appeared to depend on the preparation conditions or the "state" of the amorphous phase [6]. The so-called relaxed a-Si state [7] (in which the implanted amorphous layer is annealed to ~450 °C prior to indentation) exhibited very similar load–unload indentation curves to those of crystalline Si, including a similar hardness. However, unannealed (or unrelaxed) a-Si appeared to deform plastically at lower loads and exhibit a somewhat lower hardness. The more recent indentation measurements of Follstaedt *et al.* [8] confirmed that the hardness of relaxed ion implanted a-Si approached that of Si-I but that unrelaxed a-Si was slightly softer. Our very recent measurements [9] on ion implanted a-Si have shown that the relaxed state undergoes phase transformations under indentation whereas the unrelaxed state appears to flow plastically. In this paper, we extend such measurements to compare the indentation behaviour of both ion implanted and CVD-deposited amorphous Si.

EXPERIMENTAL DETAILS

In this current study continuous layers of *a*-Si were prepared by one of two methods: by ion-implantation of (100) Si-I wafers [9] and by PECVD deposition of a-Si on (100) Si-I. The

PECVD-deposited a-Si was pre-annealed for 5 min at a temperature of ~300 °C in a nitrogen atmosphere but this treatment is insufficient to relax the film [7,10]. The typical a-Si film thicknesses were between 600 and 900 nm. Portions of the deposited and ion-implanted samples were annealed for 30 min at a temperature of ~450 °C in an argon atmosphere to attempt to fully relax the layers [10] but not induce any crystallization.

A series of indentations were then performed on both deposited and ion-implanted (in each case on unannealed and annealed) a-Si samples using an Ultra-Micro-Indentation-System 2000 with spherical indenters of ~5.0 and 2 μm radii at room temperature. For the case of the 2 μm radius indenter, the penetration depth at maximum applied load (20 mN) for the thinnest films was less than one third of the a-Si film thickness and, hence, the underlying substrate is expected to have a minimal affect on the deformation behavior of the film.

After indentation, Raman spectra were recorded with a Renishaw 2000 Raman Imaging Microscope using the 632.8 nm excitation line of a helium-neon laser and a beam spot of ~1.0 μm radius. Residual indent impressions were prepared for cross-sectional transmission electron microscopy (XTEM) using a FEIxP200 focused ion beam system [4] and samples were analysed in a Philips CM 300 operating at an accelerating voltage of 300 kV.

RESULTS AND DISCUSSION

Figure 1(a) shows a load versus penetration plot of unrelaxed ion-implanted a-Si loaded to 50 mN. No discontinuities on either the loading or unloading sections of the curve are observed. In contrast, the nanoindentation curve taken from relaxed ion-implanted a-Si, as shown in figure 1(b), exhibits a clear discontinuity on unloading. Such an event is typically referred to as a "pop-out" event [5] and is indicative of phase transformations in Si [5,11]. Furthermore, a discontinuity on loading with a spherical indenter (called "pop-in") can be observed with crystalline Si but this is not easily observed when indenting relaxed *a*-Si.

Figure 2 displays the Raman spectra of indents in relaxed and unrelaxed ion-implanted amorphous silicon and of pristine amorphous silicon, all spectra showing clearly the broad a-Si peak at around 470 cm^{-1}. In the unrelaxed case no evidence for phase transformations could be found but for the spectrum of the relaxed sample we observed Raman shifts indicating the presence of Si-III at 383 cm^{-1} and 435 cm^{-1} and indicating Si-XII at 351 cm^{-1} and 396 cm^{-1}.

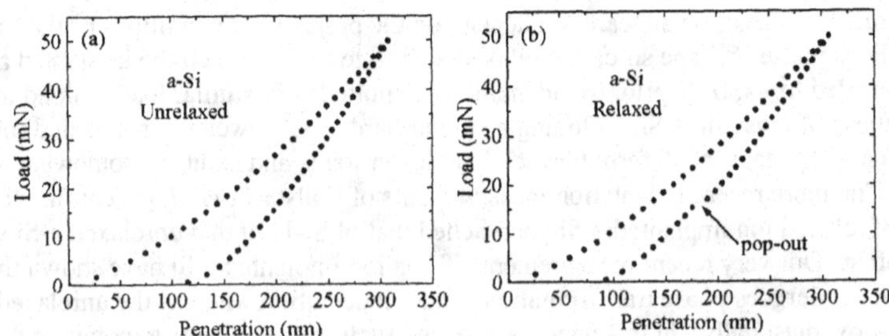

Figure 1. Load-unload curves of (a) unrelaxed and (b) relaxed 900 nm thick ion-implanted a-Si films using a continous load cycle with a spherical indenter of ~5 μm radius.

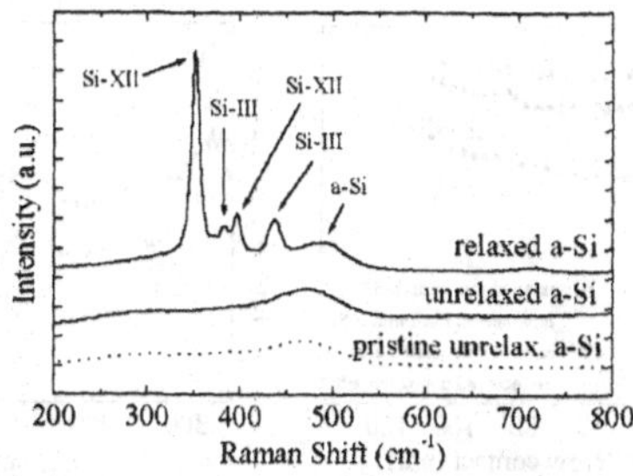

Figure 2. Raman spectra of pristine unrelaxed *a*-Si and spectra from indentations in relaxed and unrelaxed *a*-Si. Indentations made to a 50 mN load in 650 nm thick, ion-implanted a-Si with a continuous load–unload cycle using a spherical indenter of ~5 μm radius.

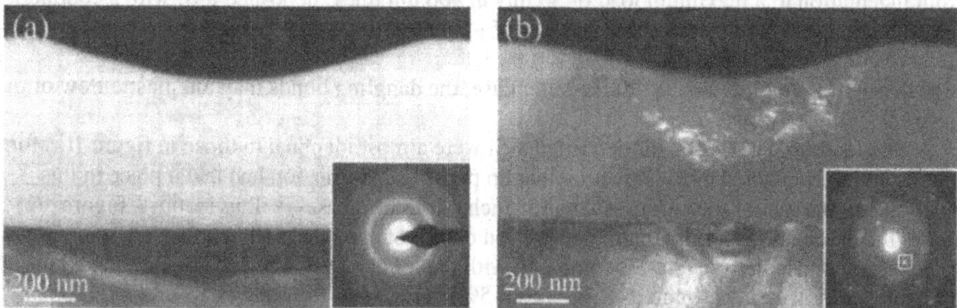

Figure 3. (a) BF XTEM image of an indent in unrelaxed 600-650 nm thick ion-implanted amorphous silicon; (b) DF XTEM image of an indent in relaxed 600-650 nm thick, ion-implanted amorphous silicon, the diffraction spot used to form the DF image is marked with a box in the inset. All indents were made using a continuous load cycle with a ~2 μm radius indenter to a maximum load of 20 mN. The SADP were taken from the region directly under the residual indent impressions.

Figure 3(a) shows a XTEM bright-field image of a 20 mN indent in unrelaxed ion-implanted a-Si made using a ~2.0 μm radius indenter. A selected area diffraction pattern (SADP) of the region immediately beneath the residual indent shows only the diffuse rings characteristic of an amorphous structure. In contrast, the dark-field (DF) image of a residual indent in relaxed ion-implanted a-Si, shown in figure 3(b), illustrates very clearly that phase transformations have occurred during loading and unloading. The DF image was taken using the boxed diffraction spot from Si-III/Si-XII shown in the inset to this figure. These extra diffraction spots arise from Si-III/XII.

The above behavior clearly indicates that phase transformations from an amorphous state to a crystalline state can occur at room temperature under indentation in relaxed a-Si. As indicated elsewhere [9], we can interpret the difference in the mechanical behavior of relaxed and unrelaxed films in terms of the different bonding configurations in these states: relaxed a-Si has near-fully-coordinated Si in a continuous random network, whereas unrelaxed a-Si contains a

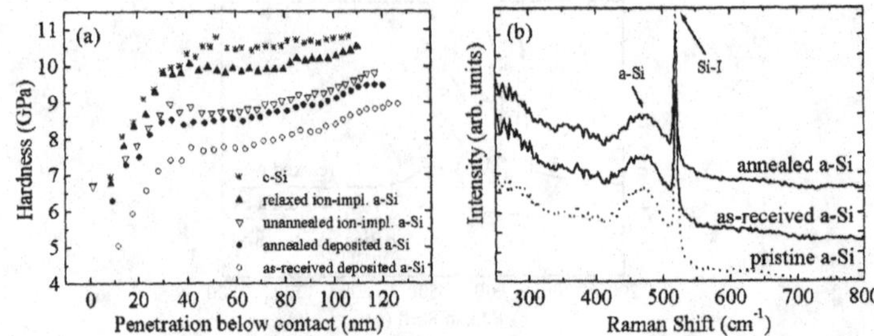

Figure 4. (a) Hardness of c-Si, compared with relaxed and unrelaxed ion-implanted and deposited a-Si films. The hardness data was taken for loading to 25 mN using a partial load-unload cycle for ~5 μm radius spherical indenter. (b) Raman spectra of pristine Si and spectra from indentation to a maximum load of 50 mN in 900 nm thick, deposited a-Si with a continuous load–unload cycle using a spherical indenter of ~5 μm radius.

high density of dangling bonds. In the latter case, the dangling bonds may aid plastic flow of the material.

The load-unload curves for deposited a-Si were almost identical to those in figure 1(a) for unrelaxed ion implanted a-Si, indicating that no phase transformation had taken place in this case. A comparison of the hardness data for each of the film cases is illuminating. Figure 4(a) shows the hardness (as a function of penetration depth) of annealed and unannealed, ion-implanted and also deposited a-Si, compared with that of crystalline Si-I. Clearly, relaxed ion-implanted a-Si is nearly as hard as Si-I (within 5%). However, the other films are considerably softer. Note that we have annealed the deposited a-Si film to 450 °C for 30 minutes in an attempt to relax the film but, although its hardness increases slightly, the material does not undergo transformation. We do not understand this behavior but suggest that the deposited films could differ from relaxed (annealed) ion implanted films in a number of respects: they may contain a significant impurity content (including hydrogen) and could exhibit porosity. As a result, the annealed deposited film may depart from the near-fully-coordinated relaxed ion implanted film and this could enhance their ability to deform plastically under indentation. We intend to further investigate this issue. The lack of phase transformation in the deposited films is confirmed by Raman analysis. Figure 4(b) shows Raman spectra of indents in as received and annealed deposited a-Si, compared with pristine deposited a-Si prior to indentation. No evidence for a phase transformation is observed, with peaks from Si-I as well as from the amorphous layer being visible. However, we would not expect Si-I peaks to be observed for 900 nm thick films and believe that pinholes may be the reason for their observation. If this is the case, then it may reinforce structural differences between the annealed ion implanted and deposited a-Si films.

CONCLUSION

In conclusion we have observed an amorphous to crystalline phase transformation at room temperature in relaxed (annealed) ion implanted a-Si during indentation, resulting in the end phases Si-III and Si-XII following unloading. Phase transformations are not the primary mode of deformation with deposited a-Si films or with unannealed (unrelaxed) ion implanted films. In the latter case, these films deform via plastic flow of the film under the indenter. Whereas the hardness values of relaxed films that undergo phase transformation are within 5% of the value for crystalline Si-I, the films that plastically deform are considerably softer, with a hardness of 75-85% of the crystalline value.

REFERENCES

1. J. Z. Hu, L. D. Merkle, C. S. Menoni, and I. L. Spain, *Phys. Rev. B* **34**, 4679 (1986).
2. D. R. Clarke, M. C. Kroll, P. D. Kirchner, R. F. Cook, and B. J. Hockey, *Phys. Rev. Lett.* **60**, 2156 (1988).
3. V. Domnich, Y. Gogotsi, and S. Dub, *Appl. Phys. Lett.* **76**, 2214 (2000).
4. J. E. Bradby, J. S. Williams, J. Wong-Leung, M. V. Swain, and P. Munroe, *Appl. Phys. Lett.* **77**, 3749 (2000).
5. J. E. Bradby, J. S. Williams, J. Wong-Leung, M. V. Swain, and P. Munroe, J. *Mater. Res.* **16**, 1500 (2001).
6. J. S. Williams, J. S. Field, and M. V. Swain, *Mater. Res. Soc. Symp. Proc.* **308**, 571 (1993).
7. S. Roorda, W. C. Sinke, J. M. Potae, D.C. Jacobson, S. Dieker, B. S. Dennis, D. J. Eaglesham, F. Spaepen, and P. Fuoss, *Phys. Rev. B* **44**, 3702 (1991).
8. D. M. Follstaedt, J. A. Knapp, and S. M. Myers, *J. Mater. Res.* **19**, 338 (2004).
9. B. Haberl, J. E. Bradby, M.V. Swain, J. S. Williams, and P. Munroe, *Appl. Phys. Lett.* **85**, 5559 (2004).
10. S. Roorda, S. Doorn, W. C. Sinke, P. M. L. O. Scholte, and E. van Loenen, *Phys. Rev. Lett.* **62**, 1880 (1989).
11. Y. G. Gogotsi, V. Domnich, S. N. Dub, A. Kailer, and K. G. Nickel, *J. Mater. Res.* **15**, 871 (2000).

[illegible] reduced. Thus the [illegible] phase transformation [illegible] within 5% of the value for crystalline Si [illegible] the film [illegible] physically [illegible] consider [illegible] with a thickness of [illegible] the crystalline Si.

REFERENCES

1. [illegible] Merkle, [illegible] and [illegible] *Phys.* [illegible] 1679 (1986).
2. [illegible] Kirchner, [illegible] and D. [illegible] *Phys. Rev. Lett.* [illegible] (1998).
3. [illegible] Appl. [illegible] (2002).
4. [illegible] *J.* [illegible] *Appl. Phys.* [illegible] (2000).
5. [illegible] Swain, [illegible] *J. Mater.* [illegible]
6. [illegible] Williams [illegible] (1991).
7. [illegible] Dennis, [illegible] *Phys. Rev.* [illegible]
8. [illegible] *Phys.* [illegible] (2000).
9. [illegible] Williams [illegible] *Appl. Phys. Lett.* [illegible] (1994).
10. [illegible] *Phys. Rev. Lett.* [illegible]
11. [illegible]

Mater. Res. Soc. Symp. Proc. Vol. 843 © 2005 Materials Research Society 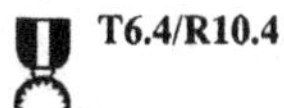 T6.4/R10.4

Cross-Sectional TEM Studies of Indentation-Induced Phase Transformations in Si: Indenter Angle Effects

Songqing Wen[1], James Bentley[2], Jae-il Jang[1], and G.M. Pharr[1,2]
[1] The University of Tennessee, Dept. of Materials Science & Engr., Knoxville TN 37996
[2] Oak Ridge National Library, Metals & Ceramics Division, Oak Ridge TN 37831

ABSTRACT

Nanoindentations were made on a (100) single crystal Si wafer at room temperature with a series of triangular pyramidal indenters having centerline-to-face angles ranging from 35° to 85°. Indentations produced at high (80 mN) and low (10 mN) loads were examined in plan-view by scanning electron microscopy and in cross-section by transmission electron microscopy. Microstructural observations were correlated with the indentation load-displacement behavior. Cracking and extrusion are more prevalent for sharp indenters with small centerline-to-face angles, regardless of the load. At low loads, the transformed material is amorphous silicon for all indenter angles. For Berkovich indentations made at high-load, the transformed material is a nanocrystalline mix of Si-I and Si-III/Si-XII, as confirmed by selected area diffraction. Extrusion of material at high loads for the cube-corner indenter reduces the volume of transformed material remaining underneath the indenter, thereby eliminating the pop-out in the unloading curve.

INTRODUCTION

The generally recognized sequence for pressure-induced phase transformation of silicon as established in diamond anvil tests is that normal diamond cubic Si (Si-I) transforms to the metallic β-tin structure (Si-II) at a hydrostatic pressure of 11GPa, or at lower pressures when aided by shear stresses [1]. Upon decompression, the β-tin structure transforms to amorphous silicon (a-Si) at high release rates, or to a variety of metastable crystalline forms at low rates 1. Nanoindentation has also proven useful in the characterization of the pressure induced phase transformations of Si [2-14]. Since the hardness of Si is dominated by the pressure needed to induce the phase transformations, metastable phases are frequently observed within and around nanoindentation hardness impressions.

Indentation-induced phase transformations in Si have been studied mostly with spherical indenters and the Berkovich indenter - a triangular pyramid with a centerline-to-face angle of 65.3°. For these indenters, there is general agreement that the "elbow" observed in unloading curves at relatively low maximum loads is caused by the formation of amorphous silicon, whereas the "pop-out" observed at higher loads results from the nucleation and growth of metastable crystalline phases such as Si-III and Si-XII. Numerous studies have documented the post indentation structures of the deformed region of spherical, Vickers and Berkovich indentations using plan-view and cross-sectional TEM [5-13].

Recently, a systematic study by Jang et al. revealed that pop-out is absent when very sharp triangular pyramidal indenters are unloaded from high loads (figure 1a), even though the Raman spectra provide evidence for the existence of crystalline Si-III and Si-XII in the indent (figure 2) 14. This new finding led us to begin a detailed TEM study of the effects of indenter angle on the deformation structures of Si. Initial results of that study are reported here.

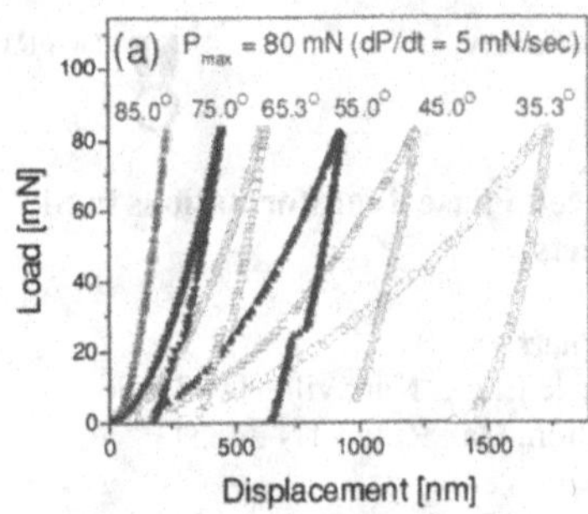

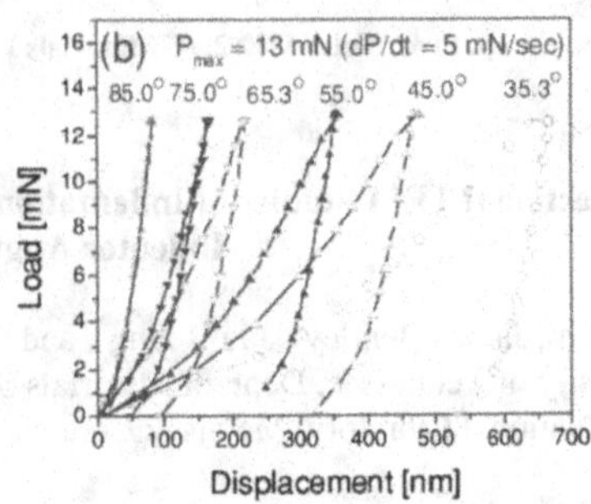

Figure 1. Load-displacement curves for nanoindentations made with triangular pyramidal diamonds with centerline-to-face angles varying from 35° to 85° at a constant loading/unloading rate of 5 mN/s: (a) P_{max} = 80 mN; (b) P_{max} = 13 mN [14].

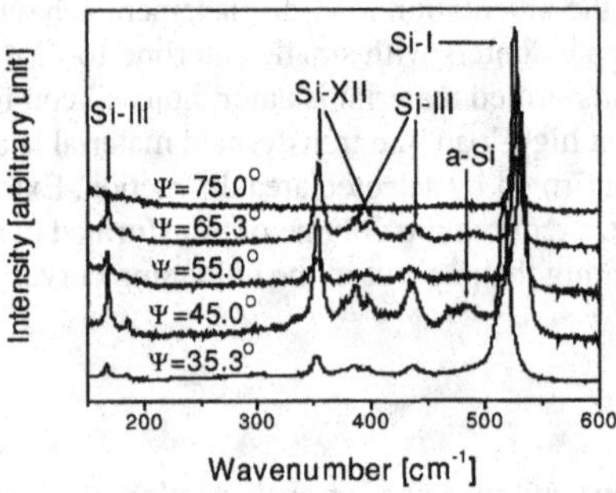

Figure 2. Raman spectra from nanoindents made with different indenters at P_{max} = 80 mN [14].

EXPERIMENTAL

Nanoindentation was conducted at the SHaRE User Center of the Oak Ridge National Laboratory using a Nanoindenter XP (MTS Systems, Oak Ridge TN). Undoped (100) single crystal Si wafer samples were indented at room temperature using a series of triangular pyramidal indenters with centerline-to-face angles of 35.3° (cube-corner), 45°, 55°, 65.3° (Berkovich), 75° and 85°. Two peak loads - 10 mN and 80 mN - were chosen to explore indenter angle effects at relative high and low loads. The loading/unloading rate was fixed at 5 mN/sec to eliminate complications brought about by rate effects. Cross-sections of the indentations were prepared utilizing a dual beam focused ion beam (FIB) mill. Cross-sectional transmission electron microscopy (TEM) and high resolution electron microscopy (HREM) were performed using FEI/Philips Tecnai20 and CM200FEG instruments at the SHaRE User Center. All the observations were made at 200kV.

RESULTS AND DISCUSSION

A. Indenter angle effects at low load

Figure 3 shows general SEM and TEM observations for indentations made at 10 mN with the 35°, 45°, 55°, and 65° indenters. Results for the 85° indenter are not included because it produced entirely elastic deformation (see figure 1). Cross-sections for the 75° indenter are currently in preparation. From the SEM plan views of the indentations, it is clear that residual depth increases as the indenter angle increases, in agreement with the indentation load-displacement curves (figure 1b). The surface radial cracking is most pronounced for the sharper indenters but is not present for indenter angles of 65° and above. The most notable feature is that the 35° cube-corner indenter shows not only the deepest hardness impression, but the transformed material also extrudes from underneath it to form a thin wispy layer on the surface. The cross-sectional TEM images reveal that the shape of the transformed zone is approximately the same for all indentations; however, the amount of transformed material in the zone decreases for the sharper indenters. For the Berkovich indenter (65°), the transformed material remains entirely beneath the indenter because the transformed zone does not extend beyond the edge of contact. The Berkovich indent shows no evidence of radial or median cracking, either in plan view or in cross-section. Cracks are observed for all other indenter angles, and material from the transformed zone flows into the median cracks. For the cube-corner indentation, the region of transformed material under the hardness impression is smaller than the others due to the extrusion of material to the surface.

Although the Berkovich indenter is frequently used for studies of this kind, indentations at low loads, e.g. 10 mN, have never been documented using cross-sectional TEM. The bright field image of a 10 mN Berkovich indent in figure 4a exhibits a large transformed zone (gray) with a few bright features, bend contours outside the transformed zone, and defects at the bottom of the indentation. Selected area diffraction (SAD) from the transformed zone (figure 4b) indicates

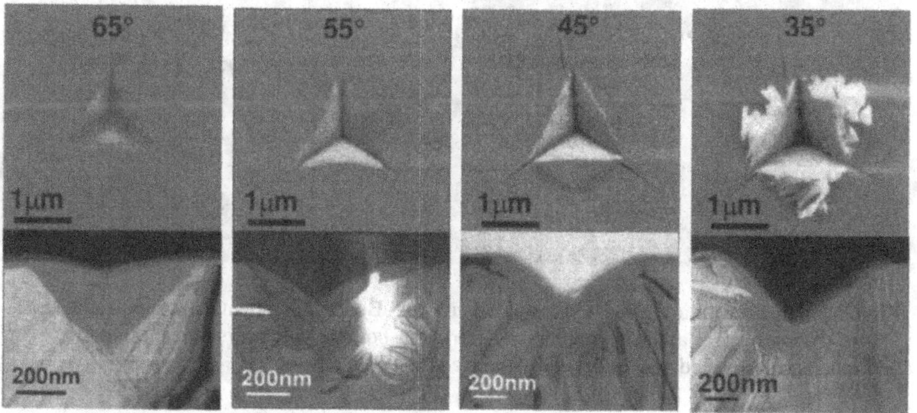

Figure 3. SEM plan-view and TEM cross-sectional images of low load indentations (10 mN) made with various triangular pyramidal indenters.

that material in the zone is mostly amorphous. The spot patterns in the SAD are produced by the surrounding untransformed Si-I matrix. There is dislocation activity underneath the transformed zone, which is further revealed by a dark field image (DF) taken with a Si-I (111) reflection (figure 4c). High resolution electron microscopy (HREM) was carried out at the crystalline-to-amorphous boundary and within the transformed zone (figure 5a). Microtwins were detected that are most extensive under the apex of the amorphous pocket and at the crystalline-to-amorphous boundary (figure 5b). Nanocrystalline grains of Si-I with typical dimensions between 5 and 20 nm are observed within the transformed region. However, observed crystallization of a-Si under the beam during HREM (figure 5c) gives rise to the concern that the original nanocrystals in the a-Si zone (figure 4a) might be artifacts caused by the ion beam during sample preparation. TEM examination of a low-load cube-corner indent is currently in progress.

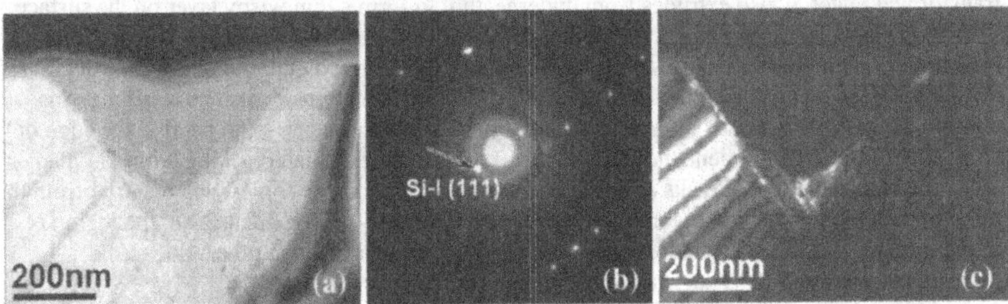

Figure 4. Cross-sectional TEM of a low load (10 mN) Berkovich indent: (a) bright field image; (b) SAD from the transformed zone; (c) dark field image from a Si-I (111) reflection.

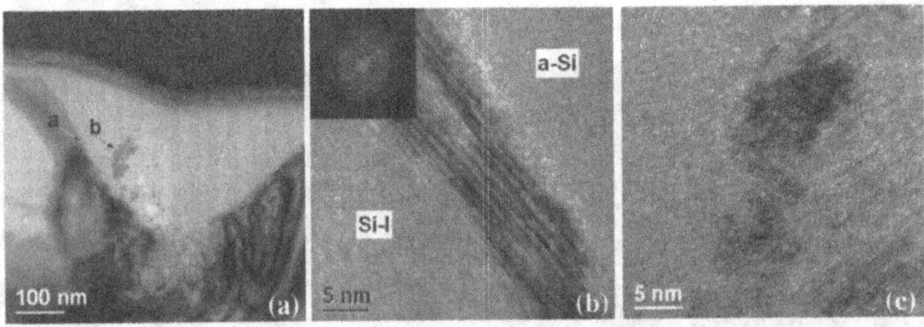

Figure 5. High resolution TEM of a low-load (10 mN) Berkovich indent: (a) bright field TEM image; (b) microtwins at the crystalline-to-amorphous boundary; (c) crystallization in the amorphous transformed zone induced by the electron beam.

B. Indenter angle effects at high load

Unlike the low-load condition, both the high-load Berkovich and cube-corner indentations made at 80 mN show radial cracks, lateral cracks, and median cracks (figure 6). The Berkovich indentation exhibits a well-capped transformed zone that is generally nanocrystalline in nature (figure 6b). The structure was confirmed by SAD to be mixture of crystalline Si-I, Si-III/Si-XII, which is in good agreement with Raman data (figure 2) and other microscopy results [9-13]. In contrast, the high-load cube-corner indentation shows extensive extrusion of the transformed material and very little remaining under the hardness impression. Since it is the volume change during phase transformation that gives rise to the pop-out event during unloading, the extrusion explains the absence of the pop-out event for the cube-corner indenter. This is a common feature for the 35° indenter and is occasionally observed for 45° indentations as well. It is noted that the appearance of the extruded material again strongly indicates the ductile and metallic-like nature of the transformed material under pressure.

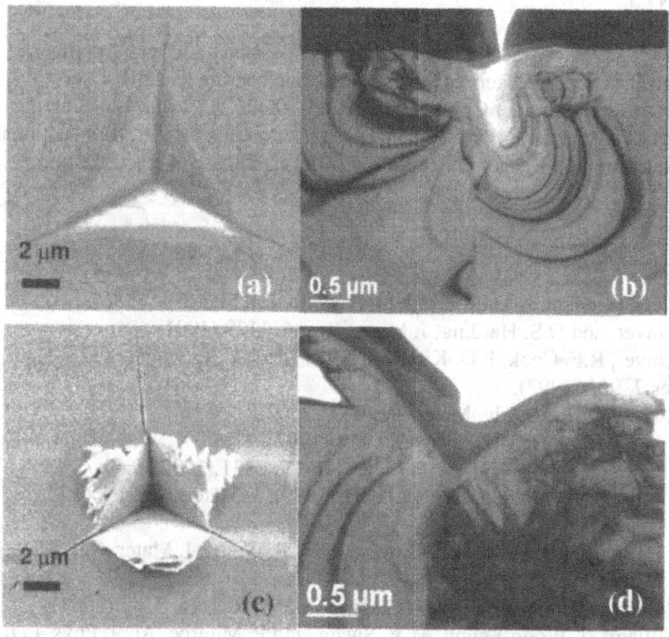

Figure 6. SEM plan-view and TEM cross-sectional images of high load indentations (80 mN): (a) and (b) Berkovich indentation; (c) and (d) cube-corner indentation.

CONCLUSIONS

Cross-sectional TEM and plan-view SEM observations show that cracking and extrusion in (100) Si are more prevalent for indenters with sharp centerline-to-face angles. At low loads (10 mN), the transformed material is confirmed to be amorphous Si for all indenter angles. Cracking is not observed when indenting with the Berkovich indenter at low loads, but at high loads (80 mN), cracking is observed for all indenters with angles in the range 35°-65°. Selected area diffraction confirms that the transformed material remaining underneath the high-load Berkovich indents is a nanocrystalline mixture of Si-I and Si-III/Si-XII. Extrusion of transformed material into the median cracks is a common feature for all indenters. Cross-sectional TEM explains the absence of pop-out for cube-corner indenters. Specifically, because the transformed zone extends beyond the edge of contact, there is extensive extrusion of transformed material and little remains under the hardness impression to reverse transform during unloading.

ACKNOWLEDGEMENTS

This research was sponsored by the National Science Foundation under grant number DMR-0203552, and by the Division of Materials Sciences and Engineering (SHaRE User Center), U. S. Department of Energy, under Contract DE-AC05-00OR22725 with UT-Battelle, LLC. The authors would like to thank Dr. Ting Tsui at Texas Instruments for the cross-sectional TEM specimen preparation, and Dr. Michael Lance for micro-Raman phase characterization.

REFERENCES

1. J.Z. Hu, L.D. Merkle, C.S. Menoni, and I.L. Spain, Phys. Rev. **B34**, 4679 (1986).
2. G.M. Pharr, W.C. Oliver, and D.S. Harding, J. Mater. Res. **6**, 1129 (1991).
3. G.M. Pharr, W.C. Oliver, R.F. Cook, P.D. Kirchner, M.C. Kroll, T.R. Dinger, and D.R. Clarke, J. Mater. Res.**7**, 961 (1992).
4. V. Domnich and Y. Gogotsi, Rev. Adv. Mater. Sci. **3**, 1 (2002).
5. D. L. Callahan and J. C. Morris, J. Mater. Res. **7**, 1614 (1992).
6. T.F. Page, W.C. Oliver, and C.J. McHargue, J. Mater. Res. **7,** 2431 (1992).
7. Y.Q. Wu, X.Y. Yang, Y. B. Xu, Acta Mater. **47**, 2431 (1999); J. Mater. Res. **14,** 682 (1999).
8. A.B. Mann, D. van Heerden, J.B. Pethica, P. Bowes, and T.P. Weihs, J. Mater. Res. **15,** 1754 (2000).
9. S.J. Lloyd., J.M. Molana-Aldareguia, and W.J. Clegg, J. Mater. Res. **16,** 3347 (2001).
10. J.E. Bradby, J.S. Williams, J. Wong-Leung, M.V. Swain, and P. Munroe, Appl. Phys. Lett. **77**, 3749 (2000); J. Mater. Res. **16**, 1500 (2001).
11. H. Saka, A.S. Himatani, M. Suganuma, and Suprija, Phil. Mag. **A82**, 1971 (2002).
12. D.B. Ge, V. Domnich, and Y. Gogotsi, J. Appl. Phys. **93**, 2418 (2003).
13. I. Zarudi and L.C. Zhang, Appl. Phys. Lett. **82**, 874 (2003); J. Mater. Res. **19**, 332 (2004).
14. J-I. Jang, M.J. Lance, S.Q. Wen, T.Y. Tsui, and G.M. Pharr, Acta Mater. (in press)

Mater. Res. Soc. Symp. Proc. Vol. 843 © 2005 Materials Research Society T6.5/R10.5

In-Situ Infrared (IR) Detection and Heating of the High Pressure Phase of Silicon during Scratching Test

Lei Dong[2], John A. Patten[1] and Jimmie A. Miller[2]
[1]Manufacturing Research Center, Western Michigan University,
Kalamazoo, MI 49008, U.S.A.
[2]Center for Precision Engineering, University of North Carolina at Charlotte,
Charlotte, NC 28223-0001, U.S.A.

ABSTRACT

A novel method of in-situ detection of the high pressure phase transformation of silicon during dead-load scratching is described. The method is based on the simple fact that single crystal silicon is transparent to Infrared light while metallic materials are not. Infrared heating during scratching has been performed to thermally soften and deform the transformed metallic material and some promising results were obtained. The sample material used here is silicon, but the same approach can be applied to germanium and other materials, such as ceramics (SiC), which have appropriate optical properties.

INTRODUCTION

It has been well established that single crystal silicon undergoes a phase transformation from the diamond cubic structure to the β-tin structure at a hydrostatic pressure of 10~13 GPa [1, 2]. Direct evidence of this transition includes conductivity measurement during indentation experiments in silicon [3-5]; the displacement discontinuity phenomenon in load-displacement curves during indentation serves another confirmation [6, 7].There are also ex-situ measurements such as transmission electron microscopy (TEM) analysis, Raman spectroscopy and X-ray diffraction analysis conducted after indentation of silicon [8-10]. In the current work, a different approach than any other method mentioned above was shown to provide an easy and valuable way to identify the occurrence of the high pressure phase transformation based upon in-situ measurements. The objective of the work is to develop a new concept for semiconductor material machining. It involves preferentially heating and softening, with an IR laser, the high pressure metallic zone to achieve tool wear reduction as well as enhanced ductile machining. Laser-assisted machining has been applied to material processing in which high power UV (watts to kilowatt) lasers are used to thermally soften the bulk material right before machining [11]. Such methods inevitably heat the bulk material. In this paper, a microscopic (μm) laser heating system rather than a macroscopic (mm) system is employed. Also the microscopic laser heating system has been designed to avoid the unnecessary heating of the surrounding material. The experimental method preferentially heats only the contact zone, between the silicon and the diamond tool, instead of a large area around the tool.

IR IN-SITU DETECTION OF METALLIC HIGH PRESSURE PHASE TRANSFORMATION IN SILICON

Due to the nature of high pressure (GPa) phase transformation in brittle materials and small size scale (micrometers), in-situ study of this phenomenon has been very difficult. Developing a new instrument or modifying a current instrument to make such an in-situ measurement would take considerable time and effort. In this paper an IR laser-detector system is used in conjunction with a pre-loaded (dead weight) scratching device (surface profilometer) to detect the metallic high pressure phase of Silicon in-situ. The phenomenon behind this simple experiment might inspire some new thoughts and novel applications.

EXPERIMENTAL DESIGN

The experimental setup is shown in figure 1. The setup consists of a modified profilometer (Surfcom110B) with measuring speed of 0.305mm/sec, a profilometer stylus which is modified to hold the laser ferrule. A 90 degree cone diamond tip (5μm nominal radius) mounted on the ferrule end using epoxy is used as the scratching tool, see figure 2. The single crystal silicon wafer is a p type (100) 4 inch diameter wafer with a thickness of 475~575 micrometer.

The apparatus is designed in such a way that IR light passes through the diamond tip onto the interface between the diamond tip and the silicon as the diamond tip is translated on the silicon wafer. An IR diode laser (8mW laser with a wavelength of 1330 nm) and a corresponding detector (INGAAS detector with 3000 μm DIA, 900-1700nm sensitivity range) are used as the illumination and detection devices respectively. The diamond tip is coated with gold (about 2000 Angstroms thick) to minimize stray radiation from being emitted from the ferrule end and diamond stylus. The laser power was reduced to several micro watts (from the peak of 8mW) at the silicon surface, due to absorption and scattering of the laser radiation. Figure 3 and 4 are the SEM images of the gold coated diamond tip before and after scratching test. The area which has the gold removed, at the tip of the diamond as shown in the center of figure 4, is about 4 μm in diameter. This small area represents the aperture of the illumination system; i.e. the IR radiation mostly emerges from this small area. This area also represents the contact area between the diamond tip and the silicon wafer. Loads used in the reported experimental program are 20, 30, 40 and 50 mN.

Fig. 1. Experimental setup.

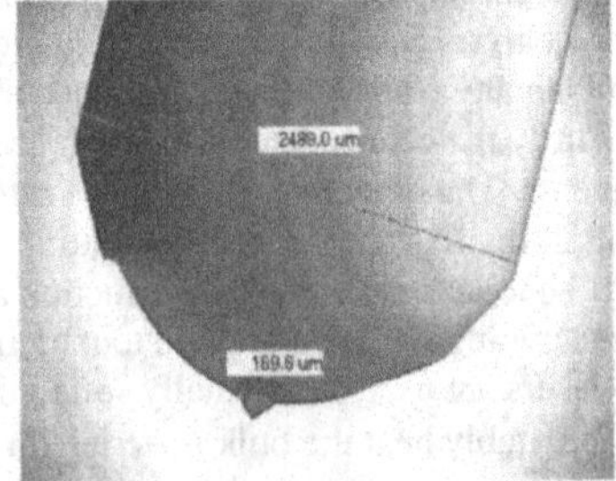

Fig. 2. Diamond tip attached on the ferrule.

EXPERIMENTAL PROCEDURES AND RESULTS

The IR detection experiment is carried out as shown in figure 5. A pre-determined length of scratch (~4mm) is made across the silicon wafer. This scratch coincides with the diameter of the

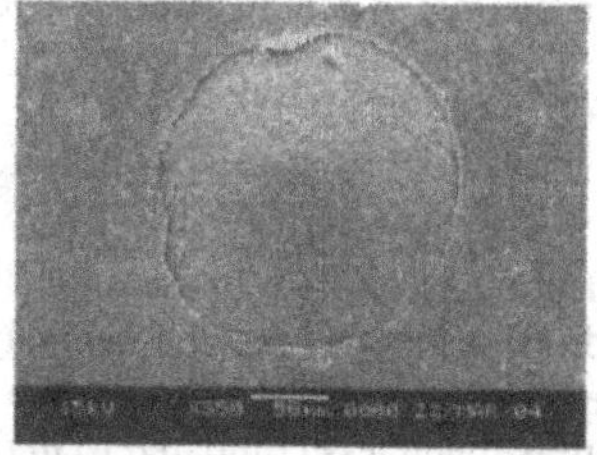
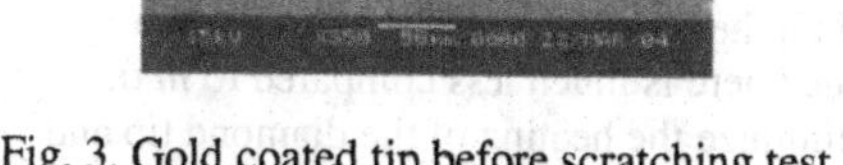

Fig. 3. Gold coated tip before scratching test.

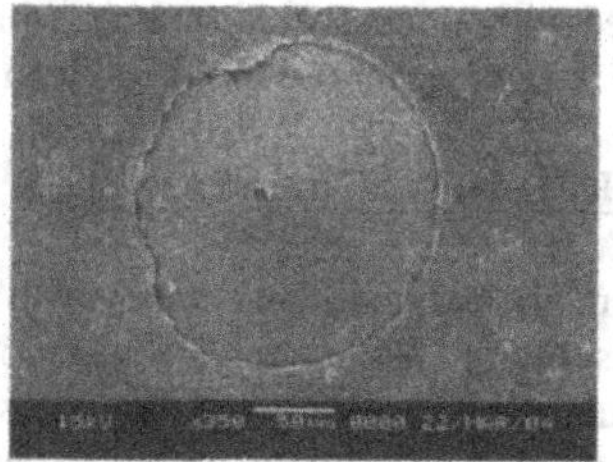

Fig. 4. Gold coated tip after scratching test.

IR detector fixed beneath the wafer. The loading direction is perpendicular to the wafer's surface. Scratching direction is [110]. The detector measures the IR laser signal transmitted through the silicon wafer and the signal is collected using Labview software.

Figure 6 shows a series of 4 scratches at 20mN load. The 0 mN scratch curve represents the nominal preload of the surface profilometer which is about 0.7 mN. The other subsequent curves correspond to scratches that are done in the same track on top of the zero weight scratch. For the various loads of 20mN, 30mN, 40mN, 50mN (only the results for the 20mN scratch are presented), the voltage of IR transmitted signal decreases after each scratch and it is observed that there is about 10% to 15% signal decrease after the 3rd scratch, compared to the zero weight scratch. The metallic Si-II is presumed to result in reduced transmission of radiation at IR wavelengths. The measurements obtained above give direct evidence of the existence of the high pressure metallic phase of silicon during scratching and this correlates well with previous reported and related work utilizing electrical detection [12].

INFRARED HEATING OF METALLIC ZONE DURING SCRATCHING TEST

The metallic deformed zone is presumed to have blocked some of the IR radiation in the experiments described above. This attenuated radiation is either absorbed or reflected by the metallic zone. If this zone can absorb the IR wavelength, then the material will be heated and possibly softened, given a sufficient amount of radiation. By measuring the depth of the scratching groove after laser heating, it is possible to determine any change in deformation

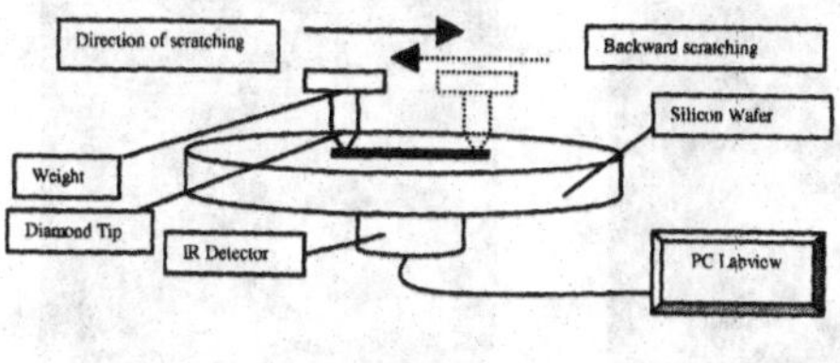

Fig. 5. Schematic of scratching procedure.

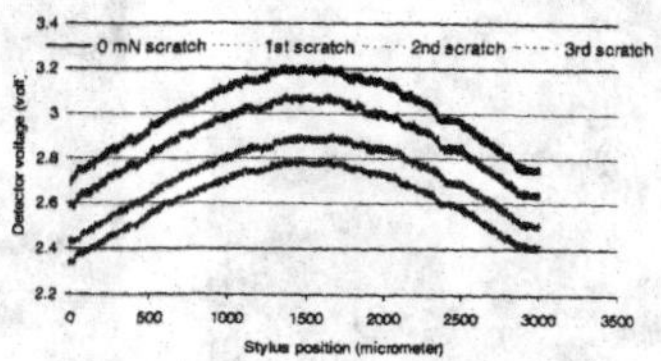

Fig. 6. Load 20mN scratch.

compared with a non-heated scratch, which could imply a thermal softening effect due to an increase in temperature of the metallized silicon.

INFRARED HEATING APARATUS

Several modifications were necessary to build the laser heating system. In this setup, a 400mw diode laser with 1480nm wavelength replaced the 8mw diode laser considered previously. A similar diamond tip as previously used is mounted on the ferrule using UV-Epoxy (80% transmittance to IR) and the amount of epoxy applied here is much less compared to in the detection system (figure 7). This is necessary to minimize the heating of the diamond tip and epoxy and maximize the IR radiation coming out of the tip. In this case the diamond tip is not coated with gold. The IR detector is also removed from the system. The scratching procedures remain the same.

BEAM SHAPE AND POWER LOSS OF THE LASER AFTER ATTACHING THE DIAMOND TIP

Figure 8 and 9 show the laser beam profile coming out of the end of the tip. Figure 8 is focused on the tip and there is an expected normal intensity distribution of the radiation, with the center having the highest value. The diameter of the central beam area measures about 10μm, which corresponds to the beam size. Figure 9 shows an image of the radiation distribution when the camera lens is out of focus, which partially explains the loss of the laser power, as some of the power is not within the well defined central core of the laser beam. Over half of the power is lost through the diamond tip attachment measured by laser power meter. UV-epoxy absorption and the high reflectivity of the diamond are suspected as the main reasons to cause the power loss mentioned above.

EXPERIMENTAL RESULTS

Two sets of experiments were designed in which only an intermediate load (25 mN) was used. Figure 10 is the results from the 'scratch and stay 'test. When a single dead-load scratching is finished, the tip "stays where it stops" for varying periods of time (30, 60,120 and 180 seconds). Different levels of laser power also acts as a variable in this test to determine if the metallic zone

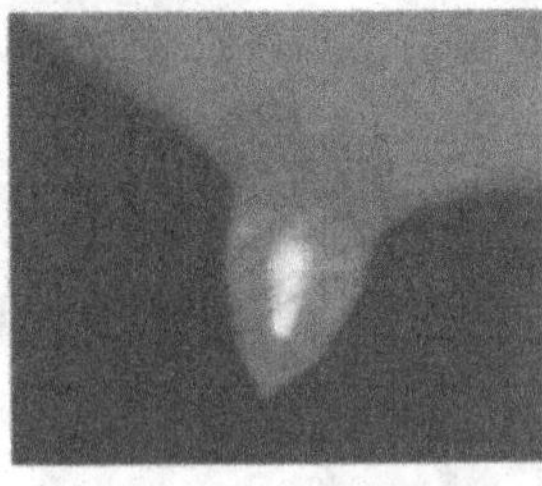

Fig. 7. Diamond tip attached on the ferrule.

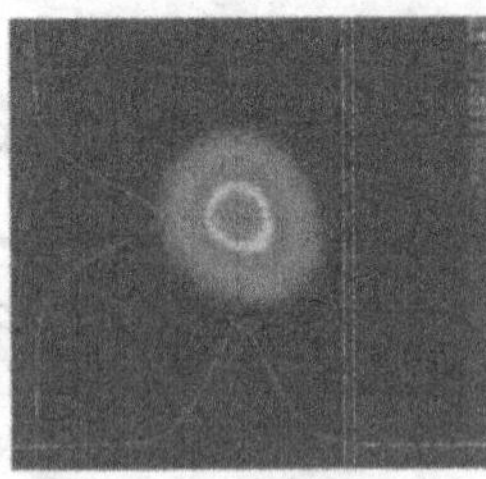

Fig. 8. Profile of the laser beam (in focus).

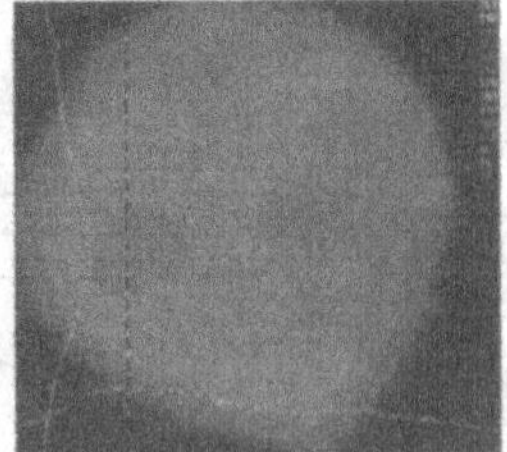

Fig. 9. Profile of the laser beam (out of focus).

absorbs more heat, leading to enhanced ductility, when given more radiation. Results in figure 10 shows that at the longest heating period the depth of the impression is the largest. It also shows that the higher the laser power, the deeper the impression. In complementary scratching speed test (figure 11), two slower speeds than the original scratching speed (0.305mm/sec) were tried. The results show significant changes especially in the case of the slowest speed. Changing either the speed or the laser driving current results in a measurable change in the thermal softening effect associated with laser heating. If the metallic zone absorbs enough heat then the thermal softening effect is measurable.

DISCUSSION AND CONCLUSION

As the loaded diamond tip scratches the silicon wafer, a layer of the high pressure metallic phase, about the thickness of the groove depth, is generated underneath the diamond tip. This metallic phase only exists during the contact, i.e. as long as the high pressure is present. As soon as the tip moves on, the material behind the diamond tip back transforms to an amorphous structure. When the next scratch occurs, the high pressure phase is reformed under the diamond tip. As the groove gets deeper and wider, the metallic zone also grows in extent (scaling with the groove size) resulting in greater attenuation of the IR laser beam. The decreasing detector signal voltage, with subsequent scratches, matches this theory. If this were not the case, the IR laser transmitted signal would actually increase with subsequent scratches as more of the gold coating would be removed (corresponding to the larger groove dimensions). This larger aperture would allow correspondingly more of the IR laser to emerge from the diamond tip. The fact that the IR signal decreases, in light of the above, clearly demonstrates the existence of the increasing size of the metallic high pressure phase of silicon under the diamond stylus tip. This also agrees with earlier work on electrical heating of the transformed metallic phase [13].

In the IR heating test, the layer of metallic phase is generated under high pressure and heated by the laser simultaneously as the scratching motion occurs. Only the material that is in contact with the tool (a diamond tip in these experiments) and a thin layer below that area is heated, which shows advantages to other laser-assisted machining processes.

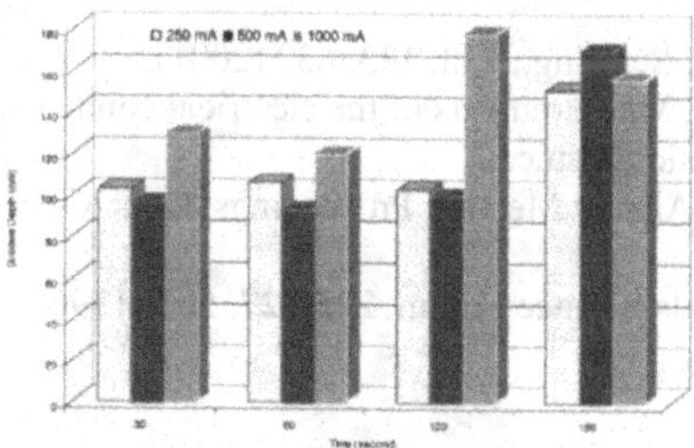

Fig. 10. Scratch and Stay Test (Scratching groove measurements on AFM).

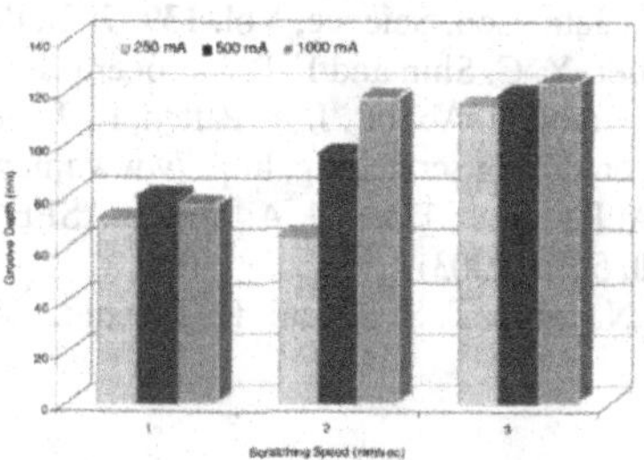

Fig. 11. Scratching Speed Test (Scratching groove measurements on AFM, Speed 1: 0.305, Speed 2: 0.02 Speed 3: 0.002).

Hardness can be calculated using the measured groove depth. For the case of scratching at 25mN load, 0.002mm/sec scratching speed and 1000 mA laser driving current, the calculated hardness is 9.5 Gpa, which corresponds to a temperature around 500 C [14]. It has been shown experimentally that at temperatures above 0.3 to 0.4 T_{ml}, the micro hardness drops abruptly with increasing temperature [3]. Temperatures in the range of 400 to 600 C would substantially soften silicon and promote enhanced ductile machining with potentially drastically reduced cutting forces and less tool wear. The laser heating system described in this paper is capable of producing these desired results. Micro Raman analysis of the deformed zone in the scratching groove will be carried out in the near future to further the characterization of the phase transformation.

ACKNOWLEDGEMENT

The authors acknowledge the support from National Science Foundation Focused Research Group program, DMR #0203552 and #0403650, Lynnette Madsen Program Director. The authors would also like to thank Jay Mathews (Digital Optics Corporation, Charlotte, NC,28262) for technical help and Scott Williams, M. Yasin Akhtar Raja (Optical Center, UNCC, Charlotte,NC,28223) for laser equipment.

REFERENCES

1. R. F. Cook and G. M. Pharr, J. Am. Ceram. Soc., Vol.**73,** 787 (1990).
2. R. J. Needs and A. Mujica, Phys. Rev. **B51**, 9652 (1996).
3. I. V. Gridneda, Yu. V. Milman, V. I. Trefilov, Phys. Stat. Sol. (a) **14**, 177 (1972).
4. D. R. Clarke, M. C. Kroll, P. D. Kirchner, R. F. Cook and B. J. Hockey, Phys. Rev. Lett., **21**, 2156 (1988).
5. G. M. Pharr, W. C. Oliver, R. F. Cook, P. D. Kirchner, M. C. Kroll, T. R. Dinger and D. R. Clarke, J. Mater. Res., Vol. **7**, No. **4**, 961 (1992).
6. G. M. Pharr, W. C. Oliver and D. R. Clarke, Scripta Metall., **23,** 1949 (1989).
7. G. M. Pharr, W. C. Oliver and D. R. Clarke, J. Elec. Mater., **19**, 881 (1990).
8. D. L. Callahan and J. C. Morris, J. Mater. Res., Vol. **7**, No. **7**, 1614 (1992).
9. A. Kailer, Y. G. Gogotsi and K. G. Nickel, J. Appl. Phys., **81**(7), 3057 (1997).
10. J. C. Jamieson, Science, Vol. **139**, 762 (1963).
11. S. Lei, Y. C. Shin and F. P. Incropera, J. Manuf. Sci. Eng., Vol. **123**, 639 (2001).
12. N. Hirata, D. Abbott, J. A. Patten, R. J. Hocken, Measurements of the electrical conductivity of Silicon during scratching, http://www.micro.physics.ncsu.edu.
13. J. A. Patten, L. Dong, J. A. Miller, ASPE 2003 Annual Meeting Proceedings, Oct. 26-31, Vol. **30**, 507 (2003).
14. M. Nagumo, T. S. Suzuki, K. Tsuchida, Materials Science Forum, **225-227**, 581 (1996).

Mater. Res. Soc. Symp. Proc. Vol. 843 © 2005 Materials Research Society T6.9/R10.9

Micro-Raman Mapping and Analysis of Indentation-Induced Phase Transformations in Germanium

Jae-il Jang [1], M.J. Lance [2], Songqing Wen [1], J.J. Huening[3], R.J. Nemanich[3], and G.M. Pharr [1,2]
[1] The University of Tennessee, Dept. of Mater. Sci. & Eng., Knoxville, TN 37996-2200, USA.
[2] Oak Ridge National Laboratory, Metals and Ceramics Division, Oak Ridge, TN 37831, USA.
[3] North Carolina State University, Dept. of Physics, Raleigh, NC 27695, USA.

ABSTRACT

Although it has been confirmed by diamond anvil cell experiments that germanium transforms under hydrostatic pressure from the normal diamond cubic phase (Ge-I) to the metallic β-tin phase (Ge-II) and re-transforms to Ge-III (ST12 structure) or Ge-IV (BC8 structure) during release of the pressure, there are still controversies about whether the same transformations occur during nanoindentation. Here, we present new evidence of indentation-induced phase transformations in germanium. Nanoindentation experiments were performed on a (100) Ge single crystal using two triangular pyramidal indenters with different tip angles - the common Berkovich and the sharper cube-corner. Although the indentation load-displacement curves do not show any of the characteristics of phase transformation that are well-known for silicon, micro-Raman spectroscopy in conjunction with scanning electron microscopy reveals that phase transformations to amorphous and metastable crystalline phases do indeed occur. However, the transformations are observed reproducibly only for the cube-corner indenter.

INTRODUCTION

Together with silicon, germanium is one of the most important materials in the electronic industry, and thus its mechanical behavior is of a considerable interest from both scientific and engineering viewpoints. Following a number of theoretical studies and diamond anvil cell experiments (see review articles [1-2]), it is now well recognized that pristine Ge-I with its diamond cubic structure transforms to the metallic β-tin phase (Ge-II) at about 10 - 11 GPa, and, upon unloading, the Ge-II transforms to Ge-III (ST12, a simple tetragonal structure with 12 atoms in the unit cell) or Ge-IV (BC8, a body-centered cubic structure with 8 atoms in the unit cell), depending on the pressure release rate. Since these transformations are broadly analogous to those occurring in silicon, one might expect the well-known indentation-induced phase transformations for silicon to also occur in Ge. The structural similarities of these two materials also support this expectation. However, whether or not a phase transformation can actually occur during the nanoindentation of Ge is still controversial. While some experimental evidence for transformation has been reported for high load indentations performed mainly using a Vickers indenter [3-6], transformed phases have very rarely been observed reproducibly in Ge nanoindentations, as reviewed by Domnich and Gogotsi [1]. Moreover, using cross-sectional transmission electron microscopy, Bradby et al. [7] have found that severe twinning, rather than phase transformation, is the primary deformation mechanism in the spherical nanoindentation. As a result, they have suggested that, unlike Si, the hardness of Ge is not determined by the

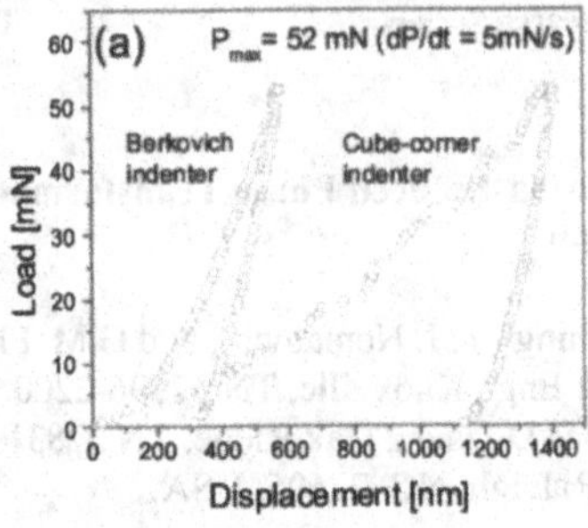

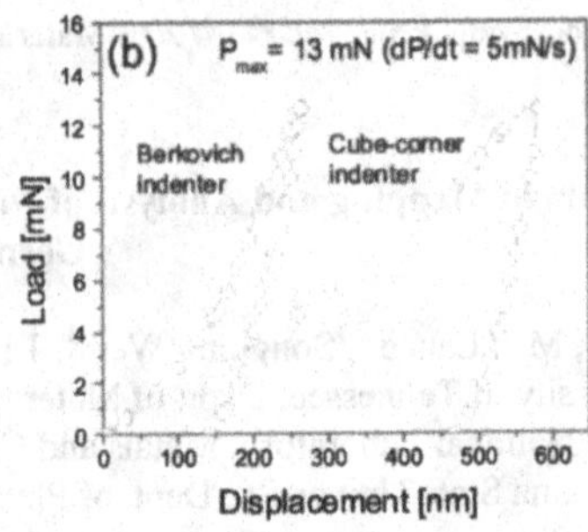

Figure 1. Nanoindentation load-displacement curves made in (100) Ge with cube-corner and Berkovich indenters at different peak loads; P_{max} = (a) 52 mN and (b) 13 mN.

phase transformation pressure. Very recently, Patriarche et al. [8] found the Ge-I and Ge-III (ST12) phases form in the residual indents of a-Ge thin films as observed in plan-view transmission electron microscopy. However, these phases result from what is probably an entirely different process involving the indentation-induced crystallization of an amorphous material. The current study was undertaken to explore other possible evidence for nanoindentation-induced phase transformation in Ge.

EXPERIMENTAL DETAILS

Nanoindentations were made on a standard (100) wafer Ge using a Nanoindenter-XP (MTS System Corp., Oak Ridge, TN). Based on the observation that a wide variety of transformation behaviors are observed in Si depending on indenter sharpness [9], two triangular pyramidal indenters were employed - the Berkovich and cube-corner indenters which have centerline-to-face angles of 65.3° and 35.3°, respectively. Indentation peak loads were varied in the range 10 to 100 mN at a fixed loading/unloading rate of 5 mN/sec. After nanoindentation testing, micro-Raman analyses were conducted within one hour using a Dilor XY800 Microprobe (JY Inc., Edison, NJ) and an Ar^+ laser operating at an excitation line of 5145 Å. To systematically characterize any new crystalline and amorphous phases that may have formed during indentation, Raman spectral maps of the indents were acquired using mapping software and a mesh size of 1 μm × 1 μm. The beam intensity was kept low to avoid possible artifacts induced by laser heating. All the hardness impressions were imaged using a Leo 1525 field-emission scanning electron microscope (SEM) (Carl Zeiss SMT Inc, Thornwood, NY) to identify important topographical features.

RESULTS AND DISCUSSION

Figure 1 shows examples of load-displacement (P-h) curves typically observed during nanoindentation with the cube-corner and Berkovich indenters. The sharper cube-corner

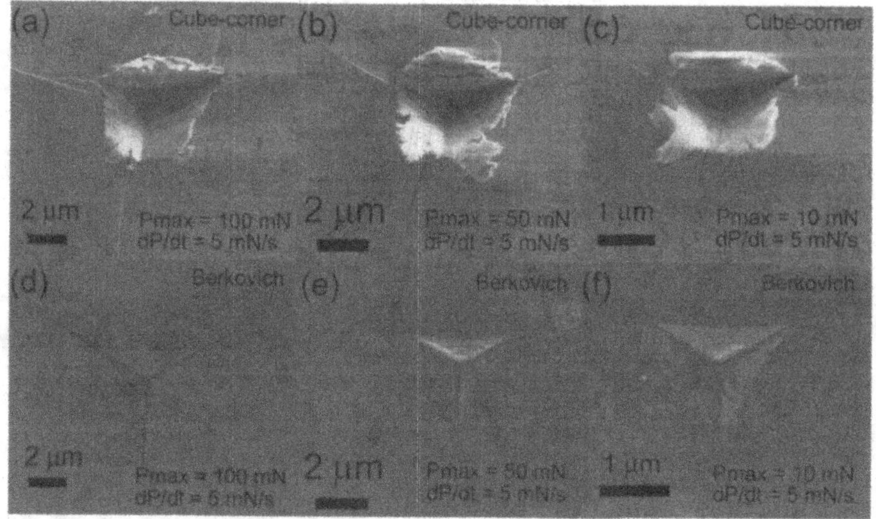

Figure 2. SEM micrographs of nanoindentation hardness impressions in (100) Ge, made at different indentation loads with (a)-(c) cube-corner indenter and (d)-(f) Berkovich indenter; Indentation peak load (Pmax): (a) and (d) 100 mN; (b) and (e) 50 mN; (c) and (f) 10 mN. Note that magnifications of the micrographs are not all the same.

indentations exhibit a higher displacement at peak load and a larger proportion of irreversible plastic deformation than the Berkovich indentations. When compared to similar curves for silicon, the most noteworthy feature is the lack of any unusual features in the unloading curve, particularly the "pop-out" and "elbow" behaviors. A number of pop-ins (discontinuities in loading sequence) are observed for the cube-corner indentations, especially at high loads. These most likely correspond to cracking and chipping rather than transformation since Ge has a very low fracture toughness, 0.6 MPa$(m)^{1/2}$, about half that of silicon. The P-h curves of indentations performed at loads lower than the first pop-in load (figure 1(b)) exhibited a large amount of plasticity, which would not be expected if the pop-ins corresponded to the initiation of phase-transformation-controlled indentation deformation.

The first solid evidence for phase transformation phenomena was found in SEM observations, as shown in figure 2. In the case of cube-corner indentations (figures 2(a) – 2(c)), thin, extruded material exhibiting metal-like plastic flow behavior is clearly seen at the contact periphery. Since such behavior is possible only when a thin layer of highly plastic material is sandwiched between the diamond indenter and the relatively hard surrounding Ge-I, the extrusions are direct evidence for the presence of a metallic-like phase such as Ge-II. Although this extrusion is well recognized in silicon since Pharr et al. first reported it in 1991 [10], it has not been observed in Ge. In contrast to cube-corner indentations, Berkovich indentations (figures 2(d) – 2(f)) showed no extruded material, although unusual deformed zones with a rough mottled appearance were observed within the hardness impressions. It has been suggested that these deformed zones represent the transformation from Ge-I to Ge-II [11], but it is difficult to state conclusively

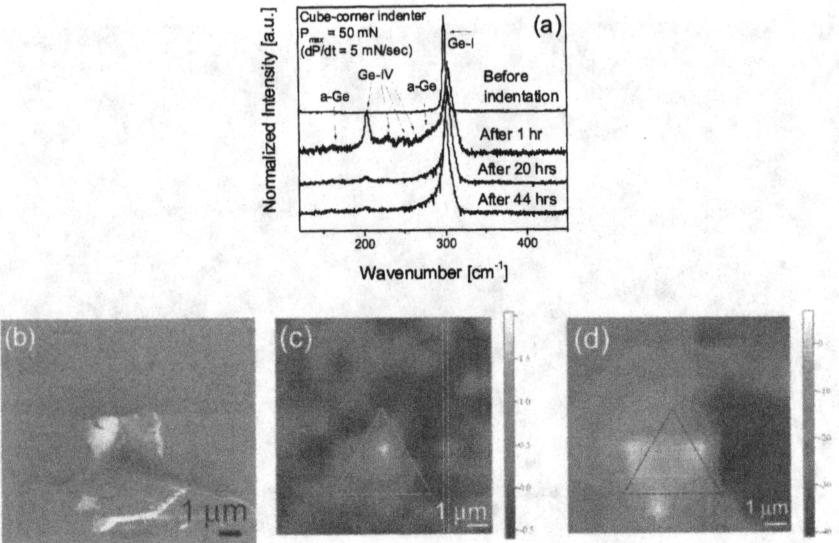

Figure 3. Results from cube-corner indentations made at P_{max} = 50 mN and dP/dt = 5 mN/sec: (a) typical Raman spectra; (b) SEM image; (c) micro-Raman map identifying the location of the transformed crystalline phase Ge-IV; (d) micro-Raman map identifying the location of the amorphous phase. In (c) and (d), the lighter regions correspond to the phase used to create the map. Creating these maps was started from 1 hour after the indentation test. Note that, in the Raman data, the Ge-IV phase was tentatively identified.

that the zones were formed by transformation rather than other deformation mechanisms such as twinning- or dislocation-mediated plasticity under highly constrained deformation conditions.

Additional strong evidence for the transformation was obtained through micro-Raman spectroscopy, as shown in figure 3(a). While pristine Ge-I exhibits only one sharp peak at 300 cm^{-1}, the micro-Raman spectra measured within one hour after cube-corner indentation show many narrow peaks at 205, 230, 250, and 264 cm^{-1} and broad bands around 150 and 270 cm^{-1}. The broad bands can be identified as amorphous Ge [6, 8], but the narrow peaks between 200 and 270 cm^{-1} represent a new crystalline phase. The peaks are very similar to ones observed during diamond anvil experiments by Hanfland and Syassen [12], who pointed out that they are strikingly similar to those for BC8-Si (Si-III). For this reason, we tentatively identify them as Ge-IV, for which the Raman spectrum is not currently known. Further examination of figure 3(a) shows that the peaks diminish substantially within 20 hours and are virtually absent after 44 hours. This is consistent with the observations of Nelmes et al. [13], who reported using synchrotron x-ray diffraction that the Ge-IV phase produced in diamond anvil cell experiments vanishes within 17 hours at room temperature and ambient pressure. By making micro-Raman maps with either the amorphous peak or the Ge-IV peak, the spatial locations of each of these phases are shown in figures 3(c) and 3(d). Comparing these maps with the SEM micrograph of

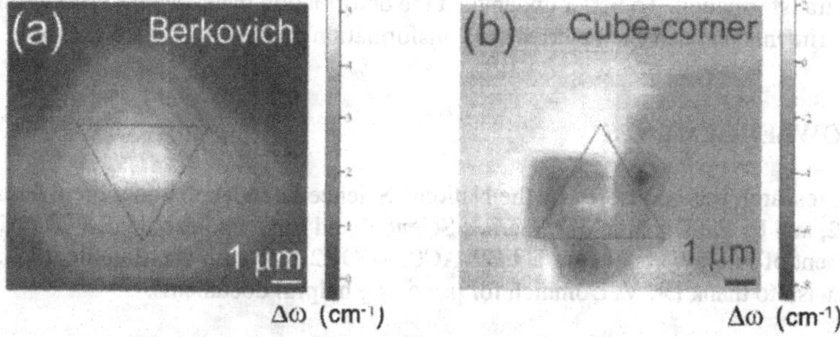

Figure 4. Micro-Raman maps showing the shift of the Ge-I peak: (a) Berkovich indentation and (b) cube-corner indentation. Both indentations were performed to a peak load of 50 mN at a loading/unloading rate of 5 mN/sec.

figure 3(b) shows that the extruded material is predominantly a-Ge, and the Ge-IV resides near the center of the hardness impression.

For Berkovich indentations made under similar loading conditions (not shown here), no crystalline peaks except for Ge-I were typically observed, although occasionally a peak corresponding to a-Ge or the Ge-IV phase was identified. These results are in agreement with those of Gogotsi et al. [14], who sometimes, but not reproducibly, could find crystalline Raman peaks for metastable phases in Berkovich indentations made at a high loading rate.

Another interesting observation is shown in the micro-Raman maps of Berkovich and cube-corner indentations in figure 4, which were constructed based on the frequency shift ($\Delta\omega$) of the primary Ge-I peak. This shift is directly related to the local residual stress; shifts to higher frequencies correspond to compressive stresses and lower frequencies to tensile stresses. The map for the Berkovich indentation in figure 4(a) reveals that the residual stresses within the indent are primarily compressive. Curiously, however, the image of the cube-corner indentation in figure 4(b) suggests that there are large tensile stresses near the contact periphery where the extrusions are observed. Since the extruded material is expected to be almost free of stress due to its thin geometry, the shift to lower frequencies is probably not an indication of tensile residual stress but rather that the extrusions are comprised of nanocrystalline Ge-I, which may crystallize from the amorphous phase, either on its own at room temperature or as activated by Raman laser heating [6]. By combining this result with the map shown in figure 3(d), it appears that the extruded material consists of a mixture of a-Ge and nanocrystalline Ge-I. Detailed studies of this and other related phenomena are currently underway.

CONCLUSIONS

Results presented here show that pressure-induced phase transformations do indeed occur during nanoindentation of Ge, but the transformations take place reproducibly only when sharper indenters like the cube-corner are used. This suggests that the larger volume displaced by the cube-corner indenter is important in the transformation process, either by increasing the extent and magnitude of the hydrostatic pressures, or by increasing the shear stresses in a manner that aids the transformation. To better understand the deformation mechanisms, cross-sectional TEM and experiments that can characterize the transformation process in-situ are desirable.

ACKNOWLEDGEMENTS

This research was sponsored by the National Science Foundation under grant number DMR-0203552, and by the Division of Materials Sciences and Engineering (SHaRE User Center), U. S. Department of Energy, under contract DE-AC05-00OR22725 with UT-Battelle, LLC. The authors wish to thank Dr. V. Domnich for providing helpful documents.

REFERENCES

1. V. Domnich and Y. Gogotsi, in *Handbook of Surfaces and Interfaces of Materials Vol. 2*, edited by H. S. Halwa (Academic Press, New York, 2001), p. 355.
2. A. Mujica, A. Rubio, A. Munoz, and R.J. Needs, *Rev. Modern Phys.* **75**, 863 (2003).
3. I. V. Gridneva, Y. V. Milman, and V. I. Trefilov, *phys. Stat. sol.* (a) **14**, 177 (1972).
4. O. Shinomura, S. Minomura, N. Sakai, and K. Asaumi, *Phil. Mag.* **29**, 549 (1974).
5. D. R. Clarke, M. C. Kroll, P. D. Kirchner, and R. F. Cook, *Phys. Rev. Lett.* **60**, 2156 (1988).
6. A. Kailer, K.G. Nickel, and Y.G. Gogotsi, *J. Raman. Spcetrosc.* **30**, 939 (1999).
7. J.E. Bradby, J.S. Williams, J. Wong-Leung, M.V. Swain, and P. Munroe, *Appl. Phys. Lett.* **80**, 2651 (2002).
8. G. Patriarche, E. Le Bourhis, M.M.O. Khayyat, and M.M. Chaudhri, *J. Appl. Phys.* **96**, 1464 (2004).
9. J.-I. Jang, M.J. Lance, S. Wen, T.Y. Tsui, and G.M. Pharr, *Acta Mater.* (in press).
10. G.M. Pharr, W.C. Oliver, and D.S. Harding, *J. Mater. Res.* **6**, 1129 (1991).
11. S.V. Hainsworth, A.J. Whithead, and T.F. Page, in *Plastic Deformation of Ceramics*, edited by R. C. Bradt, C. A. Brookes, and J. L. Routbort (Plenum, New York, 1995), p. 173.
12. M. Hanfland and K. Syassen, *High Pressure Res.* **3**, 242 (1990).
13. R.J. Nelmes, M.I. McMahon, N.G. Wright, D.R. Allan, and J.S. Loveday, *Phys. Rev. B* **48**, 9883 (1993).
14. Y.G. Gogotsi, V. Domnich, S.N. Dub, A. Kailer, and K.G. Nickel, *J. Mater. Res.* **15**, 871 (2000).

Mater. Res. Soc. Symp. Proc. Vol. 843 © 2005 Materials Research Society T6.11/R10.11

Mechanisms Underlying Hardness Numbers

John J. Gilman
Materials Science and Engineering
University of California at Los Angeles
Los Angeles, CA 90095

ABSTRACT

Relationships of indentation hardness numbers to to other physical properties are demonstrated. They differ depending on the type of chemical bonding: metals, alloys ionic, covalent, and metal-metalloid. The properties are: shear modulus; ionic charge; band-gap density; polarizability; and formation energy, respectively. In each case the rationale is provided. The concept of a "bonding Modulus" is introduced. It is concluded that the conventional wisdom that hardness is a purely empirical property does not hold. Phase transformations and indentation hardness are connected broadly.

INTRODUCTION

A purpose of this paper is to show that hardness is not a chaotic empirical property. Instead, hardness is determined by the energy densities of chemical bonding. That is, by bond moduli. These are not universal parameters, but vary depending on the type of bonding. They are closely related to chemical hardness.

Two standard methods are used for measuring the hardnesses of structural materials. One is indentation, and the other is scratching. They both depend on the type of chemical bonding in the material: alloy, covalent, ionic, metallic, metal-metalloid, or dispersion. The inelastic deformation during indentation, or scratching, involves either: the motion of glide dislocations; or a phase transformation (twinning is taken to be a monomolecular phase transformation); or both.

The atomic forces that determine hardness are electrodynamic just as they are for the elastic stiffnesses. The strongest forces are of the electron exchange type, but weaker photon exchange forces are sometimes important. They act at the atomic scale of dimensions so quantum mechanics is required to understand hardness properly. Continuum mechanics can provide an operational description of hardness measurements, but not an understanding of what underlies the numbers.

This paper describes mostly quantitative theories based on chemical behavior, including the quantity known as "chemical hardness" which is the second derivative of the chemical energy per unit volume taken with respect to a change in the number density of valence electrons. It has been found empirically that the chemical hardness and the physical hardness are proportional to one another. Thus, an understanding of mechanical hardness helps in developing an understanding of chemical stability and reactivity.

Predominantly Vickers Hardness Numbers are discussed. That is, numbers derived from indentations made by square diamond pyramids with apex angles of about 135°. The friction coefficients between such indenters and specimens are large. Therefore, the stresses in the specimens consist of two parts: shear and hydrostatic compression. The shear strains can be expected to cause dislocation activity (and twinning), while the compression will tend to cause phase transformations. Since both of these cause electron excitation from valence to conduction states in covalent crystals, they will sometimes (or often) occur simultaneously in these crystals.

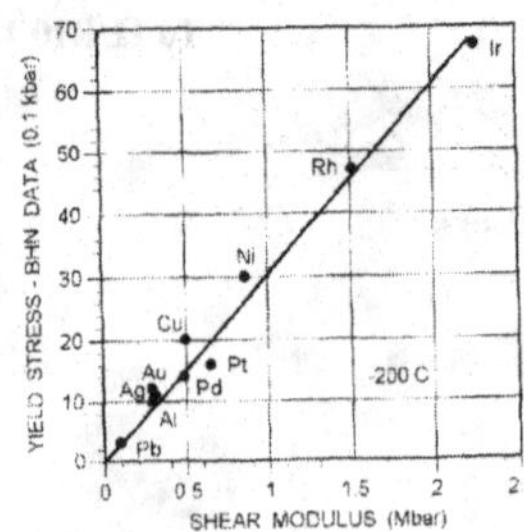

Figure 1 - Comparison of the yield stresses derived from Brinell hardnesses and the shear moduli of f.c.c. metals at -200 C.

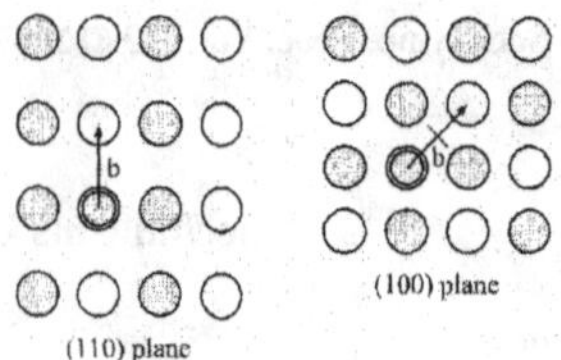

Figure 2 - Glide in rocksalt-type crystals. Comparison of glide on two planes, but the same direction. On the left, is a plan view of a (110) plane with one ion of the next plane indicated (smaller circle). The vector, b is in the glide direction. The clear circles represent ions of one sign; the cross-hatched circles represent ions of the opposite sign. As the ions in the top level glide over those in the bottom level, they always face ions of the same sign.

METALS

In pure simple metals, the mobilities of individual dislocations on the primary glide systems are indefinitely large. This results because the bonding is non-local. Therefore, the energies of the cores of dislocations are independent of their positions, so no quasi-static forces resist their motion (only electron and phonon viscosities). As a result, the deformation during indentation is only resisted by strain-hardening, by dislocation motion on secondary glide systems, or by extrinsic factors such as grain boundaries, precipitates, and the like. The principle property that determines the interactions of dislocations with these various factors is the elastic shear modulus.

Figure 1 shows that the yield stresses of a variety of pure f.c.c. metals (derived from Brinell indentation hardnesses measured at -200 C) are proportional to their shear moduli [1].

ALKALI HALIDES

High purity ionic crystals, such as the alkali halides, behave differently from metals, and crystals with other types of bonding. In them the bonding is of long range, with charge being localized at the ion positions. A result is that dislocations move with great ease in high-purity halide crystals on the primary glide planes. (110) in the rocksalt structure). Here the motion is along rows of ions of the same charge, and across the glide plane ions on one side always face ions of the opposite sign. However, dislocation motion on secondary planes (such as (100)) cuts through the rows of ions of like charge. This brings ions of the same charge into repulsive juxtaposition which is a very high energy configuration (Figure 2). On the right, a plan view of a (100) plane is shown. Now when ions in the top plane glide in the direction, b and they reach the half-way position (short dotted line), they face ions of the same sign which inhibit their motion.

The deformation during indentation requires dislocation motion on both primary and secondary planes. The ion repulsions on the secondary planes cause substantial resistance to dislocation motion; even in high purity crystals. The latter deform readily in simple compression, but not at indentations. A simple theory based on this idea yields an expression for the indentation hardness, H (kg / mm^2):

$$H = q^2 / 10\epsilon b^4 = 1.2 \times 10^{-2}\, C_{44}\ (d / cm^2) \quad (1)$$

where: q = electron charge, ϵ = dielectric constant, b = Burgers displacement, and C_{44} = cubic shear modulus [2]. Figure 3 shows that measured data agree very well with the theory which contains no disposable parameters. If impurities are present, especially di- or trivalent ions, Equation (1) must be modified.

Note that Equation (1) indicates a connection between C_{44} and ϵ. The latter depends on the polarizability of each alkali halide through the Clausius-Mossotti equation. Polarization by a one-dimensional electric field converts a spherical atom into a prolate ellipsoid. Mechanical shear causes the same shape change. The magnitude of the shape change is determined by α = polarizability in the case of the electric field, while it is determined by G = shear modulus in the mechanical case. The connection between them is [3]:

$$G = (3/4\pi)(q^2 / \alpha r) \quad (2)$$

where, r = atomic or molecular radius. Since α is proportional to r^3, the second term in parentheses is: (q^2 / r^4) which is the well-known Keyes parameter [4]. It has the dimensions of a modulus (energy / volume). C_{44} for the alkali halides is proportional to the Keyes parameter [5]. Since α is proportional to r^3, the term $(1 / \alpha r)$ may also be written: $(1 / \alpha^{4/3})$.

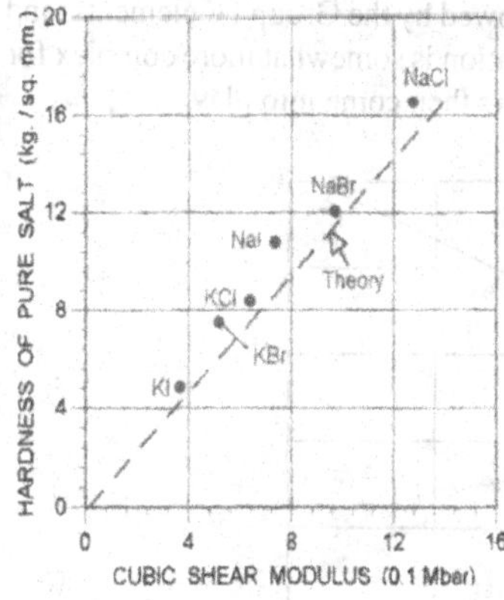

Figure 3 - Comparison of theoretical and measured values (solid circles) of the hardnesses of pure alkali halides (extrapolated from known compositions).

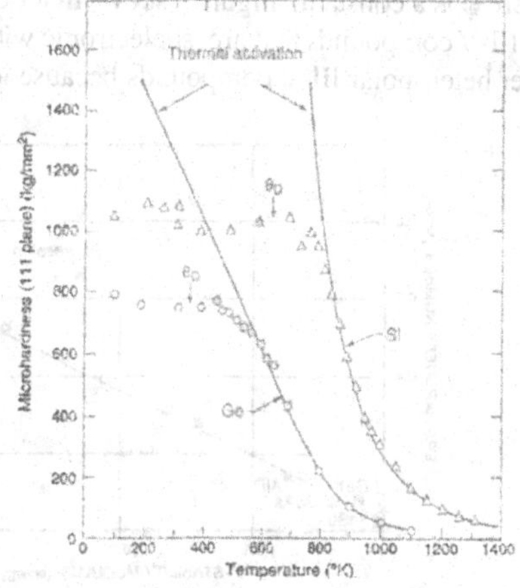

Figure 4 - Temperature dependence of hardness of covalently bonded crystals showing distinct low and high temperature regimes.

COVALENT CRYSTALS

The bonding is localized in this case with the bond charge lying between the atoms, and the structures are relatively open, so the behaviour is quite different from metals or ionics. Dislocation motion in covalent crystals (via kinks) requires the promotion of electrons from the valence bands into the conduction bands. Thus, the band-gap energy must be supplied. At temperatures above the Debye temperatures of these crystals, thermal energy excites the valence electrons, and the hardness decreases exponentially as the temperature rises. Conversely, the hardness rises rapidly with decreasing temperature until the hardness number (units of pressure) approximately equals the compressive phase transformation pressure. This interpretation was originally suggested by Gridneva, Millman, and Trefilov [6]. Then the hardness remains constant (approx.) as the temperature decreases further (Figure 4).

In the low temperature region (below the Debye temperature) both parts of the stress distribution come into play. The shear part continues to move dislocations, and the hydrostatic pressure part causes crystal structure transformations [7].

It was suggested by Jamieson [8] that structural transformations in these crystals occur when the pressure work, $P\Delta V$ (where P = applied pressure, ΔV = volume change) equals the band-gap energy, E_g. But the volume change is approximately a constant fraction of the molecular volume, V_m [9], so the work condition may be written $P(fV_m) = E_g$ where f is the constant fraction (about 20%). To equalize the units on both sides, divide by fV_m. Then, P (erg cm^{-3}) = E_g / fV_m (erg cm^{-3}). The hardness, H represents both pressure and shear stresses, but they are proportional to one another, so the critical condition might be written:

$$H = \phi (E_g / V_m) \tag{3}$$

where ϕ is a constant. Figure 5 shows that Equation (3) is followed by the Group IV elements, and the III-V compounds that are isoelectronic with them. The situation is somewhat more complex for other heteropolar III-V compounds because ionic charge centers then come into play.

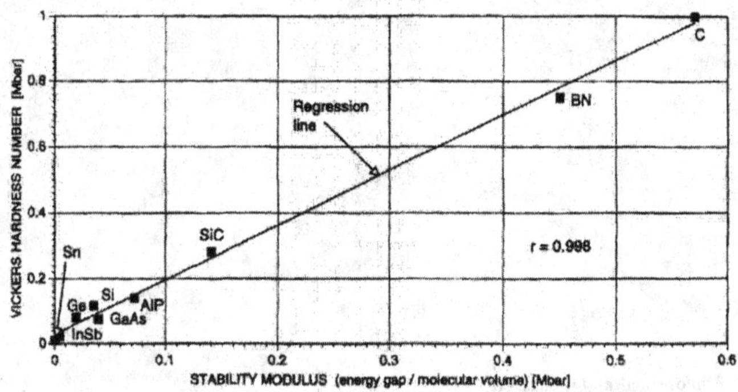

Figure 5 - Hardness numbers of the Group IV elements and the corresponding isoelec tronic III-V compounds plotted vs. the "stability modulus" (E_g / V_m). The dependence (room temperature) is consistent with the hardness numbers being determined by phase transformations.

Notice that (E_g / V_m) has units of energy density (ergs cm^{-3}), so the name "bond modulus" has been coined for it. Band gaps are related to polarizabilities as are shear moduli, so there are some additional links between hardness and other properties of dielectrics. The band gap aslo limits dislocation motion in these crystals [10]. Therefore, indentation typically involves both types of activity: phase transitions, and dislocation motion. This accounts for the form of Figure 5.

POLARIZABILITY

One of the links of hardness to dielectric properties is through the polarizability which can be obtained from the dielectric constant using the Clausius-Mossotti equation. Polarization requires excitation of a valence electron in order to change its symmetry from a spherical to a uniaxial type. Thus, it is related inversely to the energy band-gap. The dependence of indentation hardness on the reciprocal polarizability for the Group IV elements is shown in Figure 6; and for the III-V compounds isoelectronic with the Group IV elements in Figure 7.

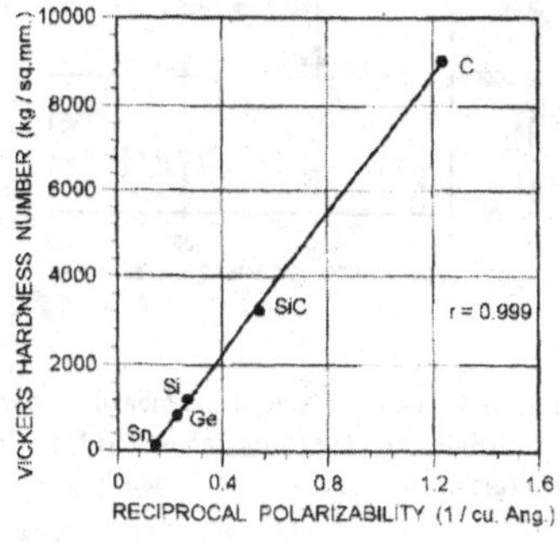

Figures 6 - Hardness vs. reciprocal polarizability for Group IV elements.

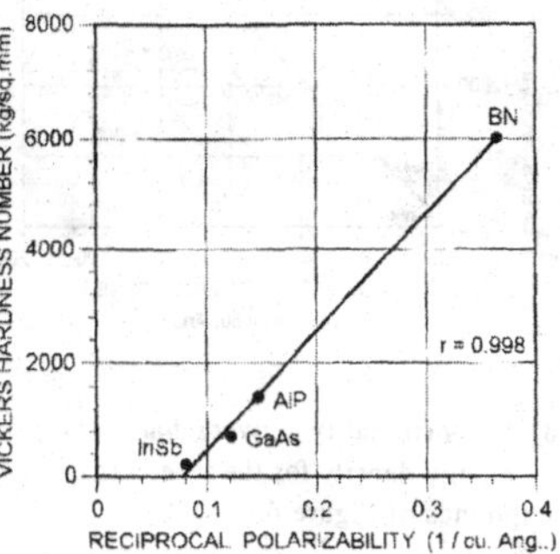

Figure 7 - Same as Figure 6 for the isoelec-electronic III-V compounds.

HEAT OF FORMATION

The chemical origin of hardness is also demonstrated by its connection with the formation energies of compounds. The finite deformation associated with indentation temporarily dissociates a compound at the centers of dislocation cores, so the work of indentation, $P\Delta V$ is proportional to to the formation energy: $\Delta U_f = -\,\text{const.}(P\Delta V_m)$. Dividing both sides by ΔV, the pressure needed to make an indentation is:

$$P^* \sim H \sim (-\,\Delta U_f / \Delta V_m)^* \qquad (4)$$

That is, the hardness of a compound is proportional to its heat of formation density. This proportionality is shown in Figure 8. Hardness is also related to electronegativity difference density, but this connection will not be pursued here.

CARBIDES AND OTHER "HARD-METALS"

The bonding in carbides, and similar compounds, is a mixture of of metallic and covalent types. There is a gap in the density of states distribution, but it is not well-defined. However, the energies of formation of carbides are definite, so the proportionality of Figure 8 suggests how their hardnesses are determined [11]. At the b/2 (mid-glide) position in a dislocation core the local symmetry is destroyed ; and so is the binding energy. In the rock-salt structure, for example, the octahedral symmetry becomes trigonal pyramidal. The work needed to cause this change must be supplied by incremental applied stress. This can be estimated as follows. The driving force on a molecular length of dislocation line is approximately τb^2 where τ is the shear stress and b is Burgers displacement. The chemical resisting force is: $-\Delta H_f / b$, so equating the forces yields: $\tau = \Delta H_f / b^3$,

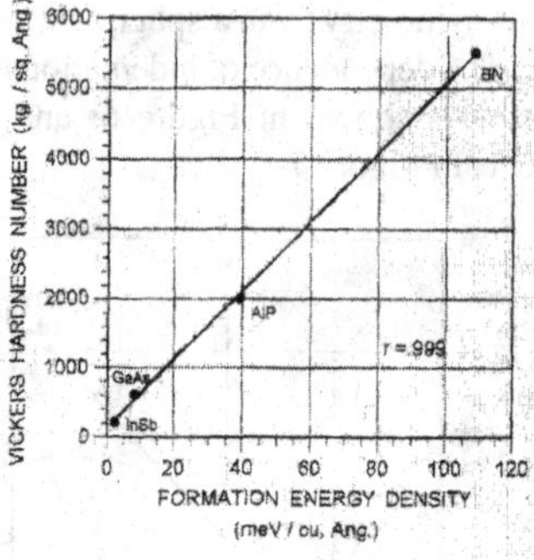

Figure 8 - Proportionality between hardness and formation energy density for the isoelectronic III-V compounds of Figure 6.

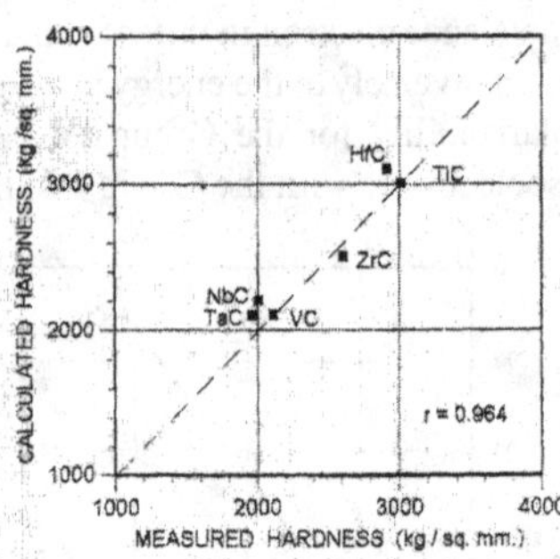

Figure 9 - Comparison of the measured hardnesses of carbides and the those calculated from formation energies.

and the yield stress is 2τ. Also, the molecular volume, $V_m \approx 3b^3$, while the hardness number is: VHN = 6τ, so it equals [12]:

$$VHN = -2\,(\Delta H_f / V_m) \tag{5}$$

This compares very well with measured values for the six prototype carbides (Figure 9).

Ag-Au ALLOYS

Because of interactions between the unlike atoms in alloys, they usually have endothermic heats of mixing. Thus, when a dislocation passes through an unlike pair its motion is resisted because the work needed to separate the pair must be supplied by the external stress.

No attempt will be made to discuss alloys generally because there are too many of them. However, one example has a simple, quantitative interpretation that illustrates how the hardness behavior can be understood. The example is the silver-gold system. It is particularly simple because the atoms are of nearly the same size, and they are isoelectronic, and they are mutually soluble. Also, yield stress measurements are available from Sachs and Weerts [13]; and a value for the heat of mixing is known ($\Delta H_m = -48$ meV / atom). The Burgers displacement for Ag is 2.88 Å, while for Au it is 2.89 Å; for an average of 2.885 Å which gives an atomic volume of 8.50×10^{-24} cm^3., and a heat of mixing density of 90.4×10^8 ergs / cm^3. The shear stress needed to disrupt a Ag-Au pair is about: $-\Delta H_m / 2V_a$, and the yield stress is twice this, and the VHN is 3 times the yield stress, so $(VHN)_{max} = H_{max} = -3\Delta H_m / V_a$. Letting H_o be the average hardness of the pure metals, an expression for the alloy hardness may be written:

$$H = H_0 + 4\,(H_{max} - H_0)[c\,(1 - c)] \tag{6}$$

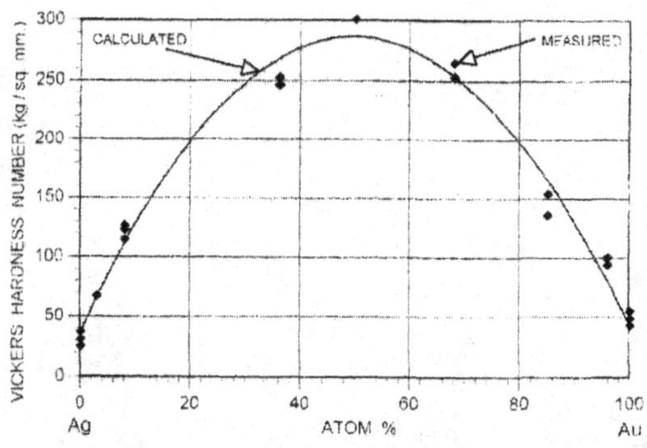

Figure 10-Comparison of Sachs and Weerts measurements with the theory described in the text.

excellent agreement between the theory and the measured values. Note that there are no disposable parameters in the theory.

SUMMARY

It is shown that indentation hardness numbers can be related quantitatively to other physical properties. The relationships depend on the type of chemical bonding. Simple relationships are presented for all of the bonding types involved in structural materials: metals, alloys, ionic crystals, covalent crystals, and "hard-metals". The hardnesses are related to elastic moduli, ionic charges, phase transformations, polarizabilities, formation energies, and molecular (atomic) volumes. Thus, the conventional wisdom that hardness is a purely empirical property is refuted.

REFERENCES

1. Gilman, J. J. (1960) Australian Jour. Phys., **13**, 327.
2. Gilman, J. J. (1973), J. Appl. Phys., **44**, 982.
3. Gilman, J. J. (1997), Mat. Res. Innovat., **1**, 71.
4. Keyes, R. W. (1962), J. Appl, Phys., 33, 3371.
5. Gilman, J. J. (1969), *Micromechanics of Flow in Solids*, p. 33, McGraw-Hill, New York.
6. Gridneva, I. V., Mil'man, Yu. V., and Trefilov, V. I. (1972), Phys Stat. Sol. A, **14**(1), 177.
7. Khayyat, M. M., Banini, G. K., Hasko, D. G., and Chaudhri, M. M. (2003), Journal of Physics D: Applied Physics, **36** (11), 1300.
8. Jamieson, J. C. (1963), Science, **139** (3557), 845.
9. Ackland, G. J. (2004), Chap. 23 in *High-pressure Surface Science and Engineering*, Ed. by Y. Gogotsi and V. Dominich, Inst. Phys. Publish., Bristol, UK, p. 120.
10. Gilman, J. J. (1993), Science, **261**, 1436.
11. Gilman, J. J. (1970), J. Appl. Phys., **41**, 1664. Data of this paper were revised for Figure 8.
12. Sachs G., and Weerts, J. (1930), Zeit. F. Phys., **62**, 473.

Synthesis and Characterization of Engineered Surfaces and Coatings

Mater. Res. Soc. Symp. Proc. Vol. 843 © 2005 Materials Research Society T7.3

Synthesis and Characterization of Carbon Nitride Thin Film with Evidence of Nanodomes

S. Chowdhury[1] and M. T. Laugier
[1]Materials and Surface Science Institute (MSSI), University of Limerick, Limerick, Ireland
Department of Physics, University of Limerick, Limerick, Ireland

ABSTRACT

We have reported the synthesis of carbon nitride thin films with evidence of formation of carbon nanodomes over a range of substrate temperature from 50 °C to 550 °C. An RF magnetron sputtering system was used for depositing carbon nitride films. The size of the nanodomes can be controlled by deposition temperature and increases from 40-80 nm at room temperature to 200-400 nm at high temperature (550 °C). Microstructural characterization was performed by AFM. Electrical characterization shows that these films have conductive behaviour with a resistivity depending on the size of the nanodomes. Resistivity values of 20 mΩ-cm were found for nanodomes of size 40-80 nm falling to 6 mΩ-cm for nanodomes of size 200-400 nm. Nanoindentation results show that the hardness and Young's modulus of these films are in the range from 9-22 GPa and 100-168 GPa respectively and these values decrease as the size of the nanodomes increases. GXRD results confirm that a crystalline graphitic carbon nitride structure has formed.

INTRODUCTION

During the last several years, many attempts have been made to grow crystalline β-C_3N_4 films because of the possibility that this hypothetical material may be even harder than diamond [1]. Generally, as-deposited carbon nitride films have been found to be either amorphous or to crystallize with a turbostratic-like microstructure [2]. In a few cases, small β-C_3N_4 second-phase crystallites have been reported [3]. In the search for crystalline carbon nitride scientists have discovered a new form of fullerene-like carbon nitride [4]. However, the amorphous form of carbon nitride is also a potential candidate for replacing diamond-like carbon (DLC) films because of its excellent mechanical and tribological properties. In this article we have reported a new form of carbon nitride with evidence of nanodomes, which are found to be hard and crystalline as well as conductive.

EXPERIMENTAL DETAILS

Carbon nitride films were deposited by RF magnetron sputtering (Leybold Lab 500) using high-purity 99.99% pyrolytic graphite disk as a target material. Base pressure of the deposition chamber was around 1.0×10^{-3} Pa. Pure argon (99.999% purity) was used as inert gas in the chamber with nitrogen (99.999% purity) gas. Coatings were deposited on Si (100) wafers at temperatures of 50, 100, 150, 350 and 550 °C using a resistance heater located behind the substrate holder with 50% argon and 50% nitrogen gas mixture. During deposition the chamber gas pressure was 0.5 Pa.

The AFM imaging on the surface of the films was carried out with a TopoMetrix Explorer surface probe microscope. The images were collected in non-contact imaging mode. The images obtained were processed by TopoMetrix SPM Lab NT Version 5.0 software supplied with the microscope. Qualitative X-Ray Powder Diffraction (XRD) was performed using a Philips X'pert PRO Multi Purpose Diffractometer. A glancing angle of 1.5 degrees has been applied. The electrical characterization of the films was performed using a Jandel four-point probe resistivity test unit at room temperature. The current applied was varied between 1 and 200 μA and at least 5 different areas were tested on each sample.

Nanoindentation testing of the samples was carried out using a CSM™ nano hardness tester (NHT) to measure the hardness (*H*) and Young's modulus (*E*) of carbon nitride films. The nanohardness tester was calibrated by using glass and fused silica samples for a range of operating conditions. The sample was indented with maximum applied loads of 1 mN in order to eliminate substrate effects. A Berkovich diamond indenter with tip radius around 50 nm was used for all the measurements. Oliver and Pharr method [5] was used in order to get the mechanical properties of carbon nitride film.

RESULTS AND DISCUSSION

Morphology study by AFM

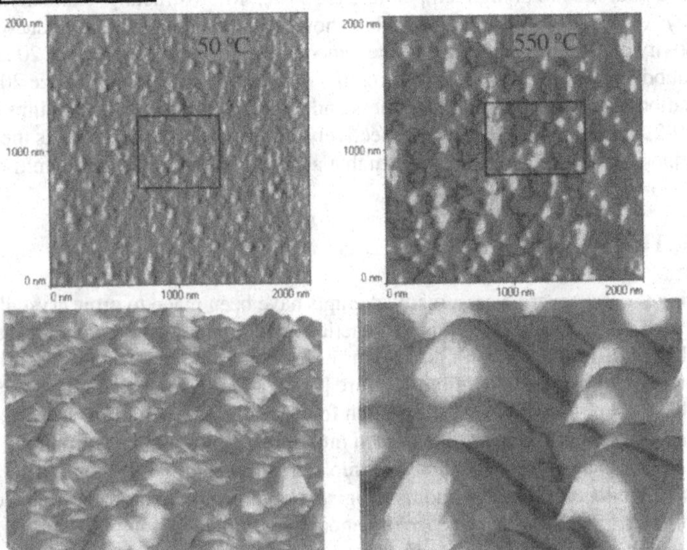

Figure 1. AFM images showing evidence of nanodomes in carbon nitride film deposited at different substrate temperature. Bottom images are 5 times closer views of the square box areas of the upper images.

The AFM images show the detailed features of the surface of the carbon nitride thin films. The carbon nitride films exhibit an interesting structure with nanometer-scale domes [6] as illustrated in figure 1. Bottom images in figure. 1 are 5 times closer views of the square box areas of the upper images, which show clearly the presence of nanodomes. Nanodomes are found to vary in sizes from 40-80 nm when deposited at room temperature. There is a little variation of size of the domes up to 350 ºC deposition temperature. When the substrate temperature is increased to 550 ºC, the size of the domes increases to around 200 to 400 nm in diameter.

We believe that the domes are formed from the bending of graphite basal planes due to the presence of nitrogen atoms [6]. The curving of the basal planes is believed to be due to the incorporation of pentagons in the graphite hexagon structure, just as in the case of fullerene-like carbon nitride [4]. Nitrogen plays a vital role here. The graphite basal planes intersect through sp^3-hybridized carbon facilitating merging and/or intersecting of the planes, such that a continuous three-dimensional dome structure can be formed. This is likely to the caused by nitrogen substitution in the graphite sites. This observation would support previous quantum chemical calculations indicating that the nitrogen atoms prefer a nonplanar surrounding [7]. With further increase of the growth temperature, more disordered / amorphous regions transform into the distorted graphitic structures. Figure 1 shows the images of the surface of a 550 ºC grown CN_x film. At that temperature surfaces where the basal planes merged together, or are possibly cross-linked can be observed. Thus the nanodomes become bigger when deposited at higher temperatures. Further investigation is needed in order to fully understand the details of formation and structure of the nanodomes. The RMS area roughness of the deposited film at different substrate temperatures has been determined by AFM and found to be 0.64 nm to 9.27 nm. A significantly higher value of area roughness has been found when the films are deposited at a temperature 550 ºC. The increase of roughness results from the increase of the size of the nanodomes in carbon nitride films with temperature. Thickness values determined from AFM line analysis were around 550 nm.

Raman analysis

Figure 2 shows Raman spectra from the carbon nitride films grown at 50, 350, 550 °C. The carbon spectra are composed of two broad bands centred on 1570 cm^{-1} (G band) and 1360 cm^{-1} (D band). There is also a small peak at ~2210 cm^{-1} corresponding to C-N triple bond in stretching mode. It can be seen from the figure that both the widths and intensities of the G and

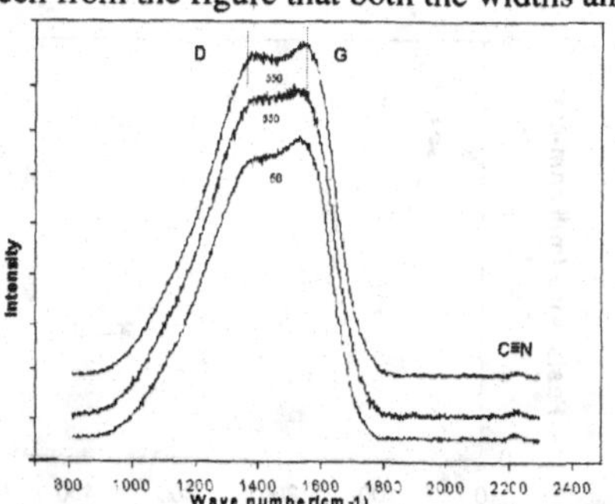

Figure 2. Raman spectra of carbon nitride films deposited at different substrate temperature.

D peaks change depending on temperature. It has been found that with increasing of temperature a shoulder band at 1380 cm^{-1} is clearly observed and becomes more pronounced with increasing deposition temperature. The relative Raman intensities of the D and G bands are therefore a measure of the degree of order in the samples. The I_d/I_g ratios are calculated by fitting the spectra by two Gaussion peaks shapes. The I_d/I_g ratio is found to increase from ~ 4.25 at 50 ºC to ~ 5.19 at 350 ºC and finally to ~ 5.54 at 550 ºC.

GXRD study

The GXRD pattern of carbon nitride film deposited at 50 °C provides evidence of the crystalline graphitic carbon nitride in the deposited film [2] (Figure 3). The peaks are matched with simulated results found in the literature. The GXRD patterns of these films clearly show the reflections from the planes corresponding to [002], [102], [200], [202] and [220] planes of graphitic carbon nitride.

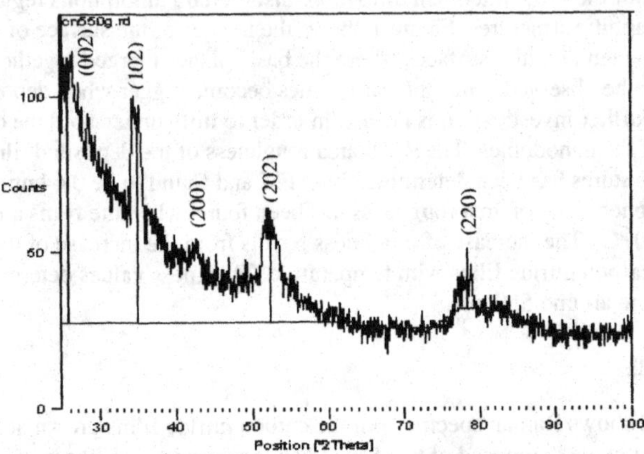

Figure 3. Glancing angle XRD pattern for carbon nitride thin film deposited at 50 °C, at glancing angle 1.5 °

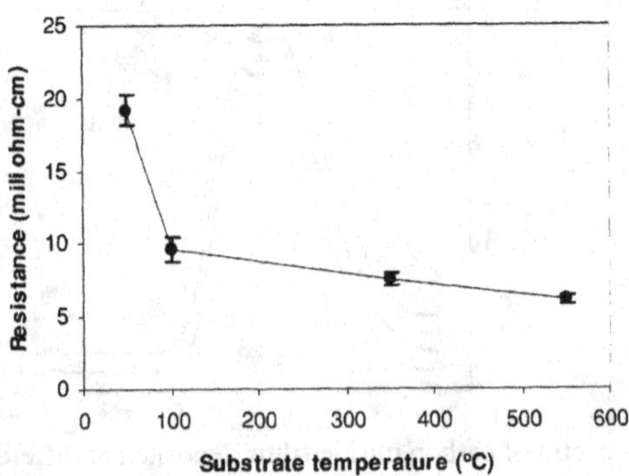

Figure 4. The variation of resistivity of carbon nitride thin film deposited at different substrate temperatures

Electrical characterization

Electrical resistivity values are found in the range of 6-19 mΩ-cm [8]. Resistivity results reported by other authors are in the range 5×10^6 to 1×10^{14} mΩ-cm for carbon nitride thin films [9]. For comparison the most conductive material, silver has resistivity around 1.5×10^{-3} mΩ-cm. We have also found that the resistivity of the deposited films decreases as the size of the nanodomes increases with temperature (Figure 4). This indicates that the nanodomes, which contain the graphite basal planes connected with nitrogen atoms, are fairly conductive. The conductivity is due to the graphite basal planes.

Nanoindentation measurement

Nanoindentation results show that the *H* and *E* values of these films are in the ranges 9-22 GPa and 100-168 GPa respectively. The *H* and *E* values of the films decreased as the size of the nanodomes increased. Percentage elastic recovery *R* -values around 78 % are found for CN_x film deposited at temperature 550 °C and around 91 % is observed for samples deposited at 350 °C. In this structure nitrogen atoms are bonded to both sp^2 and sp^3 coordinated carbon and thus the proposed structure of the film is fullerene-like where graphite sp^2 rings are cross-linked by sp^3-coordinated carbon. However, the incorporation of nitrogen in the film promotes formation of pentagons in the hexagonal graphite-like planes, and thus induces buckling of the basal planes, which facilitates cross-linking between the planes through sp^3-coordinated carbon. Due to this cross–linking, the films become harder. When load is applied to the basal planes the structure is compressed and elastic recovery occurs once the load is removed and thus the films are highly elastic.

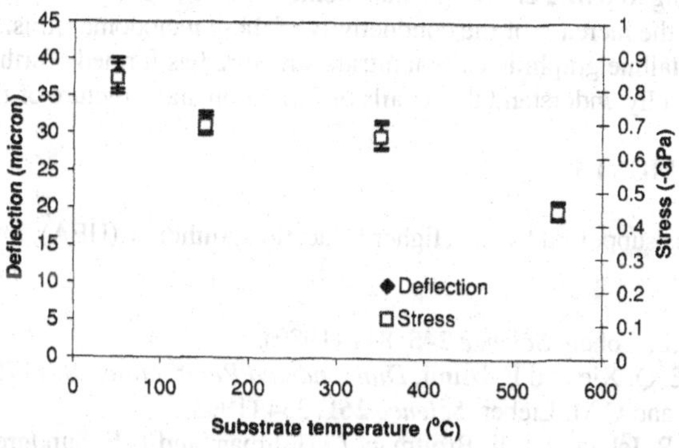

Figure 5. Variation of free edge deflections as well as compressive stresses with substrate temperature for carbon nitride thin films on silicon (100) measured by the optical fibre displacement sensor stress measurement set-up

Stress calculation

Stress values are calculated for CN_x thin films deposited at different substrate temperatures using the edge deflection values in the Stoney's formula [10] and shown in figure 5.

Negative deflections are observed on the coated substrate system and thus calculated stress values are found to be compressive. The effect of substrate temperature on stress in CN_x films is shown in figure 5. CN_x films deposited at low temperatures have higher compressive stress but with increasing temperature from 50 ºC to 150 ºC stress values gradually decrease and there is no significant increase of stress at temperatures between 150 ºC to 350 ºC. After that stress decreases until the temperature reaches 550 ºC. It has been found that, as the size of the nanodomes increase with temperature the stress in the film decreases. We obtained the lowest stress values when the nanodomes are largest at a substrate temperature of around 550 ºC. So stress relaxation occurs as the size of the nanodomes increases. The decrease of stress at higher temperature may also be due to the increase of sp^2-bonded carbon in CN_x films.

CONCLUSIONS

The present results indicate the synthesis of conductive as well as hard carbon nitride thin films with evidence of formation of nanodomes with a graphitic carbon nitride structure, over a range of substrate temperatures from 50 ºC to 550 ºC. The results also support previous indications that incorporation of nitrogen into the graphitic carbon structure leads to a distortion of the basal planes and thereby facilitates cross links between the planes as they come closer together and form nanodomes. The size of the nanodomes increases from 40-80 nm at room temperature to 200-400 nm at high temperature (550 ºC). Nanoindentation results show that the hardness and Young's modulus of these films are in the range from 9-22 GPa and 100-168 GPa respectively and these values decrease as the size of the nanodomes increases. Electrical characterization shows that these films have conductive behaviour with a resistivity depending on the size of the nanodomes. Resistivity values of 19 mΩ-cm are found for the nanodomes of size 40-80 nm falling to 6 mΩ-cm for the nanodomes of size 200-400 nm. Graphite basal planes are responsible for the increase of the conductivity of these nanodome films. GXRD results confirm that a crystalline graphitic carbon nitride structure has formed. Further investigation is needed in order to fully understand the details of formation and structure of the nanodomes.

ACKNOWLEDGMENTS

This work is supported by the Higher Education Authority (HEA), Ireland

REFERENCES

1. A.Y. Liu and M.L. Cohen, *Science* **245**, 841 (1989).
2. S. Matsumoto, E. Q. Xie and F. Izumi, *Diamond and Relat, Mater.* 8, 1175 (1999).
3. C. Niu, Y.Z. Lu and C.M. Lieber, *Science* **261**, 334 (1993).
4. N. Hellgren, M. P. Johansson, E. Broitman, L. Hultman, and J.-E. Sundgren, *Phys. Rev. B.* **59**, 5162 (1999).
5. W.C. Oliver and G.M. Pharr, *J. Mater. Res.* **7**, 1564-1583 (1992).
6. S. Chowdhury and M.T. Laugier, *physica status solidi* (a) **201**, R1-R4 (2004).
7. H. Sjostrom, S. Stafstrom, M. Boman and J.-E. Sundgren, *Phys. Rev. Lett.* **75**, 1336 (1995) .
8. S. Chowdhury and M.T. Laugier, *J. Vac. Sci. and Technol. B*: **22**, L28-L31 (2004).
9. E. Broitman, N. Hellgren, K. Ja¨rrendahl, M. P. Johansson, S. Olafsson, G. Radnoczi, J.-E. Sundgren and L. Hultman, *J. Appl. Phys* **89**, 1184 (2001).
10. G.G. Stoney, *Proceedings of the Royal Society, London*, **A82**, 172 (1909).

Mater. Res. Soc. Symp. Proc. Vol. 843 © 2005 Materials Research Society T7.4

Electron Microscopy Analysis on the Worn Surface of a High-Chromium White Iron During Dry Sliding Contact

A. Bedolla-Jacuinde[1] and W.M. Rainforth[2]
[1]Instituto de Investigaciones Metalúrgicas, Universidad Michoacana de San Nicolás de Hidalgo, Morelia, Michoacán. México.
[2]Department of Engineering Materials, The University of Sheffield, Sheffield, UK.

ABSTRACT

A series of microstructural phenomena within a thickness of material (tribolayer) below the worn surface, have been developed during dry sliding wear of a high-chromium cast iron. The overall wear behavior of the alloy is determined by the properties of this tribolayer. From the present work, a transmission electron microscopy analysis has been undertaken on the different features developing at different distances below the worn surface following wear sliding tests of a 17%Cr white cast iron alloy, whose microstructure is composed by 25% eutectic M_7C_3 carbides within an austenitic matrix. The observed phenomena is an increase in the dislocation density, plastic deformation by twinning followed by severe shear banding along with carbides fracture, a mechanical mixture formed by iron oxide and carbide particles produced from large carbides comminution, and finally a flat iron oxide layer. Wear debris was apparently created from the oxide film detaching from the outermost surface where equivalent strain is maximum. No evidence of strain induced martensite was observed from the present work, which has been reported in some austenitic materials. The implications of the microstructural evolution are discussed in terms of the wear theories and behavior of metals at high strains levels.

INTRODUCTION

Microstructural evolution during sliding contact between two surfaces is a very important factor in determining the wear resistance of a material, since sliding contact between the surfaces of ductile materials is always accompanied by severe plastic deformation localized to a small volume of material adjacent to the surface [1-7]. However, not too much is known about the work hardening behavior at the worn surface to be able to predict the ductility limit and model flow stress as a function of depth, both of which determine wear rate [2]. Currently, there is no definitive data to predict the deformation behavior at a worn surface in terms of slip systems, work hardening law, texture generated, strain distribution, flow stress, critical strain at which wear debris forms and depth below the surface at wear which debris forms [1]. Texture is one of these variables that have provided important evidence of the strain path at the worn surface. Rainforth [2] has underlined that crystallographic texture arising from the deformation can provide an important information on the strain path since some observations indicate that simple shear process occurs while some others suggest that pure shear is more appropriated. However, more data are required in this area.

It is very difficult to predict work hardening behavior at a surface even for a single phase metal, where the only surface microstructural changes taking place are associated with deformation [2]. Furthermore, when a multiphase alloy is under analysis, this prediction becomes much more complicated. However, it is common that some form of mechanical mixing with the counterface occurs or that there is some reaction with the environment, such as oxidation, which significantly alters surface deformation behavior.

In this paper the microstructural evolution resulting from sliding wear of a high-chromium white cast iron is reviewed by TEM. Previous work on the effect of silicon as a modifier element of the eutectic structure in a 17%Cr white cast iron[8] refers to the wear behavior of the alloy under dry sliding conditions by using a load range 42-238N for a distance of 70km against a hardened M2 steel counterface. The experimentation was undertaken for the alloy in the as-cast conditions as well as after a destabilization heat treatment. Results from this study reported the formation of an oxide film which resulted to be Fe_2O_3 for low loads and a mixture of Fe_2O_3 and Fe_3O_4 along with high depths of deformation for higher loads. Carbide fracture below the worn surface along with a mechanical mixture formed by iron oxide and carbide particles produced from large carbides comminution were observed. Details of the results are reported elsewhere [8].

EXPERIMENTAL PROCEDURE

Transmission electron microscopy (TEM) was carried out in a Philips 400T TEM operating at 120 kV. High resolution was also undertaken by using a Philips Tecnai F20 operated at 200 kV. The thin foils were prepared by cutting thin slices (~1 mm thickness) from the ingots. These slices were cut using an oil lubricated diamond wheel in a Vari/Cut –LECO- machine, with the wheel spinning at a speed of 12 m/min. The slices were then thinned on grinding paper to 0.5 mm and discs of 3 mm diameter were cut by spark erosion. Each disc was further thinned on grinding paper to ~ 80 μm using a special holder, then dimpled to ~20 μm in a Dimpler Model D 500 with 0.25 μm finish. Ion beam milling was carried out to perforation in a GATAN model 691 Precision Ion Polishing System (PIPS) operating at 5 kV, ~18 μ A and 4° incidence angle. Prior to analysis of a sample in the TEM, the sample was cleaned in the PIPS for 20 min using a special double side holder to permit the incidence of the beam on both faces of the disc. The conditions for this cleaning step were 3 kV, ~10 μA, and 3 deg incidence angle. Cross section- as well as back thinned- sample techniques were used for the present study. Samples were prepared from the close worn surface of samples tested at loads of 189 and 238 N, only from the as-cast alloy (austenitic matrix).

RESULTS AND DISCUSSION

Because of the large surface deformation and extensive cracking induced by wear the key areas tended to fall out before they were electron transparent. Thus the results only come from few samples and were dominated by the regions where the strain was comparatively small.

Figure 1 shows two SEM micrographs of the cross-section of the experimental iron after wear testing, where a wide deformed area along with carbide fracture as well as the formation of an oxide layer can be observed. Depth of deformation extended 40 μm for the higher applied loads. Microstructural evolution below the worn surface starts with an increase in the dislocation density as shown from Figure 2. At this stage, eutectic carbides are intact and well bonded to the highly dislocated austenitic matrix. For this level of deformation, austenite seems to support properly the carbides against spallation. However, closer to the worn surface, carbide fracture starts to occur. Figure 3 shows carbide cracking at some areas about 20 μm below the surface; such cracks will eventually destabilize the surface resulting in wear debris formation.

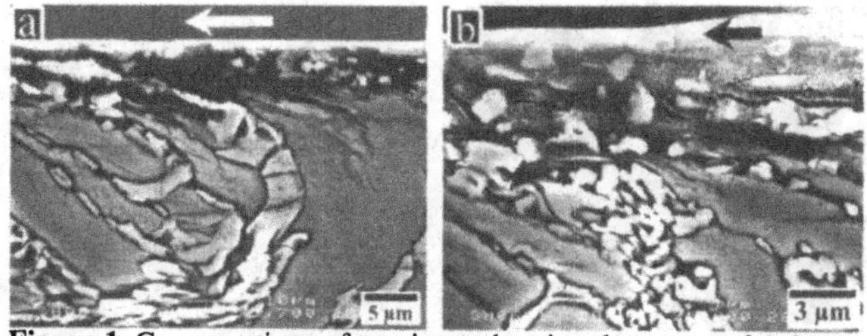

Figure 1. Cross-sections of two irons showing the extent of deformation below the worn surface (a) and the presence of carbide particles in the oxide layer (b); the arrows indicate the sliding direction. Scanning electron microscope.

Closer to the surface (less than 10 μm), a strongly twinned matrix was observed at the deformed zones, Figure 4, along with a high accumulation of dislocations. Twinning has been observed to be the initial step for deformation in an austenitic stainless steel tested under dry sliding conditions at 55N against a Mg-partially stabilized zirconia's counterface [1] and reflects the low stacking fault energy of the matrix in the as-cast condition. Closer to the worn surface (~2 μm), the strain accumulation resulted in increasing dislocation density, but no strain induced martensite (SIM) formation and austenite crystals were observed, Figure 5. Even closer to the worn surface (~300 nm) shear bands parallel to the sliding direction with no preferred orientation, as seen from the electron diffraction pattern, were observed in some carbide free zones, Figure 5. The higher strain accumulation at these zones and the resulting increase in dislocation density may have promoted the formation of shear bands instead of twin formation.

The work hardening of austenite is thought to occur by strain induced martensite formation as well as dislocation interactions. However, very little evidence of the work hardening mechanisms in white irons has been obtained, particularly under dry sliding wear conditions. Evidence of SIM formation in white irons due to three-body high stress abrasion [9-11] and in scratch test studies [12] has been published.

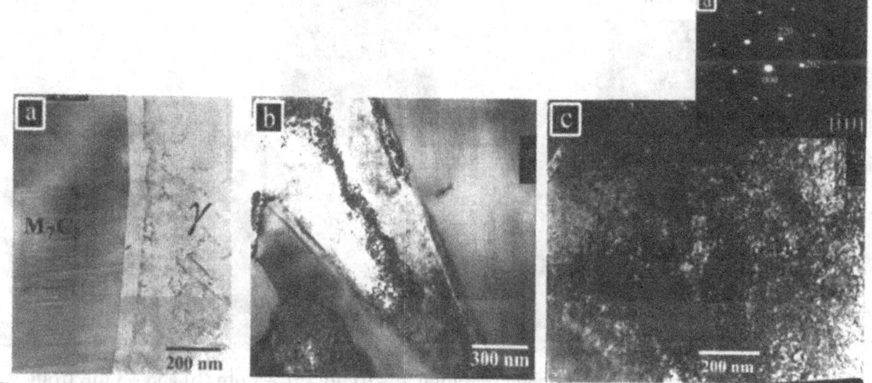

Figure 2. Dislocations density below the worn surface. (a) remote from surface, (b) ~30μm from surface, and (c) ~5μm from surface. Bright field TEM.

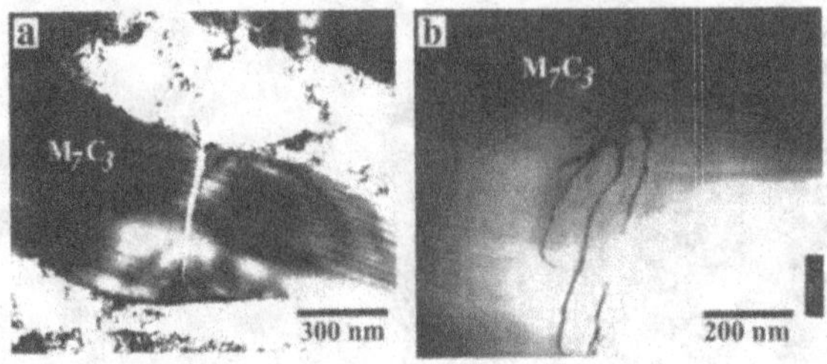

Figure 3. Bright field TEM micrographs showing details of eutectic carbides cracking at about 10 μm below the worn surface.

Pearce [9] recorded austenite hardness values up to 900 HV close to the worn surface after three-body high abrasion tests. He noticed that the extent of hardening was limited to the immediate surface regions and at a depth of 50 μm the hardness was less than 500 HV. Such hardening was attributed to the presence of thin crystals of hcp martensite together with regions of highly dislocated matrix believed to be α martensite (bcc) in the primary austenite remote from the eutectic carbides. In the current work, TEM results did not evidence the presence of SIM at any region of the deformed zone below the worn surface in the as-cast studied alloys.

Interestingly, in some highly deformed zones closer to the worn surface and at the oxide/substrate interface inclusive, the presence of nanometric carbide particles produced by the comminution of eutectic carbides were observed, Figure 6. These particles formed a kind of composite with the austenitic matrix, and then, as the sliding progressed, they should have passed to form part of the oxide layer, Figure 7.

It is quite unlikely that the presence of this nanocomposite in some areas of the worn surface could contribute to increase the wear resistance, since the fracture of large eutectic carbides in deeper zones produced large cracks which destabilised the surface. Therefore, it is suggested that eutectic carbide cracking is the controlling factor of the matrix properties and the wear resistance of the irons.

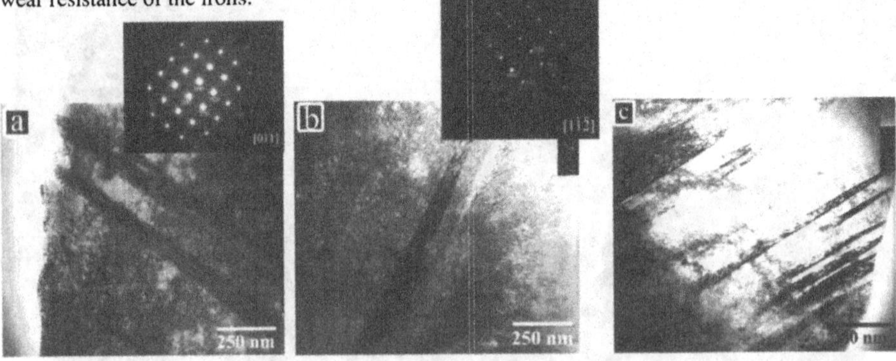

Figure 4. BF TEM micrographs showing twinning at ~10 μm (a), ~5 μm (b) and ~3 μm from surface. From the diffraction patterns it is possible to observe spots at 1/3 distance from the reflected spots of austenite, typical of fcc twinning.

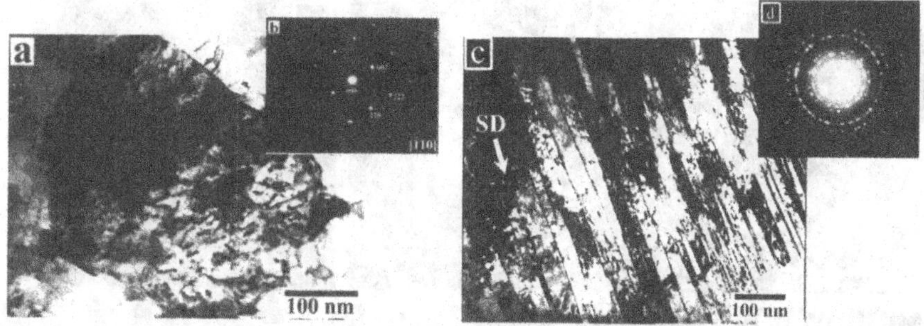

Figure 5. a) Austenite crystal ~2 μm from the worn surface; b) selected area diffraction pattern of a) from [$\bar{1}$10]. (b) shear bands parallel to the sliding direction (SD) at ~300 nm from the surface, and (d) ring electron diffraction pattern showing no preferred orientation of bands.

The frictional heating may increase the temperature locally to the extent of structural changes during wear. According to Pearce [9] it is unlikely that the formation of martensite was assisted by such heating since there was no evidence of any secondary carbide precipitation which would be needed to impoverish the austenite in chromium and carbon and hence to promote transformation to martensite. Without such precipitation, austenite in Hi-Cr irons has a very low Ms temperature and in absence of plastic deformation can only undergo some transformation to martensite by exposure to subzero conditions. Frictional heating may restrict martensite formation by rising the temperature in the surface regions above the martensite start temperature through deformation (Md) value; this is the highest temperature at which martensite may be formed by plastic deformation of austenite (Md temperatures for austenite in Hi-Cr irons are not available but are likely to be below 100°C [9]).

High resolution TEM was undertaken to characterize the nanometric particles embedded in the oxide layer forming a kind of nanocomposite by a mechanical mixture, Figure 7. Such a phenomenon is likely to occur as well at the metallic substrate just below the worn surface but no evidence of this is shown from the present work. However, the formation of nanocomposites of oxide particles embedded in a metallic matrix during wear several alloys have been well documented [1-3].

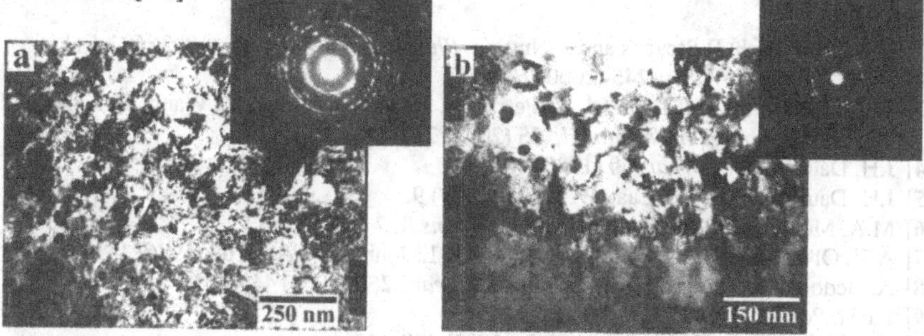

Figure 6. TEM bright field micrographs showing the presence of carbide particles within the Fe_2O_3, Fe_3O_4 oxide layer. (a) sample tested at 189 N and (b) tested at 238 N.

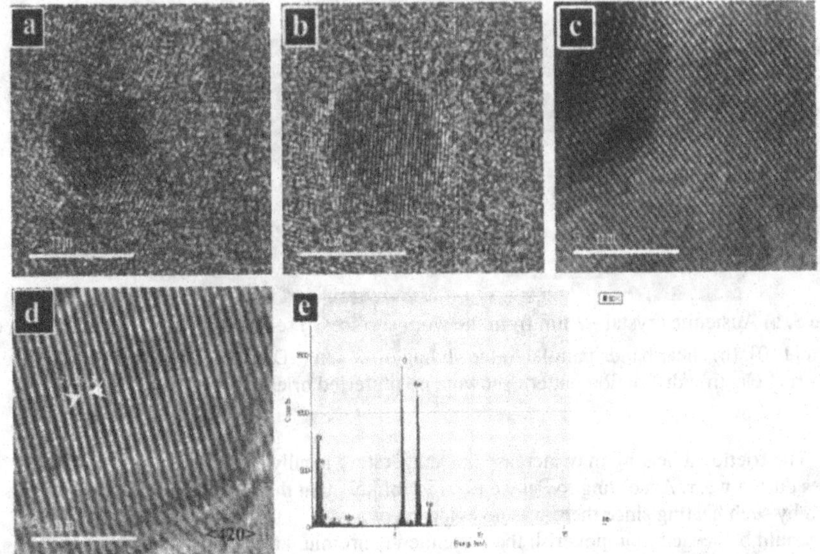

Figure 7. High resolution micrographs showing carbide particles embedded in an oxide matrix which resulted to be partially amorphous as in (a) and (b) and crystalline as in (c). (d) shows a particle where planes <420> were identified, (e) EDS taken from a spot on the particle in (d) showing to be $(Fe,Cr)_7C_3$.

GENERAL CONCLUSIONS

TEM analysis provides evidence of a highly deformed matrix induced by sliding wear that occurred by extensive dislocation flow, twinning, and shear banding. In the thin areas carbides appeared well bonded to the matrix. Thus, substantial strain occurred prior to carbide fracture; although over a very small depth and in areas where the matrix was deformed there was still a high degree of attachment between carbide and matrix. No evidence of the strain induced martensite (SIM) formation was found in the analyzed samples.

REFERENCES

[1] W.M. Rainforth, R.Stevens and J. Nutting, *Philos. Magazine A*, **66** (1992) 621-641.
[2] W.M. Rainforth, *Wear*, **245** (2000) 162-177.
[3] W.M. Rainforth, A.J. Leonard, C. Perrin, A. Bedolla-Jacuinde, Y. Wang, H. Jones and Q. Luo, *Tribology International*, **35** (2002) 731-748.
[4] J.H. Dautzemberg, *Wear*, **60** (1980) 401.
[5] J.H. Dautzemberg, J.H. Zaat, *Wear*, **23** (1973) 9.
[6] M.A. Moore and R.M. Douthwaite, *Met. Trans.A*, **7** (1976) 1833.
[7] A.V. Olver, H.A. Spikes, A.F. Bower, and K.L. Johnson, *Wear*, **107** (1986) 151.
[8] A. Bedolla-Jacuinde and W.M. Rainforth, *Wear*, **250** (2001) 449-461.
[9] J.T.H. Pearce, *Wear*, **89** (1983) 333-344.
[10] J.T.H.Pearce, *J. of Mat. Sci. Lett.*, **2**, (1983) 428-432.
[11] O.N. Dogan, J.A. Hawk, and G. Laird II, *Met. and Mat. Trans. A*, **28** (1997)1315-1327.
[12] A. Sinatora, M. Pohl and E.U. Waldherr, Scripta Met. et Mat. **35** (1995) 857-861.

200 mm Silicon Wafer-to-Wafer Bonding with Thin Ti Layer under BEOL-Compatible Process Conditions

J. Yu, J.J. McMahon, J.-Q. Lu, and R.J. Gutmann
Focus Center – New York, Rensselaer: Interconnections for Hyperintegration
Rensselaer Polytechnic Institute, Troy, New York 12180

ABSTRACT

Wafer level monolithic three-dimensional (3D) integration is an emerging technology to realize enhanced performance and functionality with reduced form-factor and manufacturing cost. The cornerstone for this 3D processing technology is full-wafer bonding under back-end-of-the-line (BEOL) compatible process conditions. For the first time to our knowledge, we demonstrate nearly void-free 200 mm wafer-to-wafer bonding with an ultra-thin Ti adhesive coating, annealed at BEOL-compatible temperature (400 °C) in vacuum with external pressure applied. Mechanical integrity test showed that bonded wafer pair survived after a stringent three-step thinning process (grinding/polishing/wet-etching) with complete removal of top Si wafer, while allowing optical inspection of bonding interface. Mechanisms contributing to the strong bonding at Ti/Si interface are briefly discussed.

INTRODUCTION

Wafer level three-dimensional (3D) integration is currently receiving wide spread attention from semiconductor industry [1, 2, 3, 4] and academia [5, 6], with International SEMATECH having identified 3D interconnect as a major technical challenge for 2005 [7]. The 3D structure offers opportunities for enhanced performance with reduced form-factor (e.g., for two-wafer stack, roughly 50% shorter global interconnect clock distribution and 50% smaller footprint compared to 2D architecture). Besides, it enables the extra degree of flexibility to integrate wafers fabricated by incompatible processes (for example, logic to memory, CMOS to BJT, CMOS to MEMS, and Si to III-V compound). The foreseeable drivers for the technology will be memory stacks, memory-intensive processors, smart imagers and mixed signal applications.

Wafer-level 3D integration generally involves four major processing steps: 1) wafer-to-wafer alignment, 2) wafer-to-wafer bonding, 3) backside thinning, and 4) inter-wafer interconnection. A review on wafer-to-wafer bonding techniques could be found elsewhere [8]. Due to the constraints of compatibility with BEOL (≤450 °C) as well as concerns on contamination (e.g., no Na^+ or Au^+), the choices of wafer-to-wafer bonding are limited to metal-based wafer bond (for instance, copper-to-copper) [1, 4, 5], dielectric adhesive bond [6], and modified fusion bond (for example oxide-to-oxide) [2]. In addition to wafer-to-wafer mechanical attachment, metal-based wafer bonding offers better thermal management and direct electrical connection. Moreover, the metal bonding layer can be incorporated as an extra ground plane or power plane.

In this work, we explore an alternative metal-based wafer bonding using ultra-thin titanium (Ti) as adhesive. For the first time to our knowledge, we demonstrate nearly void-free wafer-to-wafer bonding with blanket sputtered Ti, at 400 °C with external pressure applied. The bonded pairs evaluated with scanning acoustic microscopy (SAM) and a series of mechanical integrity tests indicate uniform bonding over 200 mm diameter wafers with sufficient bond strength to survive downstream back-side thinning processes. Mechanisms contributing to the strong bonding are briefly discussed.

EXPERIMENTAL

Blanket Ti films were deposited on 200 mm thermally oxidized silicon wafers using Endura Vectra™ IMP chamber. The Ti film thickness was 30 nm calibrated with Rutherford backscattering spectrometry (RBS). Bare silicon wafers used for bonding experiments were boron-doped P-type (100)-oriented wafers with a resistivity of 15 to 20 Ω-cm and a nominal thickness of 725 μm. Prior to bonding, a Ti wafer ($Ti/SiO_2/Si$ stack marked as "wafer I" in Figure 1) was dipped in 1:1 (by volume) $HCl:H_2O$ solution for 30 sec, followed by deionized (DI) water rinse and spin-rinse-dry. The Si wafer ("wafer II" denoted in Figure 1) was cleaned with Piranha solution for 20 min, followed by deionized (DI) water rinse, then treated in a diluted HF for 30 sec, and finally spin-rinse-dried. Once the pre-bond cleaning was completed, wafers were loaded immediately to Electronic Vision EV501 bonder and clamped face-to-face. After N_2 purge, the chamber was pumped down to $2x10^{-4}$ mbar base pressure and bonded at 400 °C for 2 hours with external down force applied.

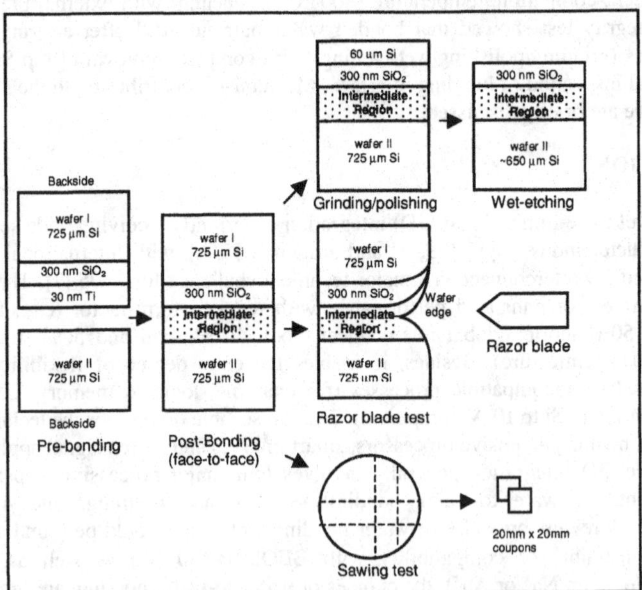

Figure 1. Schematic diagram of wafer bonding and mechanical integrity test flow (dimension not in scale).

The bonding quality was first inspected with SAM, operated at 200 MHz with 6 mm focal length. Then, three mechanical integrity tests were routinely performed, including backside thinning test (grinding, polishing and wet-etching), razor blade test [8], and sawing test (see Figure 1). Sawing test was performed using Disco dicing saw DAD-2H/5, with the spindle rotation speed set at 10 krpm. To complete these three tests, typically one or two bonded pairs processed under the same conditions were needed.

Omnimap Rs50/e four-point probe was used to measure sheet resistance, while Dektak was employed for topographical analysis of bonded surfaces after forced separation. Glancing-angle X-ray diffraction (GIXRD) was performed with incident angle of Cu Kα beam fixed at 0.7°.

RESULTS AND DISCUSSION

Figure 2(a) is the SAM image of the Ti/Si pair bonded at 400 °C. As shown, nearly void-free bonding was realized over the entire 200 mm diameter of the wafer, except for a few voids near the wafer edge, which were probably initiated by particles and surface contaminations prior to bonding. Mechanical integrity test results show that the bonding between Ti and Si wafers is sufficient to pass grinding and polishing steps. Figure 2(b) shows a photograph of the thinned wafer pair, in which, the backside of "wafer I" was ground and polished from the starting thickness of 725 μm to about 60 μm Si remaining. Minor edge chipping was observed, mostly caused by a notch-to-notch offset introduced during manual clamping of two blanket wafers prior to wafer bonding. Figure 2(c) shows a typical contour map of the remaining Si thickness (RST). The thickness non-uniformity appears in a "potato-chip-pattern" with 1-sigma standard deviation of 8.5%. Accurate thickness control is critical to guarantee the success of inter-wafer via etch and metallization. Such surface non-planarity may also cause voids during subsequent metal-based wafer bonding to next wafer level. This requirement could be relaxed for advanced devices built on silicon-on-insulator (SOI) substrate. In our case, 300 nm oxide served as an etch stop. 25% Tetramethyl-ammonium hydroxide (TMAH) kept at 85 °C was used as an etchant to completely strip the remaining Si. The bonded pair after complete strip of top Si wafer is shown in Figure 2(d), which reveals similar void distribution shown in the SAM image.

Sawing test was applied to evaluate possible die chipping and edge cracking, as part of packaging compatibility test. The bonded wafer was sawn into square coupons with nominal dimensions of 20 mm by 20 mm. Smooth edges were obtained, without any sign of delamination. In razor blade test, no evidence of wafer delamination was observed; in particular, the bonded wafer pairs could not be split without damaging them. By applying very aggressive force with razor blade wedging into the beveled edge, we started to chip off small pieces (5~6 mm wide from the edge). However, the diced coupons could never be separated regardless the force applied, indicating strong bonding.

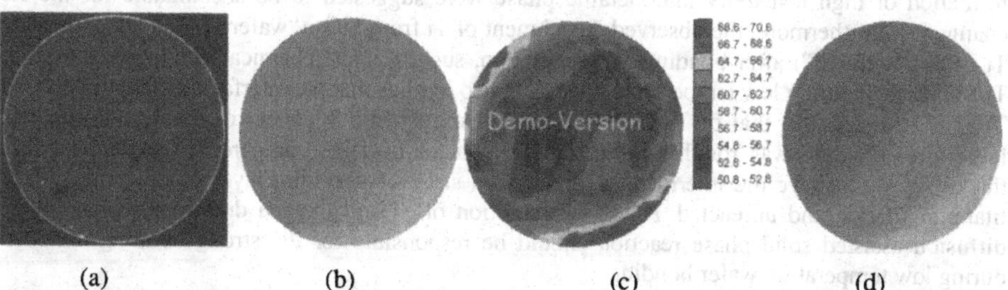

Figure 2. (a) SAM image of 200 mm Ti/Si wafers bonded at 400 °C; (b) Optical microscopy of Ti/Si bonded wafer pair after grinding/polishing; (c) Remaining Si thickness (RST) contour map after grinding/polishing; (d) Optical microscopy of Ti/Si bonded pair after wet-etching.

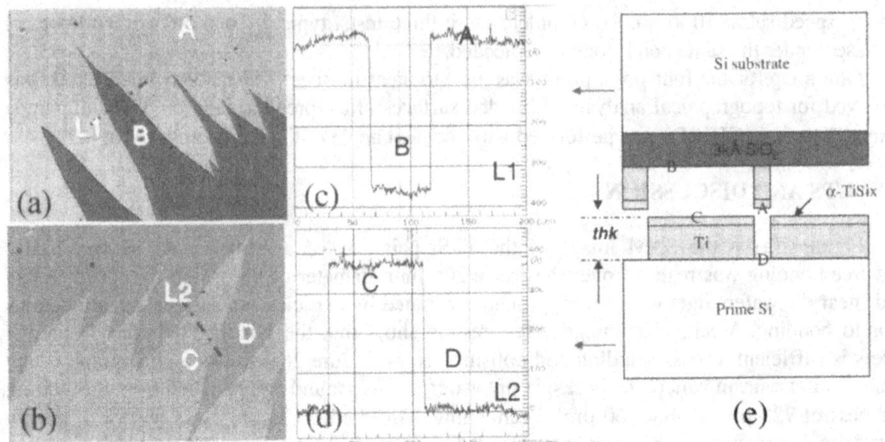

Figure 3. (a-b) Photographs of a matched pair of chipped pieces after a forced separation; (c-d) Dektak depth profile along the lines marked in (a-b); (e) Proposed cross-section schematic after razor blade separation

Figure 3(a) and (b) show a matched pair of chipped pieces after a forced separation. The majority of the surface from "wafer I" side was now in blue color, except at the wafer edge and a few discrete spots in the middle. Note that thermally oxidized Si wafer with 300 nm SiO_2 typically looks in blue color. After deposition of 30 nm Ti, "wafer I" changes to metallic silver color. It should be mentioned that Figure 3(a) and (b) are close-up views of interesting features for further analysis, which are not intended to reflect the relative coverage of two regions in different colors. With nanospec and 4-point-probe, we confirmed that the blue region is thermal oxide with an average thickness of 315 nm and is electrically insulating. Further, the blue area is below the yellow finger-like features with a step of 36 nm deep, measured by Dektak profiler, as shown in Figure 3(c). The measured average sheet resistance (Rs) of the yellow fingers over oxide was 82 Ω/sq, giving an overall resistivity of 300 μΩ-cm, about three or four times that of the as-sputtered Ti (75~100 μΩ-cm for thin Ti films).

Our results are in agreement with Murarka's report during sintering study of Ti/Si couple at low-temperature range (400~500 °C) [9]. Interdiffusion of Si and oxygen into Ti, and a formation of high resistivity intermetallic phase were suggested to be accountable for the Rs change [9]. Furthermore, the observed detachment of Ti from SiO_2 ("wafer I") and attachment of Ti to Si ("wafer II") after bonding and separation, suggest some chemical bonding occurred at Ti/Si interface during low temperature anneal. It also implies that the interfacial strength between Ti/Si is stronger than that of Ti/SiO_2. More interestingly, the thickness of the Ti intermediate layer after bonding was roughly increased by a factor of 1.25, compared to as-deposited Ti thickness. We believe the intermediate region is a stack of very thin layer of low-temperature titanium silicide and unreacted Ti in solid solution of Si and oxygen due to diffusion. Inter-diffusion assisted solid phase reaction should be responsible for the strong bonding obtained during low temperature wafer bonding.

On the opposite side of the matched pieces, reverse features were observed as shown in Figure 3(b), with two distinct regions appeared in yellow and gray color respectively. The yellow area covered most of the surface on "wafer II" side, and was about 38 nm higher than the gray area (Figure 3(d)), with a measured Rs of 54 Ω/sq. The calculated effective sheet resistance of 68 Ω/sq was reasonably close to that of the yellow fingers found on the other side on "wafer I". Therefore the yellow area should be the slightly swollen intermediate film transferred from "wafer I" to "wafer II". The gray area had a much larger Rs of 250 Ω/sq, which could be basically identified as bare Si substrate. A proposed cross-section schematic after razor blade separation is illustrated in Figure 3 (e).

Glancing angle XRD was employed to check whether there was any crystalline silicide phase formed after bonding. Figure 4 shows the spectrum of bonded wafer pair after complete strip of Si from backside of "wafer I". The result shows the presence of (110) α-Ti (hcp), with two other broad peaks around 34.6 and 48.7 degrees (2-theta). These spread peaks suggest an amorphous phase might exist at the interface, however their identification is difficult, considering the large full width at half maximum (FWHM) of these peaks. Ti-Si system has five silicide compounds, Ti_3Si, Ti_5Si_3, Ti_5Si_4, TiSi, and $TiSi_2$, in which, $TiSi_2$ has two polymorphs: a metastable C49-$TiSi_2$ (base-centered orthorhombic structure) and a stable C54-$TiSi_2$ phase (face-centered orthorhombic structure). Among them, the diffraction lines from Ti_5Si_3, Ti_5Si_4, and TiSi are crowded at the angles shown in Figure 4.

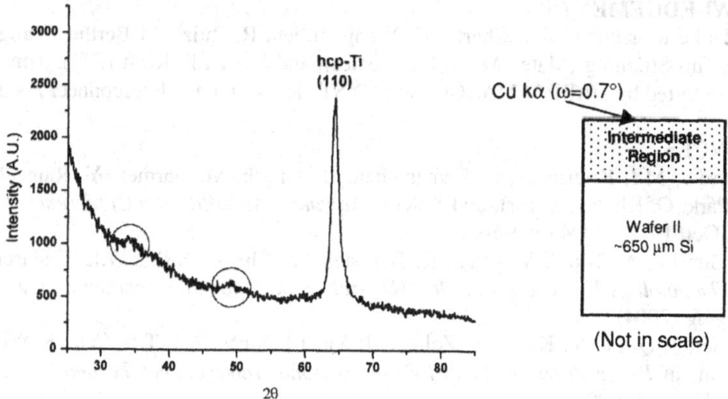

Figure 4. Glancing angle XRD spectrum of Ti/Si bonded pair after complete strip of Si from backside of "wafer I".

Ti has been extensively applied in front-end-of-the-line (FEOL) processes to form C54-$TiSi_2$ (15~20 μΩ-cm), one of the most commonly used low resistivity silicide/salicide materials. Therefore, most of the research interest has been focused on the growth kinetics of $TiSi_2$ and polymorphic phase transition from C49 to C54. The annealing behavior at low temperature range (below 500 °C) is relatively not well known. The limited work conducted in this area was mostly published in the early 1980's [9, 10]. Ti has strong affinity to oxygen and Si and can reduce native oxide to form Ti-O bond at low temperature [10]. Strong Ti-Si intermixing was reported at temperature as low as 300 °C [10], about 100~200 °C lower than the onset temperature normally expected for crystalline titanium silicide formation. Our demonstration of strong Ti/Si wafer bonding at 400 °C provides new evidence of the solid phase reaction occurring at such low temperature. More interestingly, the bonding and separation method described in this paper offers a new perspective to further explore the subject. More detailed study assisted with high-resolution structural and compositional analytical tools are under way to understand the Ti/Si bonding mechanism at low temperature.

SUMMARY AND CONCLUSIONS

We demonstrate, for the first time to our knowledge, nearly void-free 200 mm Si wafer-to-wafer bonding using an ultra-thin Ti adhesive, bonded at BEOL-compatible conditions (≤400 °C). The bonded pairs evaluated with SAM and a series of mechanical integrity tests indicate uniform bonding over entire wafer with sufficient bond strength to survive downstream backside thinning processes. After bonding, the thickness of intermediate layer was increased by approximately 25%, while the overall stack resistivity was increased to 300 μΩ-cm, about three to four times that of the as-sputtered Ti. In addition, after wafer bonding and forced separation, most Ti detached from SiO_2 side (Ti/SiO_2 deposition interface), and adhered to Si side (new bonding interface). We attribute the strong bonding achieved at low temperatures to inter-diffusion assisted solid phase reaction.

ACKNOWLEDGEMENTS

We would like to acknowledge Charles C. Wang, Julieta R. Ruiz and Bertha Chang (Applied Materials), Jim Stradling (Matec Micro Electronics), and Russel P. Kraft (RPI). This work was partially supported by DARPA, MARCO, and NYSTAR through the Interconnect Focus Center.

REFERENCES

1. P. Morrow, M.J. Kobrinsky, S. Ramanathan, C.M Park, M. Harmes, Y. Ramachandrarao, H.M. Park, G. Kloster, S. List, and S. Kim, *Advanced Metallization Conference 2004 (AMC 2004)*, Oct. 19-21, 2004 (in press).
2. K.W. Guarini, A. Topol, V. Chan, K. Bernstein, L. Shi, R. Joshi, W.E. Haensch, and M. Ieong, *Technology Venture Forum for 3D Architectures for Semiconductor Integration and Packaging* (2004).
3. S. Pozder, J.-Q. Lu, Y. Kwon, S. Zollner, J. Yu, J.J. McMahon, T.S. Cale, K. Yu, and R.J. Gutmann, in *Proceedings of the IEEE International Interconnect Technology Conference (IITC)*, 102-104 (2004).
4. http://www.tezzaron.com
5. A. Fan, A. Rahman, and R. Reif, *Electrochemical and Solid-State Letters*, **2** (10), 534-536 (1999).
6. J.-Q. Lu, Y. Kwon, J. J. McMahon, A. Jindal, B. Altemus, D. Cheng, E. Eisenbraun, T.S. Cale, and R.J. Gutmann, in *Proceedings of 20th International VLSI Multilevel Interconnection Conference*, T. Wade, Editor, 227-236 (2003).
7. http://www.sematech.org/corporate/news/releases/20040610a.htm
8. M.A. Schmidt, *Proceedings of the IEEE*, **86** (8), 1575-1585 (1998).
9. S.P. Murarka and D.B. Fraser, *J. Appl. Phys.* **51** (1), 342-349 (1980).
10. R. Butz, G.W. Rubloff, and P.S. Ho, *J. Vac. Sci. Technol.* A **1** (2), 771-775 (1983).

Mater. Res. Soc. Symp. Proc. Vol. 843 © 2005 Materials Research Society T7.8

Structural factors determining the nanomechanical performance of transition metal nitride films

K. Sarakinos, S. Kassavetis, P. Patsalas and S. Logothetidis

Aristotle University of Thessaloniki, Department of Physics, GR-54124, Thessaloniki, Greece

ABSTRACT

Chromium nitride (CrN) and Titanium nitride (TiN) thin films were deposited employing unbalanced magnetron sputtering (UBMS) for various values of substrate bias voltage (V_b). The structural characterization of the films in terms of phase identification and the density determination was achieved utilizing X-Ray techniques (XRD and XRR respectively), while the internal stresses were calculated by the change of the substrate's curvature using Stoney's equation. The nanomecahnical properties of the films (hardness and elastic modulus) were investigated using the Nanoindentation (NI) technique in the Continuous Stiffness Measurement (CSM) configuration. According to the NI results the hardness of the films ranges between between 15-25 GPa while the elastic modulus between 180-250 GPa. The analysis revealed that the hardness of the films is maximum when their orientation is pure (either [111] or [100]), while it is minimized under mixed orientation regime. Furthermore, the hardness of the films increases as the internal compressive stresses and the mass density increase. The latter can be validated through the comparison of the results concerning the above films to reported results for TiN films prepared by balanced magnetron sputtering (BMS)

INTRODUCTION

Transition metal nitrides have been established as a category of very important technological materials since they exhibit exceptional mechanical properties and find a vast range of optical properties [1]. In specific, the mononitrides of the group IVb-VIb of the transition metals form *fcc* crystals (rocksalt type) and they exhibit or not electronic conductivity depending on the position of the Fermi level in respect to the 3-d band of the constituent metal [1]. Among them TiN is the most widely investigated due to its refractory character and gold-like color. On the other hand CrN gains scientific and research interest during the last years due to its increased resistance to severe environments where TiN is not functional [2-5].

In this work we present a thorough study of the structural properties, which determine the nanomechanical properties of unbalanced magnetron sputtered (UBMS) CrN and TiN thin films. The effect of substrate bias voltage (V_b) on the nanomechanical response (i.e. hardness and elastic modulus) of the above thin films is presented in more detail. Taking into account the established correlation between the structural features of films and V_b [6], the effect of structural factors to the nanomechanical performance can be also investigated. The results revealed that the hardness is maximized in the case of clear growth orientation (either [111] or [100]). The above results are compared to results concerning TiN grown with balanced magnetron sputtering [7] in order to support the related analysis.

EXPERIMENTAL

CrN and TiN thin films were grown on c-Si (100) substrates by reactive magnetron sputtering in unbalanced configuration using a TEER UDP 450 system with P_b ~2 x 10^{-5} Torr in a mixed Ar/N_2 (99.999% purity) ambient. Prior the loading into the deposition chamber the substrates were cleaned in an ultrasonic bath using a standard chemical procedure [8, 9]. Cr and Ti targets of purity 99.95% were used for the growth. The substrates were etched in Ar plasma applying bias voltage -500 volts in order to remove the native oxide layer prior to deposition. The deposition was carried out at constant Ar and N_2 flow ($\Phi_{Ar}=\Phi_{N2}=15$ sccm) and target current (2 A for Cr and 6 A for Ti) for various values of V_b from floating to -150 volts. The deposition time was adjusted based on previous deposition rate data [10,11] in order to deposit films with maximum thickness ~150 nm.

X-Ray Diffraction (XRD) and X-Ray Reflectivity (XRR) were employed for the structural characterization and the density determination, respectively. The XRR/XRD experiments were performed in Bragg-Brentano geometry using the K_α line of a Cu anode in a Siemens D-5000 diffractometer equipped a Goebel mirror. The residual internal stresses of the films were determined *ex-situ* from the variations of the sample curvature using Stoney's formula [12] and a relative Tencor® F2320 instrument.

The nanomechanical characterization of the CrN and TiN thin films was achieved employing the nanoindentation (NI) technique [13] in a Nanoindenter XP system in Continuous Stiffness Measurements (CSM) configuration [14]. At each sample were performed 5-10 indents on pre-selected surface area while the maximum penetration depth (d_p) was ~250 nm The results were obtained in form of hardness and elastic modulus curves versus d_p.

RESULTS AND DISCUSSION

Representative XRD patterns of CrN and TiN thin films deposited at different values of V_b are presented in Fig. 1 (a) and b) respectively). Both CrN and TiN thin films exhibit the rocksalt crystal structure and their grains grow along the [111] and [100] crystallographic orientation, which correspond to the reflections (111) and (200) respectively [15,16] and they are the most common for the d-metal nitrides [8, 9]. Moreover the angular position of the reflections is shifted towards lower values compared to the angular position of the bulk and unstrained materials [15, 16] (vertical dotted lines) implying that the films underlying compressive internal stresses; this is confirmed by the stress measurements.

The preferred growth orientation is calculated from the integrated intensity of the XRD peaks (i.e the area below the peaks), as described in more detailed in Ref. 17. The primitive cell size of each film was determined, in first approximation) from the angular position of the (111) XRD reflection [where present, otherwise the (200) reflection was used]; additional XRD patterns in ψ tilted configuration were also acquired from selected samples in order to determine the film stress and the cell size with higher accuracy [17].

The hardness (a) and the elastic modulus (b) of representative CrN thin films versus the penetration depth (d_p) of the indenter are presented in Fig 2. The nanomechanical properties of the films, as acquired by CSM, are strongly affected by substrate effects. Therefore, the measured hardness values were fitted by the Battacharya-Nix equation [18], in order to exclude the substarte contribution and determine the real film hardness. The film hardness H_f corresponds to the solution of Battacharya-Nix equation at d_p=0. On the other hand, the elastic modulus of the film E_f corresponds to the values at the initial d_p. In both figures the horizontal dotted lines correspond to the mechanical properties of the c-Si substrate [19].

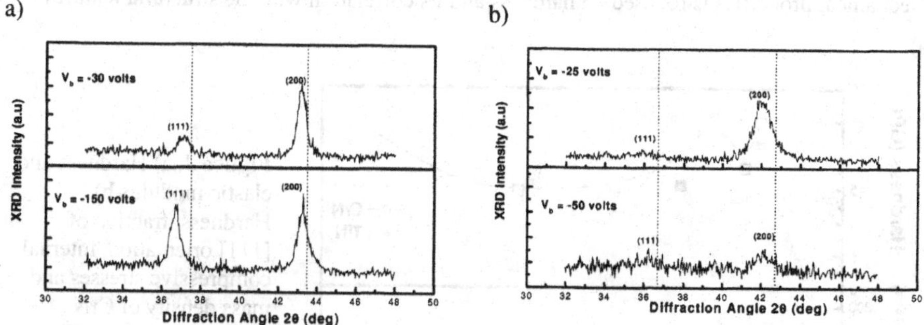

Figure 1. Representative XRD patterns of a) CrN and b) TiN thin films deposited at various values of V_b. The vertical dotted lines correspond to the angular position of the reflections in bulk and unstrained materials

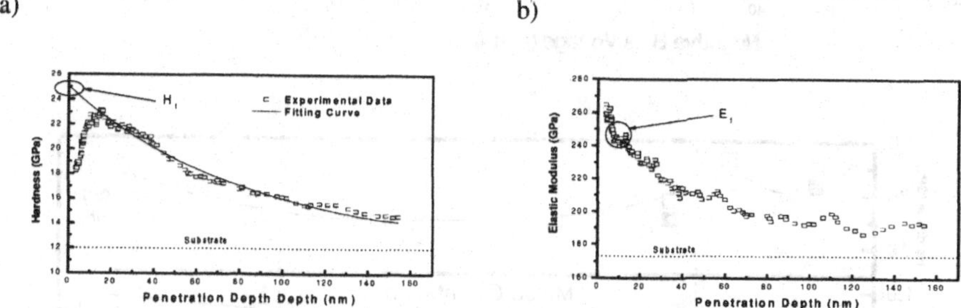

Figure 2. a) Hardness and b) Elastic modulus of representative CrN thin films versus d_p (open squares). The solid line corresponds to the fitting of the experimental data to the Battacharya–Nix model. The horizontal dotted lines correspond to the mechanical properties of c-Si.

The results about H_f and E_f of CrN (open squares) and TiN (closed squares) versus V_b are summarized in Fig. 3a. The H_f values ranges between 15-25 GPa while the E_f values between 180-250 GPa. Both quantities exhibit similar dependence on the V_b. This should not necessarily lead to the conclusion that the two quantities obey a linear law. The elastic modulus is an intrinsic characteristic and depends mainly on the bonding properties of each material [20]; take this into account, the high elastic moduli of transition metal nitrides are due to their covalent bonding [21]. On the other hand, the hardness is a macroscopic quantity, which represents the plastic deformation of a material under specific load conditions and it depends on the mobility of the dislocations across specific slip planes within the lattice. Therefore, the hardness is also affected by non-intrinsic properties of the material, such as structural defects, grain boundaries and morphological features not present in the corresponding singe-crystal. The behavior observed in Fig. 3b can be attributed to underestimation of elastic moduli, measured by CSM, which has been also reported for a-C thin films [22]. Thus, in the followings the analysis of the mechanical properties is focused on hardness and its correlation with the structural features.

a)

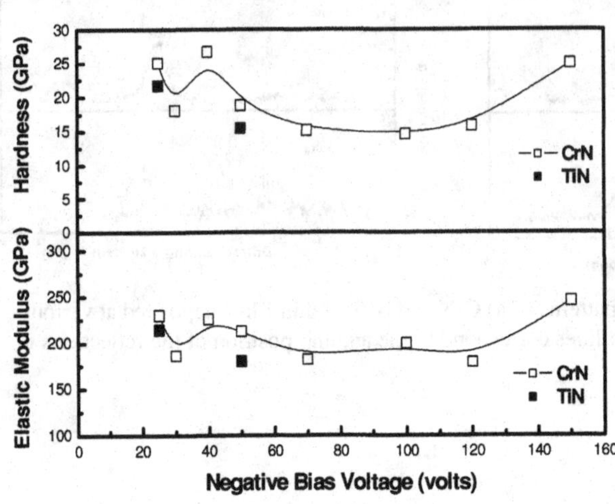

Figure 3. a) Hardness and elastic modulus b) Hardness, fraction of [111] orientation, internal compressive stresses and mass density of CrN (open squares) and TiN (closed squares) thin films versus substrate bias voltage. The solid lines are guides to the eye.

b)

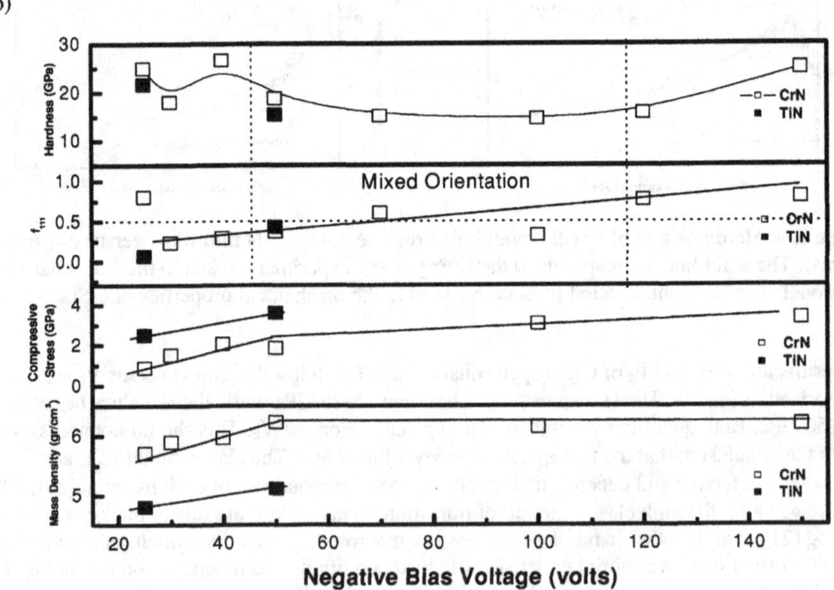

Fig. 3b shows the hardness, the fraction of the [111] growth orientation calculated as in Ref. 9, the compressive stress and the mass density of CrN and TiN (open and closed squares respectively) thin films versus V_b. Fig. 3b provides a better overview for the nanomechanical response of the above films in terms of the above mentioned structural characteristics.

In general, the nanomechanical response of the materials is affected by the growth orientation. The elastic modulus along the [100] crystallographic direction is higher than the one of the crystallographic direction [111] [22]. As a result, the increase of the fraction of grains which grow along the [100] orientation has a result an increase in elastic modulus and consequently in hardness. On the other hand, increase in hardness of TiN thin films with increase in the fraction of the [111] orientation has been reported [23]. This is attributed to the fact that TiN (and CrN similar) exhibits the crystal structure of NaCl. In the above structure the active slip system is the {111} <110> [24]. In the case which the external force is perpendicular to the (111) plane then the resolved shear stress is to all slip directions zero. As a result lower plastic deformation can be induced and the measured hardness is higher.

Besides, the internal stress level influences the hardness. Higher internal stress level results in an increased lattice deformation. Thus, the induced plastic deformation decreases and subsequently the hardness increases. Finally, increase of the mass density results in the formation of a more compact structure with fewer grain boundaries and thus higher hardness.

The above analysis is clearly illustrated in Fig. 3b. In particular, in the case of well-defined [111] grain orientation of CrN (low and high values of V_b) the hardness is maximized, while at intermediate values of V_b (mixed orientation regime in Fig. 3b) the hardness is lower. Moreover, the effect of mass density and stress is clear at low values of V_b. Thus, while the increase of negative V_b from 30 volts to 40 volts results in an increase of [111] orientation fraction, leading towards to the mixed orientation regime, the hardness increases due to the internal compressive stress and the mass density increase. Further increase of negative V_b, does not induce increase of hardness, although it results in further increase in stress and mass density, since we are in the mixed orientation regime. Consequently, in this case (CrN) the grain orientation seems to be the dominant structural factor, which determines the nanomechanical response of the films. In the case of TiN thin films the dominant effect of the preferred orientation is more pronounced. More detailed, the increase of negative V_b, results in the decrease of hardness, since it causes transition from pure [100] orientation to a mixed orientation regime, although a significant increase in compressive stress and mass density takes place.

Comparing the results for the above films with reported results for balanced magnetron sputtered (BMS) TiN thin films [7] some interesting conclusions arise. In the case of TiN films of reference [7] the hardness increases monotonically with V_b, while it is maximized at high values of [100] grain orientation and takes its minimal values at high values of [111] grain orientation. This discrepancy can be attributed to the differences between the two deposition techniques. The exhibiting [111] dominant orientation BMS TiN films were deposited at low values of V_b. This has a result lower internal stresses and mass density compared not only toBMS films deposited at higher V_b but also toUBMS films deposited at similar conditions, since the unbalanced magnetic field induces more intense bombardment of the substrate and subsequently the growing film causing an increase in density and compressive stress. For the above reason the hardness in the case of TiN films in reference [7] exhibit monotonous behavior, while in the case of the presented films high values of hardness are observed even at low values of V_b.

CONCLUSIONS

In this work was presented the effect of structure on the nanomechanical response of UBMS CrN and TiN thin films. CrN and TiN films were deposited at various values of V_b and their nanomechanical properties were investigated in terms of hardness and elastic modulus, while their structure by means of preferred growth orientation, internal stresses and mass density. The analysis showed that nanomechanical performance of the above films is mainly governed by the preferred orientation. More detailed, the hardness of the films is maximized in the case of clear either [100] or [111] orientation. This is attributed to the fact that $E_{100}>E_{111}$ for the [100] structure. In the case of [111] orientation this is due to the active slip system ({111} <110>) in the fcc crystals. For a load perpendicular to the {111} planes the resolved shear stress is zero and thus the induced plastic deformation low. Both of the above reasons for the two directions cause increase of hardness as the fraction of both [111] and [100] growth orientations increases resulting in maximized hardness under pure orientation regime, while the latter is minimized when the orientation is mixed. On the other hand the internal stresses and the density affect also the nanomechanical properties of the films. This is more pronounced in the case of BMS TiN films where the hardness increases monotonically with the fraction of [100] orientation and takes its minimal values at when [111] orientation dominates. The latter can be attributed to the fact that the [111] oriented BMS films exhibit low values of stress and density, since they are fabricated at low values of V_b.

REFERENCES

1. C. Stampfl, W. Mannstadt, R. Asahi, A.J. Freeman, Phys. Rev. B63, 155106 (2001).
2. L. Hultman, Vacuum 57, 1 (2000).
3. D.A Glocker, S.I Shat, "Handbook of thin film process technology", Institute of Physics, Wilmington, Delaware (1995).
4. G. Bertrand, C. Savall, C. Meunier, Surf. Coat. Technol. 96, 323 (1997).
5. B. Navansek, P. Panjan, I. Milosev, Surf. Coat. Technol. 97, 182 (1997).
6. K. Sarakinos, M.Sc. Thesis, Aristotle University, Thessaloniki (2004).
7. P. Patsalas, C. Charitidis, S. Logothetidis, Surf. Coat. Technol. 125, 335 (2000)
8. P. Patsalas, S.Logothetidis, J. Appl. Phys. 93, 989 (2003).
9. P. Patsalas, S. Logothetidis, J. Appl. Phys. 90, 4725 (2001).
10. S. Logothetidis, P. Patsalas, K. Sarakinos, C. Charitidis, C. Metaxa, Surf. Coat. Tech. 180-181, 637 (2004).
11. V.G. Kechagias, M. Gioti, S. Logothetidis, R. Benferhat, D. Teer, Thin Solid Films 364, 213 (2000).
12. G.G. Stoney, Proc. Roy. Soc. London A82, 172 (1909)
13. G.M. Pharr, W.C. Oliver; MRS Bulletin, XVII, 28 (1992).
14. J.B. Pethica, W.C. Oliver, Phys. Scr. T19, 61 (1987).
15. JCPDS Powder Diffraction File No. 76-2494
16. JCPDS Powder Diffraction File No. 38-1420
17. P. Patsalas, C. Gravalidis, S. Logothetidis, J. Appl. Phys. 92, 1 December (2004).
18. A.K. Bhattacharya, W.D.Nix, Int. J. Solid Structures 24, 1287 (1988).
19. H. Gao, C.H. Chiou, J.Lee, Int.; J. Solid Structures 29, 2471 (1992).
20. C. Kittel, Introduction to Solid State Physics, 4th edition, J. Willey & Sohns (1971).
21. M. Marlo, V. Milman, Phys. Rev. B62, 2899 (2000).
22. C. Mathioudakis, P.C. Kelires, Y. Panayiotatos, P. Patsalas, C. Charitidis, S. Logothetidis, Phys. Rev. B65, 205203 (2002).
23. C.T. Chen, Y.C. Song, G.-P. Yu, J.-H. Huang; J. Mater. Eng.Perform. 7, 324 (1998).
24. L. Hultman, M. Shinn, P.B. Mirkarimi, S.A. Barnett, J. Cryst. Growth 135, 309 (1994).

Mater. Res. Soc. Symp. Proc. Vol. 843 © 2005 Materials Research Society T0.0

The Tribological Behavior of MoS_2-Cr Films Sliding Against an Aluminum Counterface

James E. Krzanowski, Department of Mechanical Engineering and Materials Science Program, University of New Hampshire, Durham, NH 03824, U.S.A.

ABSTRACT

The tribological performance of MoS_2-Cr films in a pin-on-disk test sliding against an aluminum ball has been examined in this study. The MoS_2-Cr films were deposited by pulsed laser deposition (MoS_2) and simultaneous sputter deposition of Cr, giving a Cr content in the film of 10 mol. %. The results were also compared with films of MoS_2 alone. The frictional behavior of MoS_2-Cr films was not improved compared to the MoS_2 alone, so SEM/EDS studies of the ball and flat were conducted to determine the nature of the transfer films and examine any interface reactions that occurred during the pin-on-disk (POD) test. In the early stages of the POD test (500 cycles) on the MoS_2-Cr film, Al-oxide particles formed and caused cratering and scratching of the wear track, and the coefficient of friction neared 0.7. At later stages (9000 cycles), a thick oxide-based transfer film formed on the ball, but on the flat the track composition was closer to that of the original coating. For the films without Cr, after 10^4 cycles a smooth wear track was observed, and a thin transfer film of MoS_2 was found within grooves on the ball wear scar along with Al oxide, which resulted in superior tribological performance.

INTRODUCTION

The natural lubricious properties of MoS_2 have led to numerous investigations aimed at depositing it as a coating material. The early work of Spalvins [1] used sputtering methods to deposit films of MoS_2. Since then, d.c., r.f., magnetron and reactive sputtering methods have been used [2-4], in addition to ion beam enhanced deposition [5], pulsed laser deposition [6] and chemical vapor deposition [7]. The primary issues investigated were the effects of chemical purity, crystallographic texture, operating environment, and types of dopant elements on tribological properties.

The application MoS_2 in air environments has been limited by the degradation of the frictional properties of MoS_2 in humid environments. This has led to studies on the effect of oxygen on the structure properties and structure of MoS_2 films. Lince [8] found that oxygen can form solid solutions with MoS_2 by substituting for S atoms, resulting in the formation of $MoS_{2-x}O_x$ compounds. The effects of oxygen were reported as enhancing edge growth and reducing wear life. However, it has also been suggested that the presence of oxygen in $MoS_{2-x}O_x$ can reduce friction compared to pure MoS_2, and films with 5-15% oxygen often exhibit very low friction coefficients (in the 10^{-2} range) [9].

In order to improve the tribological properties of MoS_2 in humid environment applications, research has been conducted on compositional modifications to MoS_2. Elements and compounds added to MoS_2 include Au [10], PbO [11], and Fe, Ni and

Sb_2O_3 [12]. Improvements in frictional properties in humid environments have been reported in several of these studies, possibly due to protection of the edge plane by the dopant elements. The formation of composite films containing MoS_2 and other elements was investigated by Fox et al. [13], who attempted to deposit MoS_2/Ti multilayers by dc magnetron sputtering. However, a multi-layered structure was not obtained, so they proceeded to investigate co-sputtered MoS_2 and Ti films. There films demonstrated improved tribological behavior, particularly in humid environments. Dry machining tests were also conducted on tools coated with TiN and a MoS_2 + Ti topcoat. In their test, drills coated with TiN alone in dry drilling produced 141 holes before failure of the drill, while tools with MoS_2 + Ti topcoat drilled 303 holes under the same conditions.

In the present study, we examine the tribological behavior of MoS_2 films doped with approximately 10% Cr. Previous studies [14] have shown significant improvements in the frictional behavior of MoS_2-Cr films, with friction coefficients near 0.15 when riding against a steel ball. In the research presented here, we examine the behavior of MoS_2-Cr films riding against an aluminum ball. The motivation of using an Al ball is to determine the potential of MoS_2-Cr films as tool coatings for machining aluminum.

EXPERIMENTAL PROCEDURES

Thin films of MoS_2 doped with Cr were deposited on to polished 440C stainless steel substrates. The MoS_2 was deposited by pulsed laser deposition using a KrF excimer laser, and the Cr was simultaneously deposited by dc sputtering. The vacuum chamber employed had a base pressure of $3x10^{-7}$ torr, and was backfilled during deposition with argon to a pressure of 5 mtorr. Deposition was carried out without intentional heating of the substrates. The pin-on-disk test carried out in air (50% r/h) using a 1 N load, a 6.25 mm diameter Al ball, at 200 rpm. Analysis of the ball and flat was carried out using scanning electron microscopy (SEM) and energy dispersive analysis (EDS) on an AMRAY 1410 SEM. Quantification of the spectra was conducted using the IXRF EDS 2004 program (IXRF Systems, Houston, TX) and the ZAF correction routines contained therein. Analysis of the as-deposited films of MoS_2-Cr showed a Mo:Cr atomic ratio of 9:1, indicating 10 mol.% of Cr in the films.

RESULTS

<u>MoS_2-Cr Film</u>

The pin-on-disk test results for the two films riding against the Al counterface are shown in Fig. 1. Fig. 1(a) shows the friction coefficient vs. number of cycles for the MoS_2-Cr film. The initial friction coefficient is near 0.2, but then rises rapidly to a maximum of almost 0.7 after about 500 cycles. The friction coefficient then declines, and reaches an average coefficient of about 0.25. However, the noise in the scan is very high for the entire run. For the MoS_2 film shown in Fig. 1(b), there is a steady increase from 0.1 to close to 0.3, and the noise is significantly lower than in Fig. 1(a). Overall, the performance of the MoS_2-Cr film is inferior to MoS_2 alone. To understand the basis of this behavior, extensive SEM/EDS analysis was conducted and is described below.

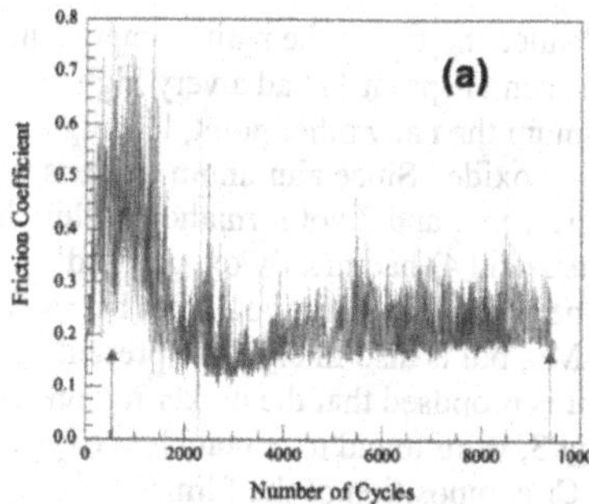

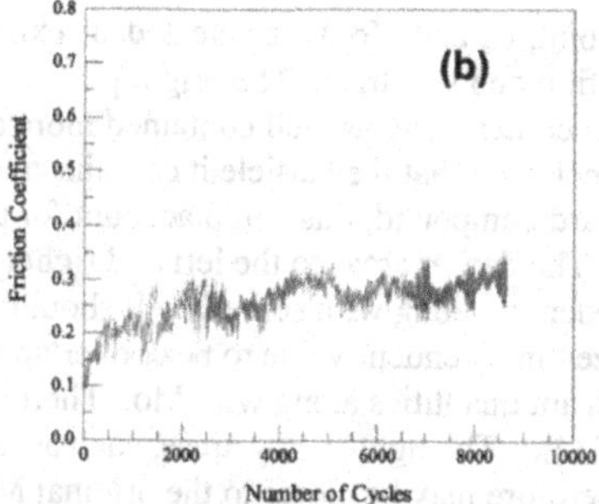

Fig. 1: Pin-on-disk test results for the (a) MoS_2-Cr and (b) MoS_2 films. In both cases the counterface was an Al ball, with a 1 N load and a speed of 200 rpm. In (a), arrows show specimens selected for analysis in this report.

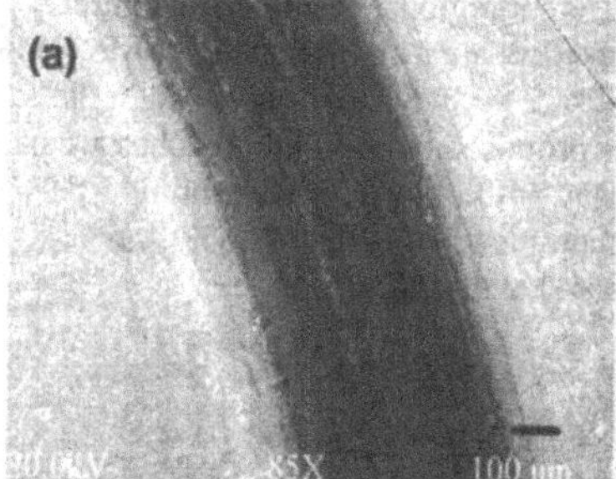

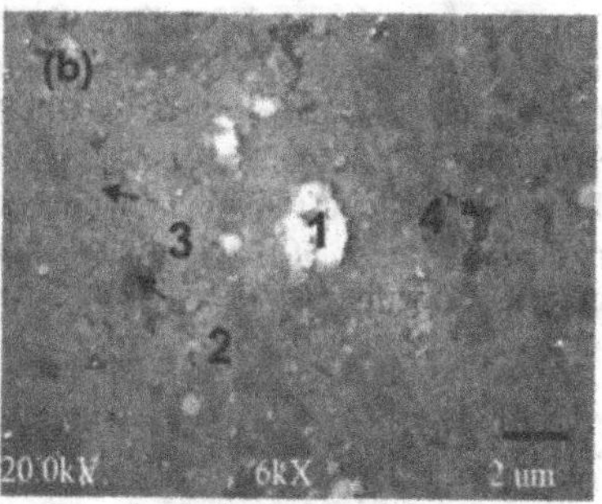

Fig. 2: Wear track for the MoS_2-Cr film after 500 cycles. Lower magnification (a) shows scratching within track; (b) higher magnification shows debris as well as points analyzed.

Table I: Analysis of points in Figure 2(b), At.%

Element	Point 1	Point 2	Point 3	Point 4
Oxygen	64.4	36	20.2	43.2
Aluminum	4.6	0.5	0.5	1.4
Chromium	6.4	9.2	10.9	8.4
Iron	0.5	0.5	1.0	0.6
Molybdenum	24	53.4	67.3	46.3

Analysis was first conducted on the MoS_2-Cr film in the wear track after 500 cycles, the point at which the friction coefficient is near a maximum. Fig. 2(a) shows the wear track at low magnification, and a significant amount of debris was noted along with scratching and the formation of divots within the track. Fig. 2(b) shows detail within the track along with EDS analysis of several points. The results of the analyses are shown in Table I. At each point, Fe and Mo were detected, as expected since they were the main components in the film and substrate. The bright particle in the center (point 1) had a very high oxygen content, and as well contained more aluminum than any other point, leading to the conclusion that the particle it contains aluminum oxide. Since aluminum oxide is a very hard compound, this could account for the scratching and divot formation within the track. The darker areas to the left and right (points 2 and 4) had mostly oxygen and molybdenum along with some Cr. It should also be noted that sulfur could not be analyzed independently due to peak overlap with Mo, but is also likely to be present in significant quantities along with Mo. Therefore, it is proposed that the darker regions are $Mo_x(S,O)_y$. The lighter grey areas, such as at point 3, were found to contain less oxygen, and therefore may be closer to the original MoS_2-Cr composition of the film.

Analysis was also conducted on the aluminum ball after 500 cycles and is shown in Fig. 3(a). Compositional analyses of the indicated points is shown in Table II. As expected, oxygen and Al are the main components, but Mo and Cr were also detected indicating a transfer of material from the film to the ball. A closer look at the ball wear scar is shown in Fig. 3(b). Here we observed two distinctly different regions, described as lighter patches on a darker background. The region marked as area 1 contained 71% Mo, 19% Cr, 8% O and 2% Al, indicating it is primarily Mo-Cr (with S also likely). With the Mo/Cr ratio close to that in the original film, it is proposed these lighter patches are parts of the film transferred to the ball but not yet reacted with oxygen or aluminum. The region marked "2" in Fig. 3(b) was found to contain mostly oxygen (46.8%) and aluminum (38.7%) with lesser amounts of Cr (6.5%) and Mo (7.9%).

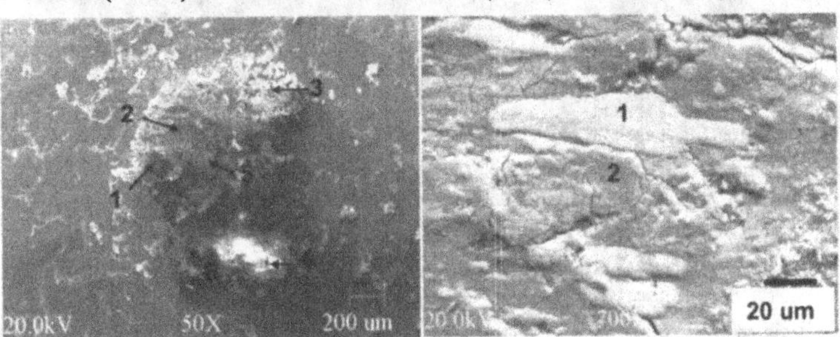

Fig. 3: Aluminum ball wear scar after 500 cycles with the MoS_2-Cr film: (a) Secondary electron image showing the overall wear scar and points analyzed; (b) Detail in the wear scar showing different phases within the transfer film.

Examination of the wear track in the MoS_2-Cr films after 9000 cycles revealed both scratched and burnished areas. Compositional analyses of both of these areas was similar, with Mo contents of over 80%, and oxygen contents of less than 5%. This suggests the track was primarily MoS_2 containing a small amount of Cr. The ball side is shown in Fig. 4(a), where a thick transfer film was observed. The primary constituents within this film (areas 1 and 2) were oxygen (>60%) followed by Al (18-20%) with lesser amounts of Mo and Cr. Therefore, near the end of the wear test, the tribological couple has become one of MoS_2(Cr) on the flat riding against an oxide-based compound containing Al, Cr and Mo (and possibly some S). It was also noted that the transfer film exhibited many fine cracks indicating it was brittle in nature.

Table II: EDS Analysis of Points in Figure 3(a), At.%

Element	Pt. 1	Pt. 2	Pt. 3	Pt. 4	Pt. 5	Pt. 6
Oxygen	61.3	55.9	30.3	59.9	64.4	36.4
Aluminum	13.7	30.2	57.1	25.7	17.6	62.4
Chromium	12.8	6.3	4.7	6.3	8.2	0.3
Molybdenum	12.2	7.6	7.9	8.9	9.7	0.9

MoS_2 Film

The MoS_2-only film was analyzed after 10^4 cycles. Fig. 4(b) shows the ball, and several analysis areas are indicated. In comparison to Fig. 4(a), the wear scar in 4(b) shows fewer cracks and a regular pattern of grooves on the surface. Analyses of the wear scar showed the grooves to be alternating in composition of Al and Mo, indicating incomplete coverage of the ball by the MoS_2 film. Outside the wear scar (point 2, Fig. 4(b)), only aluminum was detected.

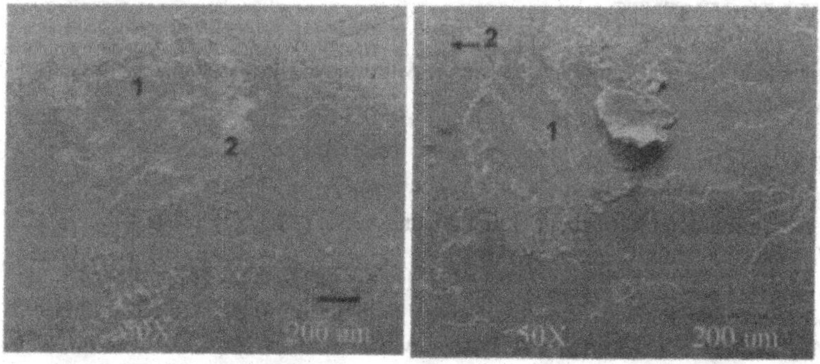

Fig. 4: (a) Al ball after 9000 cycles against the MoS_2-Cr film, showing a substantial transfer film; (b) the Al ball tested against the MoS_2 film after 10^4 cycles; here the transfer film is thinner and contains transferred MoS_2 within the grooves.

The wear track on the flat after 10^4 cycles is shown in Fig. 5. In comparison to the MoS_2-Cr film, which showed numerous divots, the wear track was smooth and free of large divots. Also, only Mo (S) was detected within the track. As a result, the final wear couple appears to be a MoS_2 film riding against a surface on the Al ball that contains grooves of MoS_2 and aluminum oxide.

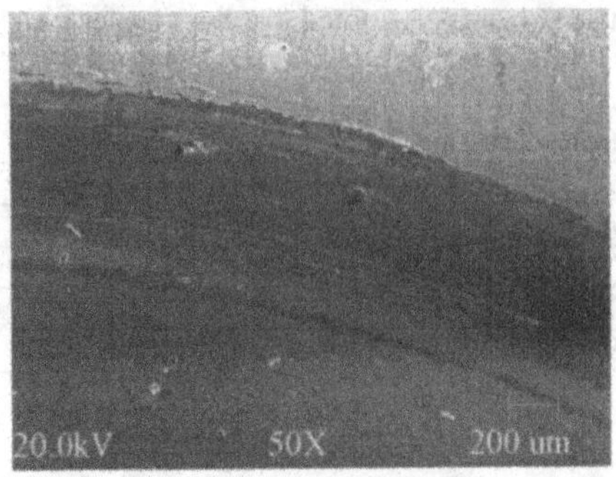

Fig. 5: Wear track of the MoS_2 coated flat after 10^4 cycles showing a smooth and uniform film of MoS_2 remaining within the track.

CONCLUSIONS

The tribological behavior of and MoS_2-Cr film has been compared to one of MoS_2 alone. The MoS_2-Cr film demonstrated poorer performance in a pin-on-disk test using and Al ball. SEM and microchemical analysis showed that the in the MoS_2-Cr film Al-oxide particles formed in the early stages of wear and caused cratering and scratching of the wear track. As the wear proceeded, a thick oxide-based transfer film formed on the ball. For the films without Cr, a smooth wear track was observed, and a thin transfer film of MoS_2 was found within grooves on the ball wear scar along with Al oxide. This gave superior tribological performance.

ACKNOWLEDGEMENTS

The author would like to acknowledge William A. Porter for his assistance with the pin-on-disk wear testing.

REFERENCES

1. Spalvins, T., Thin Solid Films, v. 96, p. 17 (1982).
2. P.D. Fleishauer and R. Bauer, Tribol. Trans. v. 31, p. 239 (1988).
3. V. Buck, Vacuum, v. 36, p. 89 (1986).
4. J. Moser, F. Levy, F. Bussy, J. Vac. Sci. Tech., A12, p. 494-500 (1994).
5. L.E. Seitzman, R. N. Bolster, I.L. Singer, Thin Solid Films, v. 260, p. 143 (1995)
6. M.S. Donley, P.T. Murray, S.A. Barber, and T.W. Haas, Surf. Coat. Tech., v. 36, pp. 329-340 (1988).
7. J. Cheon, J.E. Gozum, and G.S. Girolami, Chem. Mater. v. 9, pp. 1847-1853 (1997).
8. J.R. Lince, J. Mater. Res., v. 5, pp. 218-222 (1990).
9. T. Le Monge, C. Donnet, J.M. Martin, N. Milliard-Pinard, S. Fayeulie, and N. Moncoffre, J. Vac. Sci. Tech. A12, pp. 1998-2004 (1994).
10. M.C. Simmonds, A. Savan, E. Pfluger and H. Van Swygenhoven, J. Vac. Sci. Tech. 19A, pp. 609-613 (2001).
11. S.D. Walck, M.S. Donley, J.S. Zabinski and V.J. Dyhouse, J. Mater. Res., v. 9, pp. 236-245 (1994).
12. J.S. Zabinski, M.S. Donley, S.D. Walck, T.R. Schneider, and N.T. McDevitt, Tribology Transactions, v. 38, pp. 894-904 (1995).
13. V.C. Fox, N. Renevier, D.G. Teer, J. Hampshire, literature available from Teer Coatings Ltd.
14. J.E. Krzanowski, unpublished research, University of New Hampshire (2003).

Mater. Res. Soc. Symp. Proc. Vol. 843 © 2005 Materials Research Society T3.16

Real Time Measurement of Functional Groups Substitution on Fluorocarbon Surface by ATR FT-IR

Yuki Sato and Masataka Murahara
Depertment of Electrical and Electronic Engineering, Tokai University
1117 Kitakaname Hiratuka Kanagawa, 259-1292, JAPAN

ABSTRACT

The photochemical reaction process steps in substituting the functional groups on the surface of fluorocarbon [FEP] by irradiating an Xe_2 excimer lamp on water or formic acid and FEP placed on the attenuated total reflectance [ATR] prism, has been measured in real time. These steps include, water photo-dissociation, the defluorination of the FEP and the hydrophilic group substitution.

In case of formic acid, the absorption peaks of the -CHO in the region of 2940 cm^{-1} and the -COOH in the region of 1710 cm^{-1} decreased respectively by photo dissociation, but that of the -OH in the region of 3300cm^{-1} increased. The results indicated that the -CHO and -COOH have turned to -OH.

Furthermore, the contact angle with water was measured. When compared to the untreated sample, whose contact angle was 110 degrees, the contact angle of the sample treated with water and the Xe_2 lamp irradiation for 25 minutes became 31 degrees, and that of the sample modified with formic acid and the lamp irradiated for 25 minutes further improved to 17 degrees. This implies that the hydrophilic groups were produced more in formic acid than in water.

INTRODUCTION

Fluororesin, which is a very chemically stable material and has several excellent properties such as heat and chemical resistance and resistance to hostile environments. It is thus employed as implant material in soft tissue as artificial blood vessels, artificial ligaments and artificial corneas. Fluororesin itself, however, is weak in tissue affinity because of its repellency.

In order to improve its affinity for the tissue, Kurotobi et al. tried to incorporate oxygen atoms into expanded poly (tetrafluoroethylene) [e-PTFE] by using He^+ and Ar^+ ion-beam irradiation, and evaluated the treated samples by fourier transform infrared spectroscopy [FT-IR] and scanning electron microscope [SEM]. By this treatment, the e-PTFE surface was roughened and carbonyl groups and carbonization were also produced to promote the fibronectin and fibrinogen absorption [1, 2]. Suzuki et al. reported that hydrophilic property was temporarily generated on the PTFE surface by H_2O ion irradiation; XPS analysis and Cu adhesive test were carried out to investigate relationship between the PTFE surface and Cu adhesion. The results showed that the Cu adhesion was dependent on the quantity of the OH adsorption [3]. Park et al. modified the FEP surface to be hydrophilic by remote H_2, Ar, N_2 and O_2 plasmas, and improved the adhesion with copper [4].

We have incorporated desired functional groups or metal atoms onto various kinds of polymer surfaces using the photochemical reaction of UV-photon to generate the hydrophilic, hydrophobic or metallization surfaces [5-10]. The treated samples were investigated by XPS and FT-IR, to study the substituted hydrophilic, hydrophobic and metal atoms on the sample surface.

It was confirmed by XPS that a PTFE surface was defluorinated by ArF laser light irradiation and the F1s peak (690 eV) decreased as the number of laser shots increased and diminished after 3000 shots [7, 8]. In these surface modifications, the sample surface was studied by FT-IR or XPS before and after treatment. However, the mechanism of the modification process was still a matter of conjecture [5-10].

In this study, both the photo-dissociation of reaction liquid and the functional groups substitution onto fluororesin surface were measured in real time, when the process of substituting hydrophilic group onto fluororesin by using photochemical reaction of UV-photon, was analyzed by ATR FT-IR.

PRINCIPLE OF PHOTOCHEMICAL REACTION

The FEP was composed of $(CF_2\text{-}CF_2)_m(\underset{CF_3}{CF}\text{-}CF_2)_n$

The reaction liquid used was water or formic acid having a hydrophilic group. The FEP was irradiated with the Xe_2 excimer lamp (172nm) in a reaction liquid ambience. The water was photo-dissociated to produce -H and -OH radicals as shown in Equation (1). Using the formic acid as a reaction liquid, it was also photo-dissociated to produce -H, -CO and -OH radicals, as shown in Equation (2). The Xe_2 excimer lamp photon energy was as high as 165kcal/mol, whereas the C-F bonding energy of the FEP is 128kcal/mol. Therefore the C-F bond can be dissociated. At the same time, the H atoms photo-dissociated from reaction liquid pulled out the F atoms of FEP surface and bonded to produce the HF. Thus, the -OH functional groups are substituted to the dangling bonds of C, and the FEP is modified into hydrophilic, as shown in equations (3) and (4).

$$H_2O + h\nu \rightarrow H^* + OH^* \quad \text{------} \quad (1)$$

$$HCOOH + h\nu \rightarrow H^* + OH^* + CO^* \quad \text{------} \quad (2)$$

$$(CF_2\text{-}CF_2)_m(\underset{CF_3}{CF}\text{-}CF_2)_n + H_2O + h\nu \rightarrow (CF_2\text{-}CF_2)_m(\underset{CF_2\text{-}OH}{CF}\text{-}CF_2)_n + nHF \quad \text{------} \quad (3)$$

$$(CF_2\text{-}CF_2)_m(\underset{CF_3}{CF}\text{-}CF_2)_n + HCOOH + h\nu \rightarrow (CF_2\text{-}CF_2)_m(\underset{CF_2\text{-}OH}{CF}\text{-}CF_2)_n + nHF + CO_n \quad \text{---} \quad (4)$$

EXPERIMENTAL

The experimental setup is shown in Figure 1. First, the reaction liquid is dropped on the ATR FT-IR prism made from germanium. Next, FEP and fused silica glass are put on it to make the thin reaction liquid layer by a capillary phenomenon. In this condition, the Xe_2 excimer lamp is vertically irradiated. As the same time, the infrared spectroscopy is sequentially run.

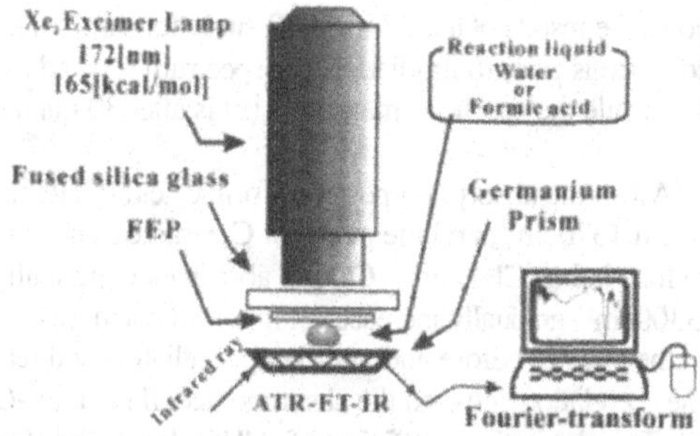

Figure 1. Schematic diagram of experimental set-up.

RESULTS AND DISCUSSION

Real-time measurement by ATR-FT-IR

Figure 2-(A) shows the result of real-time measurements by ATR FT-IR for samples modified with water. The spectrum marked as (a) is from the sample before Xe_2 lamp irradiation, the spectrum marked as (b) is after 15 minutes of Xe_2 lamp irradiation, while the spectrum marked as (c) is after 30 minutes of Xe_2 lamp irradiation.

In the case of 2-(A)(a), the absorption peaks of the -OH are located at 3300cm^{-1} and 1650cm^{-1}, while the peaks of C-F radicals are located at 1210cm^{-1} and 1150cm^{-1}. The -OH absorbance peaks (at 3300cm^{-1} and 1650cm^{-1}) gradually decreased as the lamp irradiation time increased (Figure 2-(A)(b), (c)).

The difference in the spectra before and after lamp irradiation and reaction was calculated. Figure 2-(B) shows the correlation between the absorption coefficient of the -OH at 1650cm^{-1} and 3300cm^{-1} and the lamp irradiation time. The absorption coefficient of -OH decreased (Figure 2-(B) (a), (b)) with increasing lamp irradiation time, and saturated at 25 minutes.

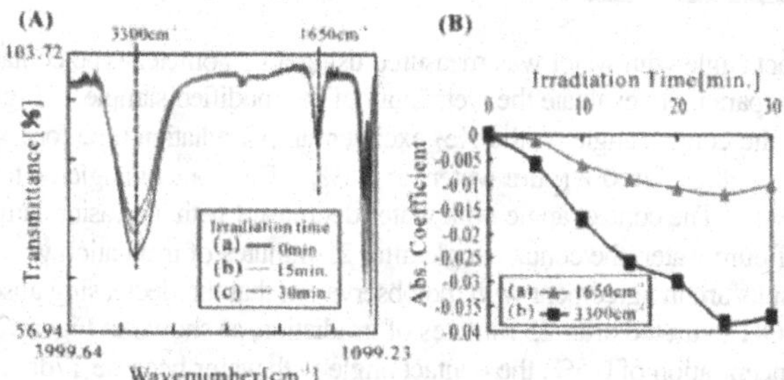

Figure 2. The spectra of real-time measurement by ATR-FT-IR in the case of the sample modified with the water. (A) ATR FT-IR spectra of photochemical reaction. (B) The correlation between the absorption coefficient of O-H at 1650cm^{-1} and 3300cm^{-1} and the lamp irradiation time.

Figure 3-(A) shows the results of the ATR FT-IR study for samples modified by formic acid. The spectrum marked as (a) is prior to irradiation, the spectrum marked as (b) is after15 minutes of Xe_2 lamp irradiation, while the spectrum marked as (c) is after 30 minutes of Xe_2 lamp irradiation.

In the case of 3-(A)(a), the absorption peaks of formic acid -CHO and -COOH are located at 2940cm^{-1}, 1710cm^{-1} and 1370cm^{-1}, and the peaks of C-F radicals are located at 1210cm^{-1} and 1150cm^{-1}. During irradiated, the -CHO and -COOH absorbance gradually decreased, while the -OH absorbance at 3300cm^{-1} gradually increased (Figure 3-(A)(b), (c)).

The difference in the spectra before and after lamp irradiation and reaction was calculated. Figure 3-(B) shows the correlation between the absorption coefficient in -OH and -CHO and the lamp irradiation time. The absorption coefficient of -CHO decreased (Figure 3-(B)(a)) with increasing lamp irradiation time, while the absorption coefficient of -OH increased and saturated (Figure 3-(B)(b)) at 25 minutes in the same manner as water.

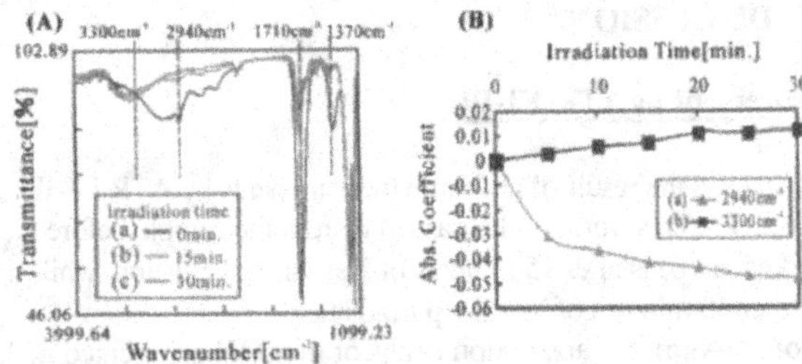

Figure 3. The spectra of real-time measurement by ATR-FT-IR in the case of the sample modified with the formic acid. (A) ATR FT-IR spectra of photochemical reaction. (B) The correlation between the absorption coefficient of -CHO at 2940cm^{-1} and of -OH at 3300cm^{-1} and the lamp irradiation time.

Contact angle measurement

The contact angle with water was measured using a goniometer-type contact angle measurement apparatus to evaluate the wettability of the modified sample. Figure 4 shows the dependence of the contact angle on the Xe_2 excimer lamp irradiation time for fixed formic acid concentrations ranging from 0% (pure water) to 98%. The contact angle on the untreated FEP was 110 degrees. The contact angle with water decreased with increasing lamp irradiation time. For the case of pure water, the contact angle after 25 minutes of irradiation was 31 degrees.

These results are in agreement with the observation that the decreasing absorption coefficient of -OH saturated after 25 minutes of irradiation, as shown in Figure 2-(B). With the formic acid concentration of 0.5%, the contact angle with water became 17degrees after 25 minutes of irradiation. In Figure 3-(B)(b), the -OH absorbance increased when the Xe_2 excimer lamp was irradiated on the FEP surface in the presence of formic acid. This result contradicts the findings of Figure 2-(B) that -OH absorbance decreased as the lamp irradiation time increased. In the case of formic acid, -COOH and –CHO produced by photo-dissociation of formic acid, were photo-dissociated again and produced -OH or CO_2. Therefore, upon irradiating with the Xe_2 lamp using formic acid as the reaction group, the OH groups were always produced to substitute on the FEP surface. The formic acid is thus able to produce the hydrophilic groups more efficiently than water; and consequently the contact angle with water decreased.

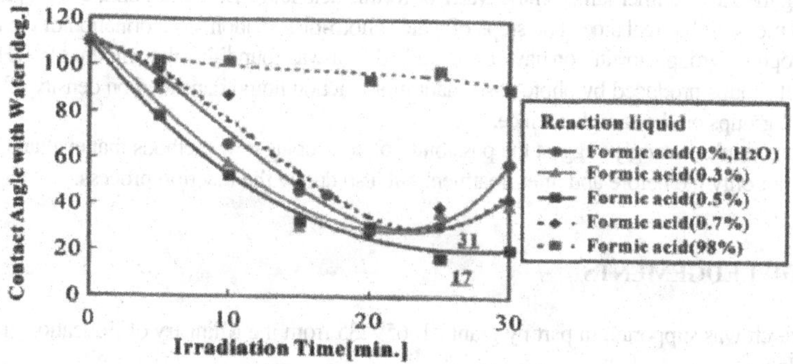

Figure 4. Contact angle with water versus lamp irradiation time.

ATR FT-IR analysis with and without UV irradiation

The FEP surfaces were evaluated before and after Xe_2 lamp irradiation by ATR FT-IR. Figure 5 (a) shows the spectrum from untreated FEP, (b) is from water treated FEP and (c) is from formic acid treated FEP. Untreated FEP has no peak of -OH of 3300cm^{-1}. After treatment of the FEP, the peak of -OH appears at 3300cm^{-1}. This is a result of -OH functional groups changing from H_2O or HCOOH and substituting onto the FEP surface. These results confirm that the -OH radicals were incorporated on the FEP by the photochemical reaction shown in Equation (1-4).

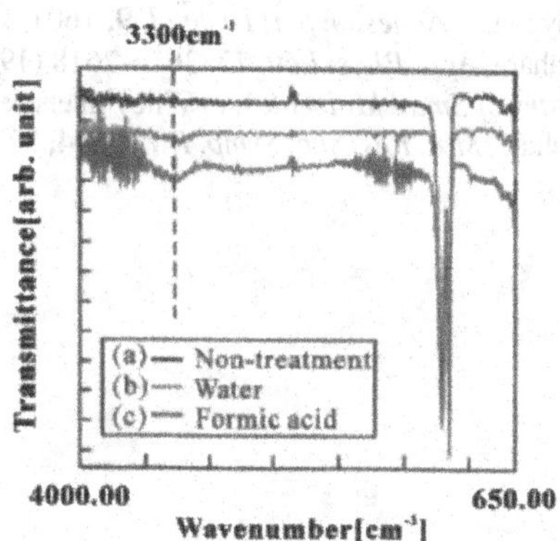

Figure 5. ATR FT-IR spectra of the fluororesin surface before and after lamp irradiation.

CONCLUSIONS

The photochemical reaction process in substituting functional groups on the FEP surface by irradiating the Xe_2 excimer lamp on the water or formic acid and FEP placed on the ATR prism has been measured in real time. The steps of water photo-dissociation, defluorination of the FEP and hydrophilic group substitution have been studied. It was found that the amount of -OH functional groups produced by photo-dissociation of reaction liquid depended on density of -OH functional groups on the sample surface.

These results, strongly suggest the possibility of developing new methods that evaluate surfaces not only the before and after treatment but also during the reaction process.

ACKNOWLEDGEMENTS

This research was supported in part by grant #16659235 from the Miinistry of Education and science Japan.

REFERENCES

1. K. Kurotobi, M. Kaibara, Y. Suzuki, M. Iwaki and H. Nakajima, *Nucl. Instr. And Meth. In Phys. Res.* **B175-177**, 791-796 (2001).
2. Y. Suzuki, N. Takahashi, M. Iwaki, H. Ujiie and T. Hori, *Technology on Adhesion and Sealing of Japan* **48 No.9**, 393-397 (2004)
3. H. Suzuki, S. Sobue, M. Nagakubo and T. Hattori, *Surface Science Society of Japan* **18 No.7**, pp.399-404 (1997)
4. Y. W. Park, S. Tasaka, N. Inagaki, *J. Appl. Polym. Sci.* **83,** 1258-1267 (2002)
5. M. Okoshi, M. Murahara and K. Toyoda, *J. Mater. Res.* **7**, 1912-1916 (1992)
6. M. Murahara and M. Okoshi, *J. Adhesion Sci. Technol.* **9**, 1593-1599 (1995)
7. M. Murahara and K. Toyoda, *J. Adhesion Sci. Technol.* **9**, 1601-1609 (1995)
8. M. Okoshi and M. Murahara, *App. Phys. Lett.* **72**, 2616-2618 (1998)
9. H. Okabe, *Photochemistry of Small Molecules*, (a Wiley Interscience, N. Y., 1978) 62-163
10. T. Ikegame and M.Murahara, *Mat. Res. Soc. Symp. Proc.* **544,** 227-232 (1999)

Mater. Res. Soc. Symp. Proc. Vol. 843 © 2005 Materials Research Society T3.28

Formation of High-Resistivity Silver-Silicon Dioxide Composite Thin Films Using Sputter Deposition.

Ann Marie Shover, James M.E. Harper, Nicholas S. Dellas, Warren J. MoberlyChan[1]
Department of Physics, University of New Hampshire
Durham, New Hampshire 03824, USA
[1]Center for Imaging and Mesoscale Structures, Harvard University
Cambridge, MA 02138, USA

ABSTRACT

Composite Ag-SiO_2 thin films were deposited to examine the stability of materials with high resistivity above 5000 μΩ-cm. We found that the resistivity increases exponentially with SiO_2 volume fractions larger than 0.50 which is consistent with a tunneling conductivity mechanism. In order to obtain a broad composition range, these films were deposited on a stationary substrate placed above Ag and SiO_2 sputtering sources. This configuration allowed compositions ranging from 0.8 to 56.3% SiO_2 to be deposited on the same sample. Resistance measurements were made using a four-point probe and a profilometer was used to measure thickness. Predicted values of thickness and composition were obtained by calibrating deposition rates from the separate sources, and were verified using Rutherford Backscattering Spectroscopy. Electron microscopy analysis revealed Ag agglomerating on the surface of the film. Because of the high mobility of Ag, the films should be capped to prevent Ag agglomeration. The results of this research demonstrate that high-resistivity thin films can be grown using Ag-SiO_2 composites.

INTRODUCTION

The aim of this research project was to determine whether composition could be consistently controlled for co-deposited metal-insulator thin films in order to achieve high resistivities. These thin films have utility in a variety of space sensor applications. In particular, cosmic ray detectors require resistivities greater than 20,000 Ω/square and this research indicates that such values can be achieved with our composition control techniques. In addition, the deposition of multi-component thin films will be used in the formation of nanoscale templates and in molecular sensors that depend on obtaining specific microstructures and compositions. To date, a number of metal-insulator combinations including Ni-SiO_2 [1] and W-Al_2O_3 [2] have been grown to examine percolation and tunneling properties for these films. Resistivities for Ag-SiO_2 thin films have been reported in the literature and the data suggests critical values for percolation between 0.4-0.6 volume fraction SiO_2[3,4].

EXPERIMENT

Our setup consists of a custom-designed vacuum chamber (Figure 1) with an approximate 1.0 m^3 volume. The chamber contains two three-inch diameter magnetron sputtering sources. The substrates were strips of polished silicon approximately 2 x 12 cm in dimension, vitreous carbon and TEM windows. These were inserted into the chamber through a load-lock. Base pressures were typically 1-2 $x10^{-8}$ Torr. The deposition pressures were 1.0 mTorr and 2.0 mTorr Argon.

Deposition rates at the center of the substrate, using DC power for Ag and RF power for SiO_2, are listed in Table I. When the runs were complete, the films were removed from the chamber through the load-lock and immediately analyzed using a step-profilometer for thickness and a four-point probe for resistivity. Additional measurements were made including Rutherford Backscattering Spectroscopy, X-Ray Diffraction and electron microscopy.

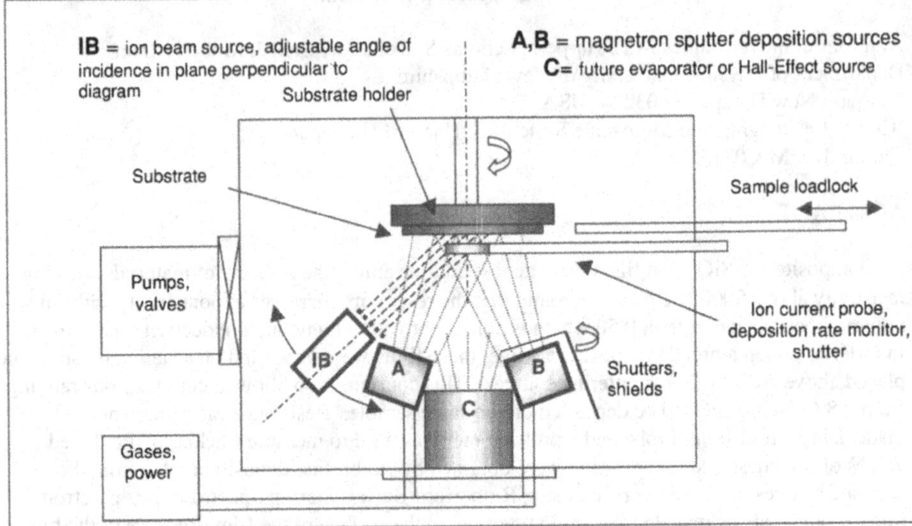

Figure 1. Experimental Chamber Layout

Table I. Deposition rates of Ag and SiO_2.

	Power (Watts)	Deposition Rate (Å /min)
Ag	37	77.4
SiO_2	100	11.25

RESULTS & DISCUSSION

Film thicknesses ranged from 4.4 to 12.2 kÅ, yielding resistivities as high as 40,000 μΩ-cm on the most insulating portions of the strips grown at 2 mTorr. To achieve the sheet resistance required for cosmic ray detectors, these films would have to be no more than 200 Å thick. The volume fractions of Ag and SiO_2 were estimated by conducting single source runs of Ag and SiO_2 under the same conditions as the dual-source runs and measuring their respective thicknesses across the sample. Some densification occurred for the compound runs compared with the single runs, possibly due to the density differences between the two materials which may enable Ag atoms to insert themselves within the SiO_2 structure. These composition estimates were verified with RBS data at three separate locations along the strip of film (Example shown in Figure 2). X-ray diffraction (not shown) confirmed the presence of crystalline Ag, with decreasing diffraction intensity as the volume fraction of SiO_2 was increased.

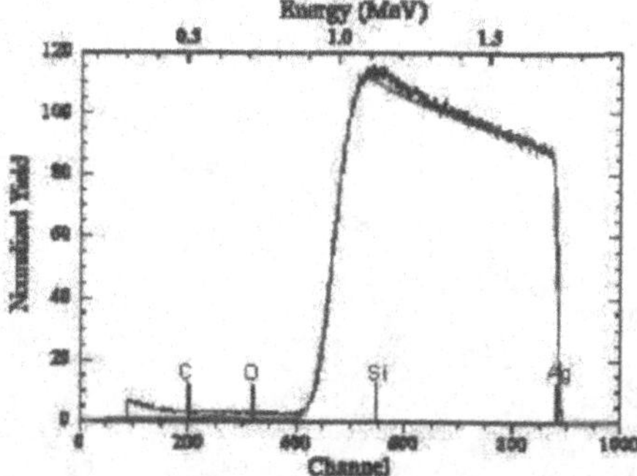

Figure 2: RBS data for Ag-SiO_2 on Carbon substrate measured 60 mm from the Ag-rich end. RBS measurements confirmed a volume composition of 81% Ag, 19% SiO_2 and thickness of 8 kÅ at this location.

The resistivity (ρ) vs. volume fraction (X) SiO_2 is plotted in Figure 3 which shows four identical runs at 1.0 mTorr. For these particular composites, the data cannot be fit strictly by a power law or exponential dependence alone. Percolation theory describes the region of thin film containing a low volume fraction of SiO_2. Above a certain volume fraction SiO_2 given by

$$X_c = (1 - x_c), \qquad (1)$$

tunneling becomes the main conduction mechanism. The value x_c is the critical value for metallic volume fraction and was calculated to be 0.42. Percolation theory predicts a simple power law dependence for resistivity of the form

$$\rho \sim (x - x_c)^{-t} \text{ [5]} \qquad (2)$$

where values around t = 2 have been reported in the literature for metal-insulator thin films [3,4]. The resistivity rose too rapidly to fit with an exponent of t = 2, therefore we observed a power fit of the form $\rho \sim (x - 0.42)^{-4.8}$. This is shown as the solid line in Figure 3. For insulator volume fractions above 0.5, the data supports tunneling by conforming to an exponential dependence of the form

$$\rho = \alpha e^{\beta x} \qquad (3)$$

as depicted by the braided line in Figure 3.

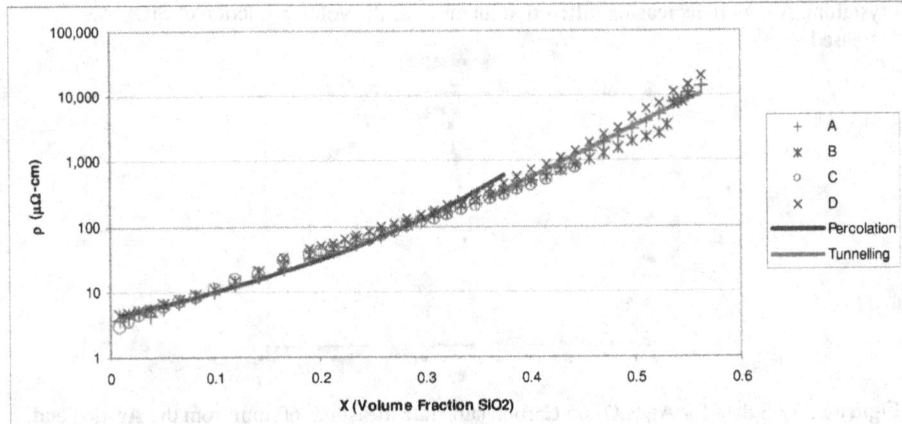

Figure 3. Resistivity versus volume fraction SiO_2 for four runs at 1.0 mTorr illustrating the comparison with percolation and tunneling theory at low and high volume fractions respectively.

We noticed the films began to develop a matte appearance several days after they had been grown. This was especially noticeable in the SiO_2-rich end of the sample. Further examination of these films using electron microscopy and confirmed by energy-dispersive X-ray spectroscopy revealed that Ag had been agglomerating on the surface of the film. We observed more surface agglomerations in regions containing less volume fraction of Ag. The higher Ag volume fraction portions of the film show more interconnected channels of conductive material than in the lesser Ag volume fraction region, potentially deterring additional Ag from diffusing out of the film and up to the surface. Figure 4a shows a portion of the film containing 0.63 Ag by volume fraction. Our measurements indicate that nearly 17% of the Ag originally in the film has diffused out of the film and onto the surface. In Figure 4b, the portion of the film containing 0.47 Ag by volume fraction has released to the surface nearly 40% of the Ag deposited into the film. This surface Ag was easily smeared during sample preparation as shown in the top portion of Figure 4b. It appears that two other mechanisms are competing in that some vaporization of Ag occurs when the sample is heated in the electron microscope, however grain growth was also observed potentially due to the films being porous in nature and thus allowing for recondensation. The close-up image of Figure 5 reveals a structure that suggests pores within the film. The bright field TEM images of Figure 6 and 7 show dark channels of conductive material that may have diffused up to the surface to form the Ag globules (Figure 8). TEM and SEM showed that the surface Ag agglomerations had multiple grains per globule. Higher magnification images indicate that the grain size is even finer than it appears. The 2-phase Ag-SiO_2 of Figure 5 has a spacing of 20-30 nm but each region is actually made up of many finer grains on the order of 5 nm.

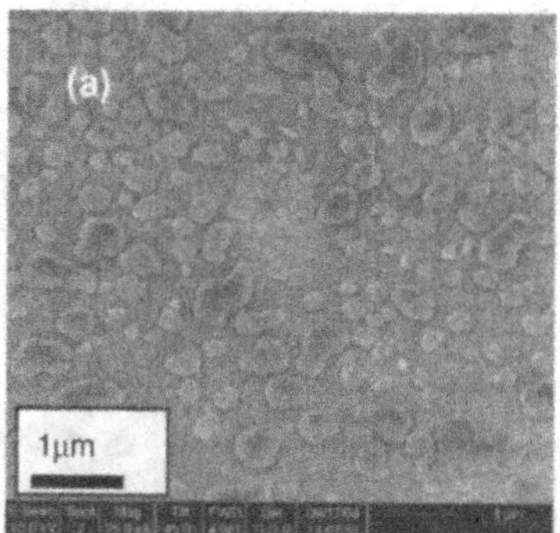

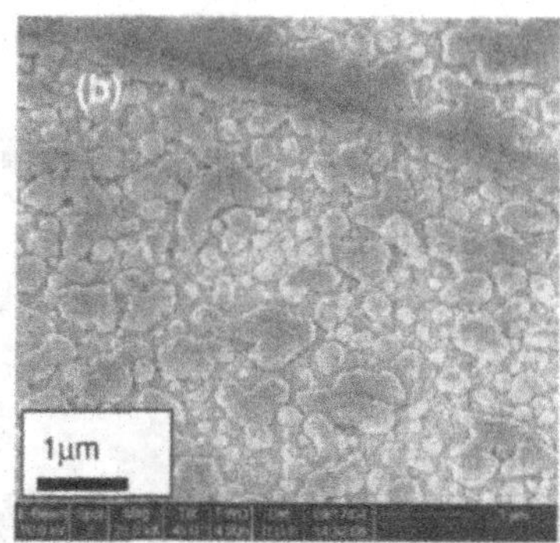

Figure 4. SEM images of the film for (a) 0.63 Ag volume fraction and (b) 0.47 Ag volume fraction showing agglomerations of Ag on the surface.

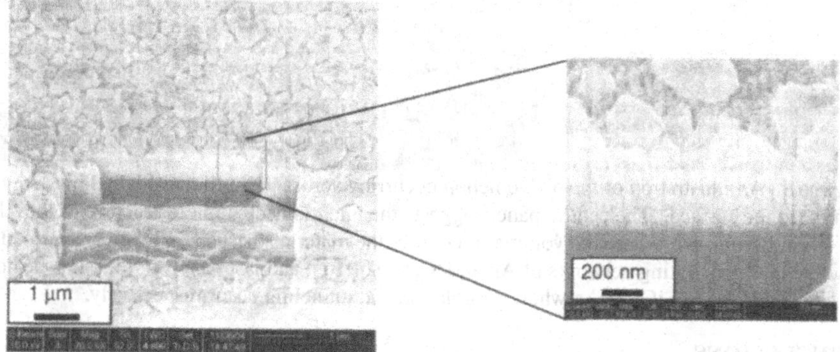

Figure 5 SEM image of ion milled portion of sample with close-up showing porous structure of the film.

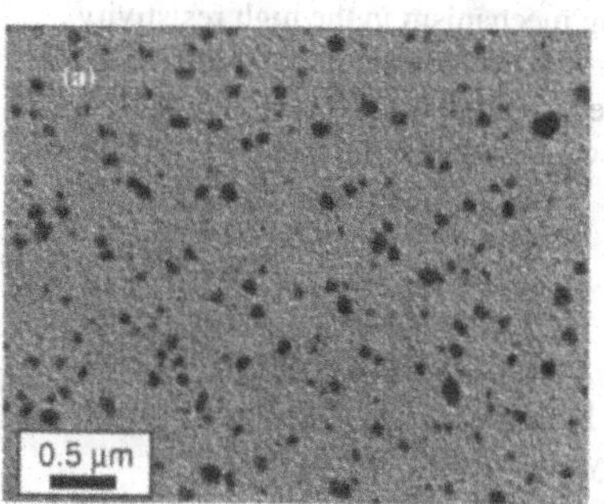

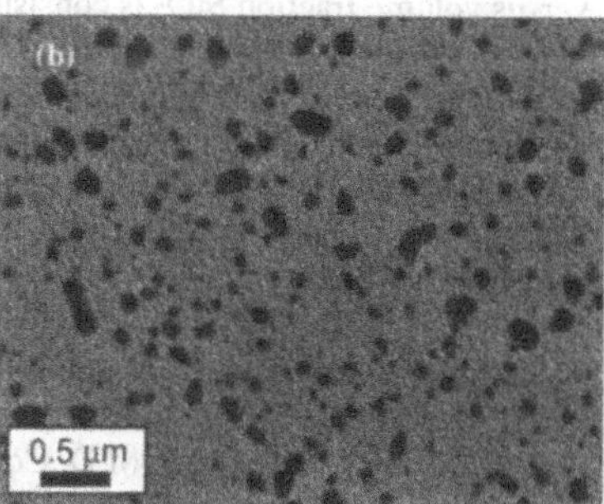

Figure 6. Bright field TEM images; dark spots are Ag islands on the film surface for (a) Ag volume fraction of 0.64 and (b) Ag volume fraction of 0.56.

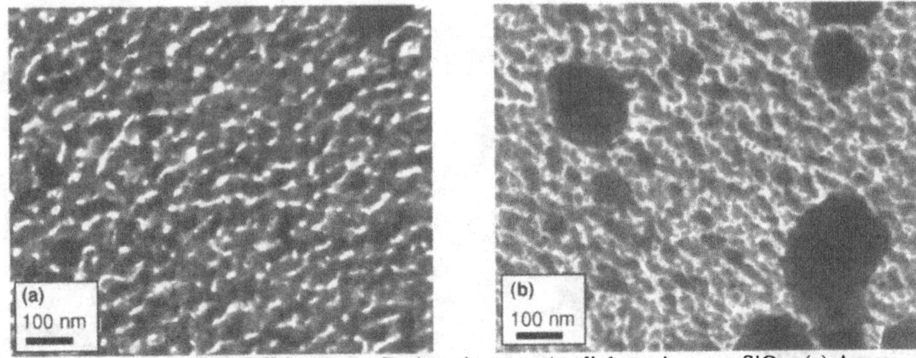

Figure 7. Bright field TEM images. Dark regions are Ag, light regions are SiO_2. (a) Ag volume fraction 0.64 and (b) Ag volume fraction 0.56.

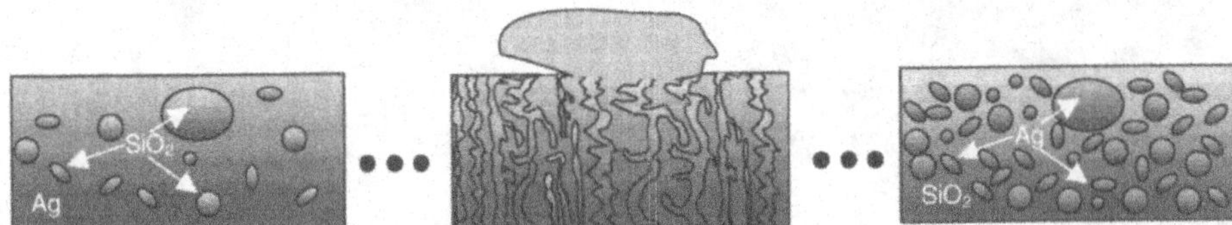

Figure 8. An illustration of the phenomenon occurring across these films which allows Ag to diffuse to the surface. The central panel suggests the interconnected microstructure of the films shown in figures 4-7, where the volume fraction of the materials is such that they are formed into discrete interpenetrating networks of Ag and SiO_2. The first and last panels of this sketch show isolated particles of SiO_2 or Ag where percolation and tunneling occur respectively.

CONCLUSIONS

High resistivity thin films were obtained by co-deposition of Ag and SiO_2. The resistivity versus volume fraction SiO_2 is consistent with a tunneling mechanism in the high resistivity region. At lower volume fraction SiO_2, the data does not follow a simple percolation model to explain the behavior in the metallic region. We also observed agglomerations of Ag at room temperature indicating that these films would have to be capped with a thin layer of SiO_2 before being used in any detector applications.

REFERENCES

1. D. Toker, D. Azulay, N. Shimoni, I. Balberg, O. Millo, Phys. Rev. B **68**, 041403 (2003).
2. B. Abeles, H.L. Pinch, and J.I. Gittleman, Phys. Rev. Lett. **35**, 247, (1975).
3. R.W. Cohen, G.D. Cody, M.D. Coutts, B. Abeles, Phys. Rev. B **8**, 3689, (1973)
4. E.B. Priestley, B. Abeles, and R.W. Cohen, Phys. Rev. B **12**, 2121 (1975)
5. D. Stauffer and A. Aharony, in *Introduction to Percolation Theory* (Taylor and Francis, London, 1992) pp. 10-56.

Mater. Res. Soc. Symp. Proc. Vol. 843 © 2005 Materials Research Society T3.32

A Surface Coating for Silicon Carbide Whiskers

Heng. Zhang[1], T.D. Xiao[1], D.E. Reisner[1], and P. Santiago[2]

1. Inframat Corporation, 74 Batterson Park Road, Farmington, CT 06032, U.S.A.
2. Central Microscopy Research Facilities, Physics Institute, UNAM A.P. 20-364, C.P. 01000 04510, Mexico D.F., Mexico

ABSTRACT

A mullite coating process has been conducted to form an electrically insulating layer on the surface of the silicon carbide whiskers for whisker reinforcement application by a chemical deposition and hydrolysis. The phase, structure, morphology and electrically insulating property of the coated silicon carbide whiskers have been investigated by using x-ray diffraction, scanning electron microscopy, transmission electron microscopy and electrical measurement. The study indicates that a crystalline mullite coating layer is formed after heat-treating the precursor coating at elevated temperature. The formed mullite coating significantly increases the electrical resistance of the silicon carbide whiskers.

INTRODUCTION

Silicon carbide in the form of whisker (SiC-w) has excellent mechanical properties, and is one of the best components for whisker toughening reinforcement for ceramic matrix composites [1]. However, as a semiconductor, it will dramatically reduce the electrical resistivity of an insulating composite system, which has critical requirement for high electrical resistivity. Hence, silicon carbide whiskers have to be coated with an insulating layer to increase its electrical resistance to fulfill the combined functions of superior mechanical and electrical insulating properties. Mullite, with a formula of $3Al_2O_3.2SiO_2$, possessing good electrical resistance and interface characteristics [1], is a good candidate for such a purpose. Recently, it was reported that mullite coated Hi-Nicalon SiC fibers exhibit higher tensile strength than uncoated SiC fibers [2] at temperatures of 1200°C.

A few methods have been used to elaborate thin films on fibers and whiskers, including chemical vapor deposition (CVD), sputtering and the sol-gel process. Among the numerous advantages of the sol-gel process compared to conventional methods [3], the low temperature and the possibility of controlling the porosity by adding various templates were of great interest. In this work, a sol-gel process combining a hydrolysis has been selected for producing mullite coated silicon carbide whisker. The process, morphology and electrical insulating properties will be reported in this presentation.

EXPERIMENTAL DETAILS

SiC whiskers with an average diameter. of 1-2 μm and a length of 15-20 μm, purchased from Alfa Aesar, were used as the substrate of mullite deposition. In order to provide the SiC whiskers with a suitable surface functionality for subsequent mullite coating treatment, the SiC whiskers were first heated at 600 °C for 0.5 h to form a thin SiO_2 layer on the surface of SiC whiskers.

A sol-gel dipping technique, combined hydrolysis, has been employed to fabricate mullite coating on the SiC-w surfaces. The starting materials for the deposition solution include 2-propanol $(CH_3)_2CHOH$ (Alfa Aesar) as the solvent, aluminum tri-sec-butoxide $Al(OBu)_3$ (Alfa Aesar) as aluminum precursor, tetraethoxysliane (TEOS) (Alfa Aesar) as silicon precursor, acetylacetone (acacH) (Alfa Aesar) as the stabilizer to balance the reaction rate between aluminum and silicon, and cetyltrimethylammonium bromide (CTAB) (Alfa Aesar) as a chelating ligand [4-5]. The solution preparation follows the procedure: (1) dissolving 0.3 mol $Al(OBu)_3$ into 2-Propanol (3 mol), (2) dissolving 0.1 mol TEOS into 2-Propanol (1 mol), (3) mixing $Al(OBu)_3$ and TEOS solution, (4) adding 0.3 mol acacH into the above mixed solution to stabilize the aluminum precursor towards hydrolysis reactions, and (5) adding 0.05 mol CTAB to 2-propanol (1mol) as a chelating ligand for porous. The solution was stirred for 1 h under a stream of nitrogen gas. This solution was finally filtered.

The desized SiC whiskers were submersed into the deposition solution contained Al and Si precursors and under stirring for 5-10 minutes. The coated SiC whiskers were filtered and dried by slow heating to 120°C for 1 h. In order to get thick mullite coating, the dip coating and drying process were repeated up to 4 cycles.

The coated SiC whiskers were heated to 1000, 1300 °C under air atmosphere for 2 h to form crystalline mullite layers. In order to identify the electrical insulating behavior of the mullite coated SiC-w, the coated SiC-w was pressed to a pellet with a diameter of 15 mm and a thickness of 1 mm under a uni-axial pressure load of 800 MPa. Then, the pellet was densified at 1850°C for 3 h under an argon atmosphere. The electrical property of the sintered mullite coated SiC-w pallet was characterized using an electric multimeter.

The phase and structure of the mullite coated SiC whiskers are characterized using x-ray diffraction (XRD). The morphology and microstructure of the mullite coated SiC-w were examined using scanning electron microscopy (SEM). The coating thickness and the bonding between the coating and substrate were characterized using a transmission electron microscopy (TEM).

RESULTS AND DISCUSSIONS

The morphology of the dip coated SiC whisker can be seen from the SEM image in figure 1. This image indicates that the coated SiC whiskers have a rod shape with a diameter of 1-2 μm and a length of 10-20 μm.

The phase and structure of the coated SiC whiskers with mullite precursors can be seen from the XRD patterns shown in figure 2 (I). The coated SiC whiskers show a mixture of crystalline SiC and amorphous mullite precursor (see the diffuse peak between 2θ 19-25°). With the increase of multiple coating cycles, the diffraction peak of the SiC becomes weaker and weaker, while the diffuse peak of the amorphous mullite precursors becomes stronger and stronger. This means that the thickness of amorphous mullite precursor coating increases with coating cycle. The XRD patterns for coated SiC whiskers after calcined at 1300°C indicate that the crystalline mullite

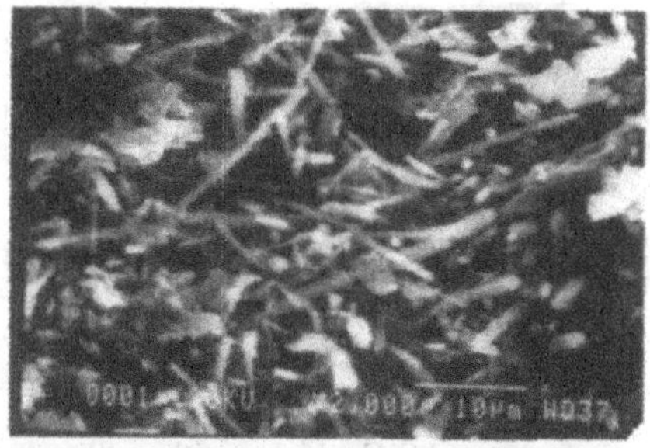

Figure 1. SEM image for the SiC whiskers with mullite coated precursors (coated twice)

phase was formed, and the fraction of mullite phase increases with the coating cycle (see figure 2 (II)).

Transmission electric microscopy (TEM) analysis demonstrated that mullite coating has been successfully formed as shown in figure 3(a). The mullite coating with a thickness of 100 to 200 nm can be prepared by controlling the coating process parameters. As shown in figure 3 (b), the interface between the mullite coating and silicon carbide whisker exhibits a firm and continuous combination.

The electrical insulating property of a sintered mullite coated SiC whisker has been examined using a multimeter. Electric property test result is listed in table 1 and excellent electrical insulating property has been achieved (over 1×10^9 Ω.cm). This demonstrates that the mullite

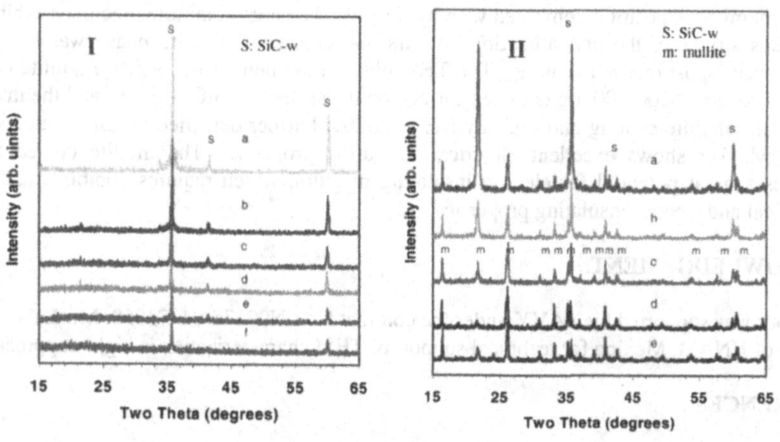

Figure 2. XRD patterns for **(I)** (a) SiC whisker, the coated SiC whisker with mullite precursors for one (b), two (c), three (d), four cycles (e), and directly evaporated (f); and **(II)** coated SiC whiskers with mullite fired at 1300°C for 2 hours for one (a), two (b), three (c), four cycles (d), and directly evaporated (e).

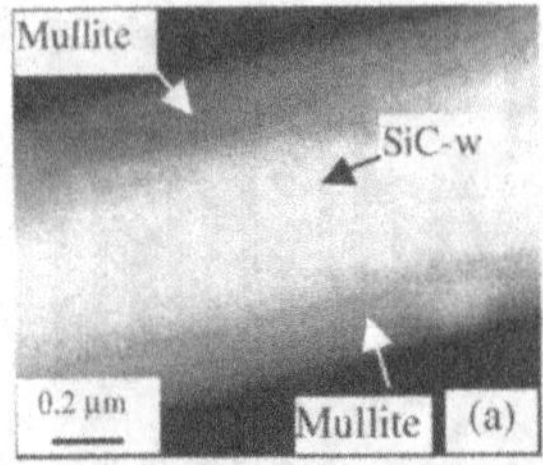

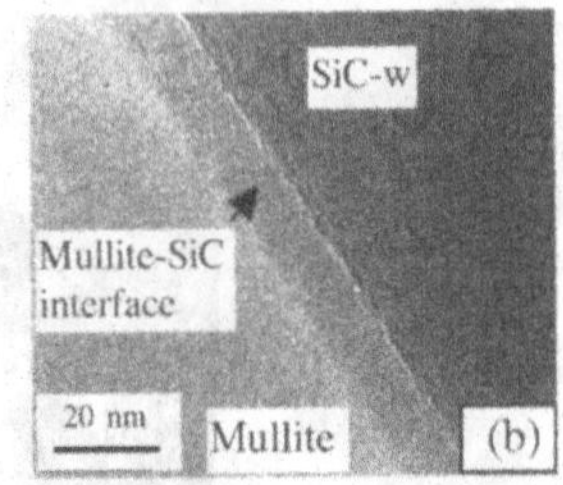

Figure 3. (**a**) TEM image for coated SiC whisker with mullite and heated at 1300°C for 2 hours and (**b**) for the interface of crystalline SiC and mullite for the coated SiC whisker with mullite and heated at 1300°C for 2 hours. The interface of mullite and SiC whisker shows firm and continuous combination.

coated SiC whiskers can offer both superior mechanical properties and electrical insulating properties, and can be served as an ideal whisker reinforcement component for a composite that requires both excellent mechanical toughness and electrical insulating properties.

Table 1. The electric insulating property for a sintered mullite coated SiC-w

Sample ID	Coating/calcination	Pressing	Sintering	Resistivity
M-SiC-w	4cycles/1300°C	800 MPa	1850 °C/Ar	$>1.0\times10^9$ Ω.cm

CONCLUSIONS

Silicon carbide whiskers have been successfully coated with an electrical insulator mullite layer using a chemical deposition combined with hydrolysis. The coated mullite precursor exhibits an amorphous structure, the crystallization appears and crystalline mullite phase was formed at 1300°C during post thermal heating. The TEM observation demonstrates that a mullite coating with a thickness of 100-200 nm has been formed on the surface of SiC whisker, and the interface between the mullite coating and SiC whisker exhibits. Further densified mullite coated silicon carbide whisker shows excellent electrical insulating property. The mullite coated SiC-w possesses a great potential for electric insulating medium, which requires combined excellent mechanical and electric insulating properties.

ACKNOWLEDGEMENTS

This work was supported by NAVY under the contract No.: N00178-04-C3058. Mr. Luis Rendon of UNAM, Mexico for technical support of TEM characterization is highly appreciated.

REFERENCES

1. R. Warren, *Ceramic-Matrix Composites,* (Blackie and Son Ltd., Bishopbriggs Glasgow, 1992) p2.
2. Martine Verdenelli, Stephane Parola, Fernand Chassagneux, Jean-Marie Letoffe, Henri Vincent, Jean-Pierre Scharff and Jean Bouix, "Sol-gel preparation and thermo-mechanical

properties of porous xAl_2O_3–$ySiO_2$ coatings on SiC Hi-Nicalon fibres", *J. Europ. Ceram. Soc.* **23** 1207–1213 (2003).

3. C. J. Brinker and G. W. Scherer, *Sol-Gel Science: The Physics and Chemistry of Sol-Gel Processing*. (Academic Press, San Diego, 1990).
4. N.Ya Turova, E. P. Turevskaya, V. G. Kessler and M. I. Yanovskaya, *The Chemistry of Metal Alkoxides*, (Kluwer Academic Publishers, Norwell MA, 2001).
5. D. C. Bradley, R. C. Mehrotra and D. P. Gaur, *Metal Alkoxides*, (Academic Press, London, 1978).

Mater. Res. Soc. Symp. Proc. Vol. 843 © 2005 Materials Research Society T3.4

Limiting Catalytic Coke Formation by the Application of Adherent SiC Coatings via Pulsed Laser Deposition to the Inner Diameter of Tube Material Traditionally Used for Ethylene Pyrolysis Service

Alok Chauhan [1], Wilton Moran [1], Elizabeth Casey [2], Weidong Si [3], Henry White [1,*]

1. Dept. of Materials Science, Stony Brook University, Stony Brook, New York 11794-2275
2. Comsewogue High School, Port Jefferson Station, New York, 11776
3. Physics Department, Brookhaven National Laboratory, Upton, New York 11973-5000

Abstract

Pulsed laser deposited coatings can be used to enhance the corrosion resistant of materials traditionally used for industrial applications. In this paper, we describe our initial results on coating HK40 (a material used for ethylene heater tubing) with silicon carbide (a carburization resistant coating) to increase tube life. A 1 um thick film of silicon carbide was successfully deposited onto a heated HK40 substrate. An array of characterization techniques (scanning electron microscopy, atomic force microscopy, and scratch tests) demonstrated that the processing conditions were suitable for good coverage and promising adhesion behavior.

Introduction

Iron based alloys such as HK40 have been used for ethylene heater tubing with operating temperatures near 1000 C [1,2]. In service carburization has reduced tube life from 100,000 to ~ 50,000 hours [3]. Nickel based materials have been suggested for higher operating temperatures with only marginal improvement in carburization resistance [1,2]. In this paper, we report on our initial efforts to utilize pulsed laser deposited (PLD) coatings to reduce carburization and increase tube life. A thin (~ 1 um) silicon carbide (SiC) layer, a candidate material for this application, was deposited on as machined and shot peened HK40 substrates. Hou et al. [4] deposited SiC on Ni-Cr surfaces and found that adhesion improved when composition graded intermediate layers were employed. Surface preparation, which is an essential variable for good adhesion, was not reported in this work. Pulsed laser deposition has been previously used to deposit SiC on smooth semi-conductive surfaces [5-7]. To our knowledge, this paper represents the first study of the effect of surface preparation on the adhesion behavior of PLD films.

Experimental Procedure

Centrifugally Cast A609 Grade HK40 [19.0-22.0% Ni, 23.0 –27.0%Cr, 0.35-0.45% C, 0.05 – 2.0% Si, 1.5% max. Mn] was obtained from Ultra-cast Inc. The target material, SiC [0.009% Al, 0.020% B, 0.031% Fe, 0.015% O and 0.017% Mn, traces of Ni and Ti], was supplied by Kurt J. Lesker Company. A KrF Excimer Laser (λ = 248 nm) was used to ablate and deposit SiC on as machined and shot peened (170/ 325 mesh glass beads) HK40 substrates. The films were deposited inside a stainless steel vacuum system pumped to a base pressure in the range of 10^{-4} to 10^{-5} torr. The KrF laser was focused to a rectangular spot 0.125 cm x 0.078 cm onto the rotating (10 - 20 rpms) 2.54 cm in diameter SiC target. The pulse energy and frequency of the laser was ~500 mJ and 5-10 hertz respectively. The laser/ target angle of incidence was 45° and the target to substrate distance was 4 - 5 cm. Depositions were performed at room temperature and while resistively heating the substrate up to 500 C. These temperatures were selected to limit sigma phase formation [7] during the deposition process. Figure 1 shows a schematic of the experimental setup.

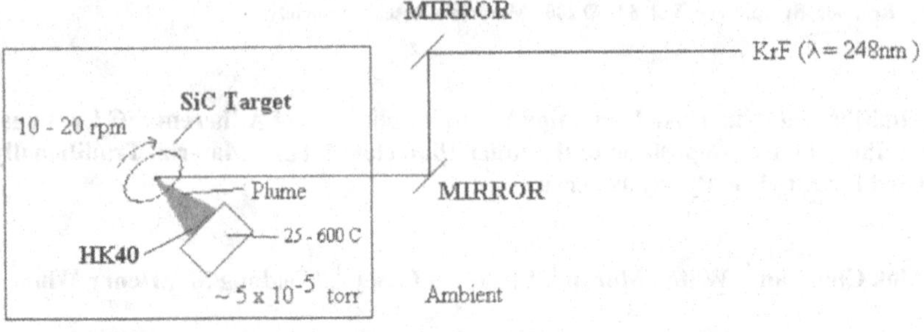

Figure 1. PLD System equipped with rotating target and substrate heating stage.

Results and Discussion

Atomic Force Microscopy

Contact mode Atomic Force Microscopy (AFM) studies were performed on the target, substrate, and coating. All specimens were scanned at 50 um x 50 um x 2 um and no correction were made during analysis. The roughness of all surfaces were comparable and varied from 300 to 500 nm. The surface morphology of the coated and uncoated substrates did not change. Shot peening introduced compressive stress on the HK40 surface and produced a "dimple like" morphology. Machining produces a "scratch like" appearance and surface tensile stresses. The surface morphology of the machined and shot peened surfaces is shown in figure 2.

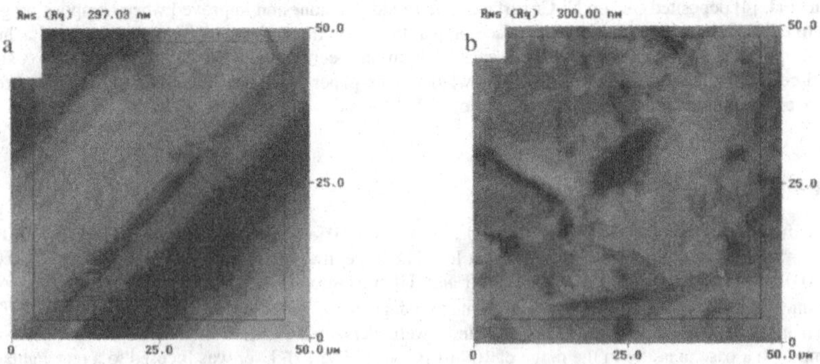

Figure 2. a) Scratch like AFM image of machined HK40 substrate. b) Dimple like image of shot peened HK40 substrate.

Scanning Electron Microscopy

Typical backscatter electron (BSE) microscopy images of as deposited SiC are shown in figure 3. During room temperature deposition on as machined "scratch like" surfaces, poor coverage and adhesion is apparent (figure 3a). Deposition at higher temperatures and on shot peened dimple like surfaces (figure 3b) showed significant coverage and improved adhesion.

a

b

Figure 3. a) BSE micrograph of PLD of SiC on HK40 at room temperature. High z-contrast areas (white) represent HK-40 (Fe-Ni-Cr) and low z-contrast areas (dark) represent SiC coating. Poor adhesion and coverage are apparent. b) BSE micrograph of PLD of SiC on HK40 at 500 C. Complete coverage of SiC on HK40 is shown.

Qualitative energy dispersive spectroscopy (EDS) studies of the target and coating shown in figure 3b showed similarities in the target and coating chemistry (see figure 4). A strong Aluminum (residual element from target) peak existed in the spectra for both the target and coated material.

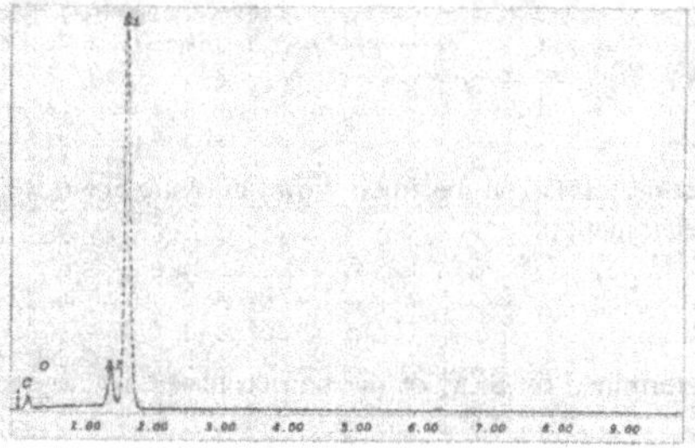

Figure 4. EDS of coated HK40 (Fe-Ni-Cr) substrate (grey) superimposed on target spectrum (black- dotted lines). In the spectrum of the coated material (grey) note the presence of a strong oxygen peak at 0.5 Kev and weak Cr and Fe peaks at 5.4 Kev and 6.4 Kev respectively.

Adhesion Test

The scratch-test method consists of generating scratches with a Rockwell C type diamond indenter (tip radius 50μm) which was drawn at a constant speed across the coating-substrate system [8]. Other test conditions are shown in table 1.

Table 1. Scratch Test Conditions

Loading range	0 to 6 N
Loading rate, dL/dt	6 N/min
Scratch Length	4 mm
Scratching speed, dx/dt	4 mm/min

A critical load (i.e. the smallest load at which a recognizable failure occurs, see inset figure 5a) was attained at 0.48 ± 0.06 N. Complete delamination (see figure 5b) occurred at higher loading 4.37 ± 0.4 N.

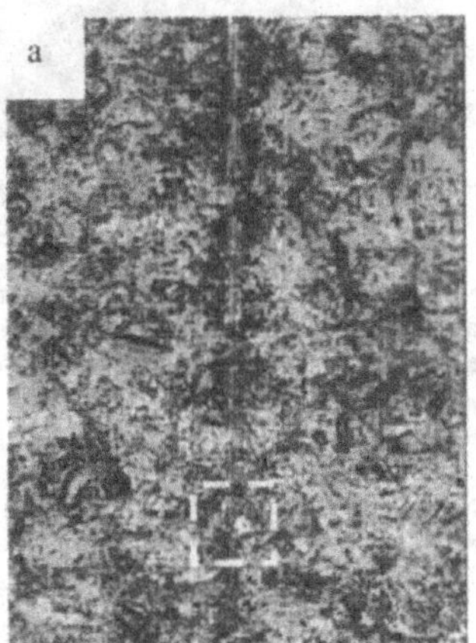

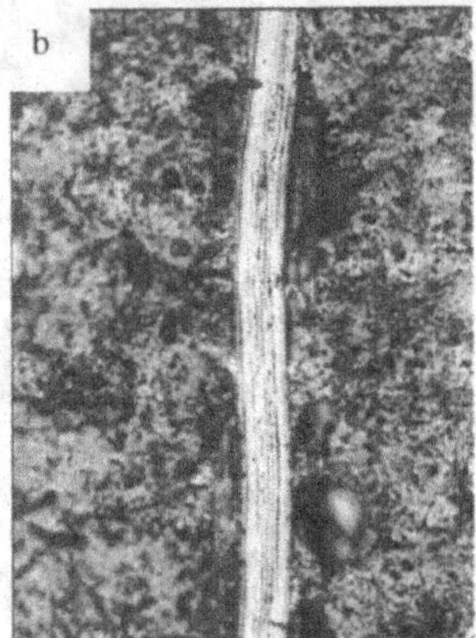

Figure 5. Adhesion test performed on specimen shown in figure 3b. a) Region of initial failure. b) Complete delamination.

The coating thickness was determined by SEM of the scratch tested specimen. Figure 6 shows that the coating thickness is ~ 1 um.

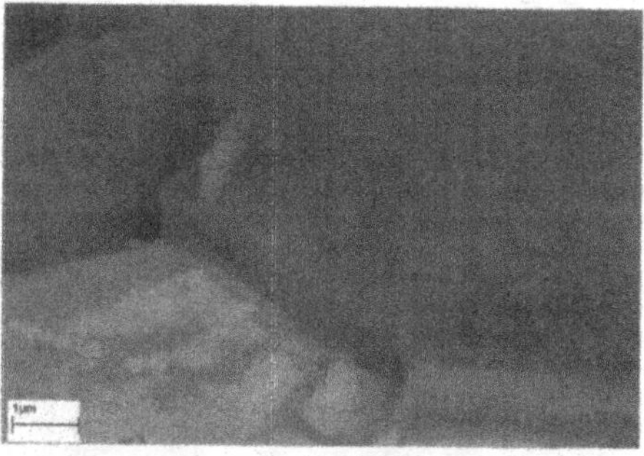

Figure 6. BSE of scratch tested specimen. The specimen was tilted 45 degree. The dark region (low z- contrast) is the SiC film, the bright region (high z- contrast) is the HK40 substrate.

Conclusions

In this paper, we presented our initial results on ablation and deposition of SiC onto HK40 substrates. It was determined that the surface condition and processing temperature play a role in the adhesion properties of the resulting films. Shot peening tended to improve coverage and adhesion properties. Substrate temperature during processing also improved coverage and adhesion, however, limitation must be imposed to avoid compromising the mechanical properties of the substrate. These efforts will be extended to determine the suitability of the coating/ substrate system for in-service testing conditions.

Acknowledgements

HJW would like to thank his colleagues Dr. Shouren Ge and Dr. James Quinn of Stony Brook University for their assistance with the collection of AFM and SEM data respectively; Dr. Weidong Si of Brookhaven National Laboratory for assistance with film depositions; Ethel Poire of Microphotonics for help with Adhesion Tests and Ultra-cast Inc. for supplying the HK40 materials. This worked is supported by the National Science Foundation under grant DMII- 0346947.

References

[1] M. W. Mucek, "Laboratory Detection of Degree of Carburization in Ethylene Pyrolysis Furnace Tubing," Materials Performance, 9, 25-28, 1983.
[2] F. Ropital, A. Sugier, M. Bisiaux, "Etude De La Restauration Des Caracteristiques De Tubes De Four De Vapocraquage En Manaurite 36XS," Revue De L'Institut Francais Du Petrole, 44, 91-101, 1989.
[3] S. Ibarra, "Materials of Construction in Ethylene Pyrolysis- Heater Service," In Pyrolysis: Theory & Industrial Practice, L. F. Albright, B. L. Crynes, W. H. Corcoran, Academic Press, 427-436, 1983.
[4] Q. R. Hou, J. Gao, S. J. Li, "Adherent SiC Coatings on Ni- Cr Alloys with a Composition- Graded Intermediate Layer," Applied Physics A, 67, 367-370, 1998.
[5] M. Y. Chen, P. T. Murray, "Deposition and Characterization of SiC and Cordierite Thin Films by Pulsed Laser Evaporation," Journal of Materials Science, 25, 4929-4932, 1990.
[6] F. Neri, F. Barreca, S. Trusso, "Excimer Laser Ablation of Silicon Carbide Ceramic Targets," Diamond and Related Materials, 11, 273-279, 2002.
[7] L. Rimai, R. Ager, E. M. Logothetis, W. H. Weber, J. Hangas, "Preparation of Oriented Silicon Carbide Films by Laser Ablation of Ceramic Silicon Carbide Targets," Applied Physics Letters, 59, 2266-2268, 1991.
[8] P. A. Steinmann, Y. Tardy, H. E. Hintermann, "Adhesion Testing by the Scratch Test Method: The Influence of Intrinsic and Extrinsic Parameters on the Critical Load," Thin Solid Films, 154, 333-349, 1987.

Mater. Res. Soc. Symp. Proc. Vol. 843 © 2005 Materials Research Society T3.43

Characterization of Electrochemically Activated Surface on Rolled Commercial AA8006 Alumnium

Yingda Yu, Øystein Sævik, Jan Halvor Nordlien[1] and Kemal Nisancioglu
Department of Materials Technology, Norwegian University of Science and Technology, N-7491 Trondheim, Norway
[1]Hydro Aluminium R&D Materials Technology, N-4256 Håvik, Norway

ABSTRACT

Commercial alloy AA8006 (nominal composition in wt%1.5Fe, 0.4Mn, 0.2Si) exhibits electrochemical activation in chloride media as a result of annealing at temperatures above 350ºC. Activation is manifested by a significant negative shift in the corrosion potential and a significant increase in the anodic current when polarized above the corrosion potential in aqueous chloride solution. This phenomenon was correlated with the enrichment of trace element Pb at the oxide-metal interface. For fixed annealing time, maximum activation and Pb-enrichment are obtained at an annealing temperature of 450 °C independent of the original surface condition. Reduced activation with increasing temperature is attributed to increasing density of the spinel crystalline phase to cause the passivity of the surface.

INTRODUCTION

Thermomechanical processing has a significant role in the determination of the surface properties of aluminum alloys. Electrochemical activation of various commercial aluminum alloys resulting from high temperature heat treatment has been a subject of attention for the past several years because of its importance in galvanic and filiform corrosion [1-9]. The phenomenon is observed by heat treatment at temperatures above 350 °C. The anodic activation was characterized by deep corrosion potential transients in slightly acidified chloride solutions, starting from highly negative potentials in relation to the usual pitting or corrosion potential of about - 0.75 V_{SCE} [10]. In addition, anodic polarization in neutral chloride solutions gave a high anodic current output with oxidation peaks corresponding to the potential arrests, as also confirmed by potentiostatic data [5-7, 11]. Recent work on commercial alloys AA8006 and AA3102 has shown the cause of activation to be the enrichment of the material surface by lead, which was present in the material as a trace element only at the ppm level [5-7]. Further work on model-binary AlPb alloys verified the role of Pb in activation [11, 12].

The present work was undertaken to obtain a more detailed chemical and microstructural information on hot-rolled commercial alloy AA8006 in the activation behavior. It is further desirable to obtain additional evidence of Pb segregation resulting from heat treatment of a commercial alloy, specifically the actual location and state of the segregated Pb, the effect of heat treatment in general on the near surface chemistry and microstructure of the alloy and the effect of these factors on the electrochemical and corrosion behavior.

EXPERIMENTAL DETAILS

The commercially rolled AA8006 material was supplied in the form of fully annealed O temper with 0.8 mm thick sheets. Samples were heat-treated at different selected temperatures in an air circulation furnace for 2 h, followed by quenching in distilled water at room temperature and degreasing in acetone. Some samples packaged inside high-purity aluminum (99.98%) foil were also heat-treated for 2 h under Ar gas protection at 600 °C followed by slow cooling. This procedure was reported to promote metallic Pb particle segregation on the surface in contrast to air annealing and subsequent water quenching [13]. For both as-received and heat-treated samples, alkaline etching was carried out in 10 wt.% NaOH at 60 °C for 10 s, removing approximately 1 μm material from the sample surface, and then desmutted for 1 min in concentrated nitric acid at 25 °C.

Electrochemical polarization measurements were performed at a potential sweep rate of 6 mV/min in a 5 wt% NaCl electrolyte exposed to ambient air at 25 ±1°C, and stirred by a mechanical stirrer at a fixed rate. In all cases, polarization curves were recorded at least twice in order to ensure reproducibility of the plotted results. In preparing TEM cross-sectional specimens, four strips of sample material, 4 mm wide, were first glued together with epoxy and the final electron transparency achieved by ion-milling [12]. Some TEM cross-sectional specimens were also generated by ultramicrotomy, with final sectioning with a diamond knife, using well-documented procedures [14]. Stripped surface oxide specimen was prepared by immersing an annealed sample into a mercuric chloride-methanol solution to remove the thermal oxide film from the aluminum substrate.

TEM observations were carried out using a Philips CM30 microscope with a point-to-point resolution of 0.23 nm. The microscope was equipped with an energy dispersive x-ray spectroscopic (EDS) detector, operated at an accelerating voltage of 300 kV. Quantitative depth profiling was performed by using a LECO GDS-750A optical emission spectrometer (GD-OES) in the direct current mode. Surface morphology was investigated by using a Hitachi Field Emission Gun SEM (FEG-SEM) equipped with an EDS detector operated at 5 kV.

RESULTS

Figure 1(a) shows the effect of 2 h annealing in the range 350 – 600 °C on the anodic behavior of as-extruded AA8006 in 5 wt.% NaCl solution. In comparison to the behavior of an etched sample as shown in Fig. 1(b), the material was already active in the as-received condition as a result of hot rolling and annealing during fabrication. Therefore, the polarization curve for the sample annealed at 350°C was nearly identical to that of the as-received condition [6]. If the surface was first etched and then annealed, the anodic behavior obtained was quite similar to those for the as-received samples, as shown in Fig. 1(b). Maximum activation, as judged by the general anodic current level, appeared to be attained by annealing at 450°C for the etched surface.

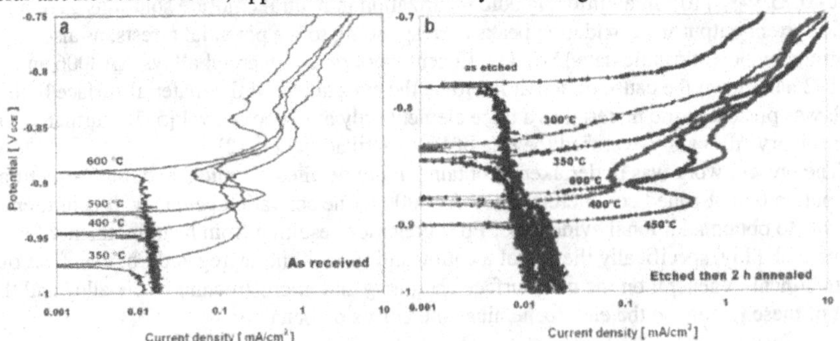

Figure 1. Effect of annealing temperatures on the polarization behavior of as received (a) and etched (b) alloy 8006 in 5% NaCl solution. The samples were annealed for 2 h annealing followed by quenching in water.

Annealing at higher temperatures gave reduced activation as shown by the 600°C annealed curve in Fig. 1(b). Increasing the annealing periods beyond 2 h did not significantly affect the shape of the polarization curves in the annealing range 350 to 600 °C. It should be reiterated that these results differ from those for the binary model AlPb alloys [11, 12] in the fact that activation of alloy 8006 occurs much easier at lower temperatures of annealing.

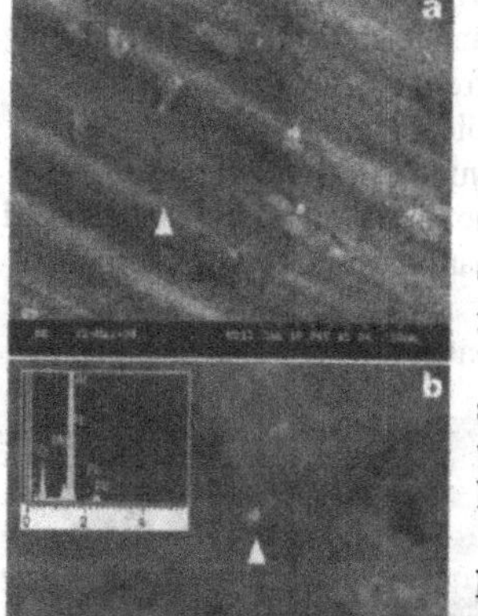

GD-OES elemental depth profiles obtained from the as-received and air-annealed AA8006 alloy verified earlier data[8] insofar as the enrichment of Pb together with significant Mg enrichment at the surface. The profiles for the etched surface indicated that Pb was removed from the metal surface.

Since the thick thermal oxide layers covered the sample surfaces after high temperature air-annealing, it was difficult to visualize the segregated Pb particles by SEM investigation [12]. However, Pb-containing particle was visible by SEM analysis, as shown in Fig 2, after promoting Pb segregation at 600 °C for 2 h annealing under oxygen-lean Ar atmosphere followed by slow cooling in air. TEM characterization of cross-sectional specimens verified the earlier results [1-5] that a grain refined surface layer (GRSL) existed on the as-extruded surface, as shown in figure 3(a). The round particles with dark contrast underneath the GRSL, indicated by the black arrows, were confirmed by EDS as the primary particles α-Al(Mn,Fe)Si and $Al_6(Mn,Fe)$. After annealing for 2 h at 600 °C, the remnants of GRSL was still present in the form of finely dispersed Al nanograins, as indicated by arrows, in the TEM cross-sectional micrograph of figure 3(b), which was prepared by ultramicrotomy. Rest of the GRSL was probably oxidized to form the thick oxide layer visible in the micrograph, rather than becoming recrystallized. The metallic aluminum microstructure of the nanograins was confirmed by TEM micro-diffraction analysis. The diffraction pattern example inserted in Fig. 3(b) was indexed as Al [321] orientation of the double-arrow marked grain in the figure.

Figure 2. Plan-view morphology of the as-received sample, annealed for 2 h at 600 °C under oxygen-lean Ar atmosphere, followed by slow cooling in air. The enlarged figure (b) shows the SEM EDS analysis of the Pb-rich particle as insert.

Significant effort to locate evidence of Pb, either in segregated form or in solid solution with the aluminum matrix, gave limited result. Specifically, a few segregated Pb-containing particles could be detected on a foil sectioned from a sample which was annealed for 2 h at 450 °C, as shown in figure 4, as

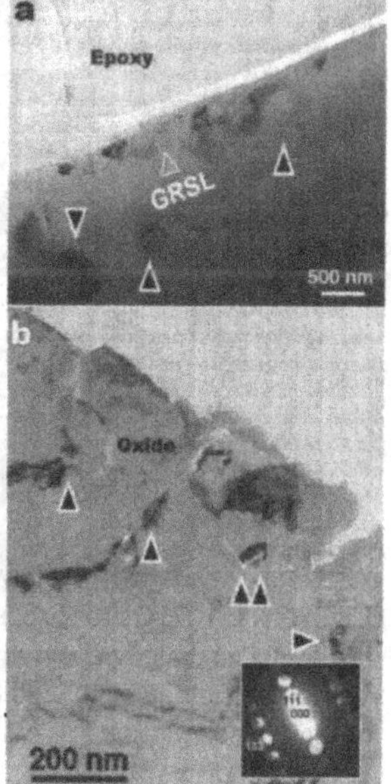

Figure 3. TEM cross-sectional micrographs of (a) as-received sample (b) and as-received sample annealed for 2 h at 600 °C, followed with water quenching. Nanocrystals of aluminum remaining from the original GRSL are indicated by arrows on the annealed sample.

confirmed by the corresponding EDS spot analysis. TEM selected area electron diffraction (SAED) pattern inserted in Fig. 4(a) was recorded from the top oxide surface region which demonstrated amorphous alumina together with spinel and γ-Al_2O_3 microcrystalline phases inside the oxide layer. As indicated by the high resolution TEM (HRTEM) image shown in figure 4(b), this particle was located at the metal-oxide interface and in contact with the Al metal surface. The inserted HRTEM micrograph in Fig. 4(b), enlarged from the arrow-marked area in the figure, indicated that the particle lattice fringes in two perpendicular directions were both 0.287 nm, as determined by using the bulk Al (111) lattice 0.234 nm as internal reference [12]. Such lattice-fringe configuration indicated that this particle was not in metallic Pb state. HRTEM data of the observed Pb-rich particle did not reveal further information about the composition and structure.

Figure 5(a) and (b) show the TEM micrographs of surface oxides stripped from sample

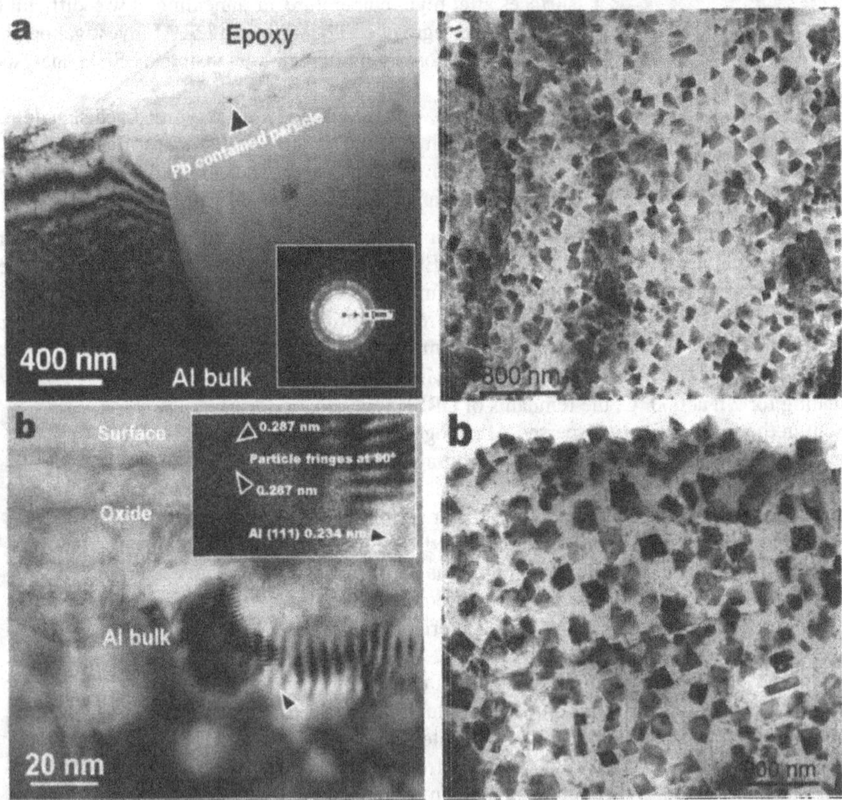

Figure 4. (a) TEM cross-sectional micrograph of sample annealed for 2 h at 450°C, showing a Pb-rich particle (arrow), and (b) HRTEM image of the Pb-rich particle.

Figure 5. TEM micrographs of the thermal oxides stripped from samples annealed for 2 h at (a) 500 °C and (b) 600 °C.

surfaces annealed for 2 h at 500 and 600 °C, respectively. TEM SAED pattern of the dark contrast crystalline phases in both figures revealed the spinel structure $MgAl_2O_4$ [15]. Only a tiny fraction of the oxide crystals were characterized as γ-Al_2O_3 crystals. Both the spinel and γ-Al_2O_3 crystals were embedded in an amorphous aluminum oxide matrix which corresponded to the well known oxide structures for Mg containing aluminum alloys discussed in the literature [16]. As shown in Figures 5(a) and (b), it demonstrated the significant increasing size of the spinel crystalline phase with increasing annealing temperature from 500 to 600 °C.

DISCUSSION

If the segregation of the trace element Pb is a normal activation controlled diffusion process, increase of both the surface enrichment of Pb and anodic activation of the surface would be expected by increasing annealing temperature. However, annealing temperature did not seem to have such an effect on alloy 8006. The results verified, in fact, that alloy 8006 became electrochemically activated at a lower annealing temperature [5] than the model binary alloy AlPb [12] for a fixed time of annealing. The present work indicated furthermore that Pb enrichment of the surface occurred most effectively at 450°C than at 600°C for alloy 8006, whereas the higher annealing temperature was more effective for Pb enrichment of the surface of the binary model AlPb alloy. Among the commercial alloys investigated earlier, alloys 3005 [7], 5754 [6] and an "Alloy X" [3] with the nominal composition 0.4% Fe, 0.2% Si, 0.2% Mn, 0.1% Mn exhibited behavior similar to alloy 8006, while alloys, 1050, [3] 1080, [3] 1200 [3] and 3102 [9] were similar to the binary AlPb.

One difference which could give this distinction between the two alloy groups is the fact that the first group contained Mg, and the second group did not. A direct role of Mg in the activation mechanism was ruled out because it existed in a stably oxidized state at the surface. However, magnesium is known to diffuse readily to the surface by heat treatment [4-9, 11, 12, 15, 16], if it is present in the alloy, probably by diffusion of the metal along the grain boundaries. The metal oxidizes readily once it becomes in contact with oxygen close to the alloy surface.

It is difficult to discuss the factors affecting the rate of Pb transport to the surface, nature of Pb segregation at the surface, and effect of these on surface activation because appreciable Pb was enriched already as a result of the thermomechanical history of the as-received sample. The present results verify again that most of the Pb near the alloy surface was removed by deep alkaline etching. The most significant enrichment in terms of the peak concentration of Pb was obtained by annealing at around 450°C. Highest anodic activation, as judged by the anodic current output, was also obtained at this temperature. The solid solution concentration of Pb in aluminum metal is limited at this temperature. Even in a supersaturated state at the metal surface, the results so far indicated a maximum possible solid solution concentration of about 1 wt%. HRTEM analysis of cross-sectional foils seemed to support this hypothesis although the present HRTEM data were not enough to allow determination of their exact microstructure and composition. Furthermore, the probability of capturing Pb-containing particles of adequate size and number in cross-sectional TEM foils is small. Whether Mg enhances Pb diffusion, since the two elements have a certain affinity to one another in terms of solid solution solubility and formation of intermetallic compounds, remains to be shown. Work is in progress by use of ternary model AlMgPb alloys.

The reduced activation resulting from annealing at temperatures higher than 450°C is difficult to explain based on the view that the Pb in solid solution with the aluminum matrix is responsible for the activation, since the solid solution concentration should increase with increasing temperature. However, this study could not ascertain the type of Pb containing phases formed by annealing at 450°C, which may be responsible for the activation, and their stability with increasing temperature. Furthermore, formation of the spinel oxide with increasing density with increasing annealing temperature may contribute to the passivity of the surface [16]. In addition, evaporation of the surface Pb with increasing temperature may become possible [12].

CONCLUSIONS

The data presented for the rolled commercial AA8006 aluminum alloy and data available for various other commercial and model alloys in the literature indicated that segregation of the trace element Pb as a result of heat treatment requires annealing at lower temperatures (450°C) on Mg-containing aluminum alloys than on Mg-free alloys (600°C) for fixed annealing time (2 h). The presence of magnesium in the alloy enhanced enrichment and segregation of Pb at the surface, thereby increased anodic activation of the alloy in chloride environment. Although the bulk concentration was at the ppm level, the trace element Pb became enriched at the surface of rolled alloy 8006, probably in solid solution with the aluminum matrix, at a supersaturated concentration of 1 wt%. Reduced activation with increasing temperature is attributed to increasing density of the spinel crystalline phase to cause the passivity of the surface.

ACKNOWLEDGMENTS

This work was supported by the Norwegian Research Council under contract no. 158545/431.

REFERENCES

1. H. Leth-Olsen, J. H. Nordlien, and K. Nisancioglu, *J. Electrochem. Soc.*, **144**, L196 (1997).
2. H. Leth-Olsen and K. Nisancioglu, *Corros. Sci.*, **40**, 1179 (1998).
3. H. Leth-Olsen, A. Afseth and K. Nisancioglu, *Corros. Sci.*, **40**, 1195 (1998).
4. H. Leth-Olsen, J. H. Nordlien, and K. Nisancioglu, *Corros. Sci.*, **40**, 2051 (1998).
5. Y. W. Keuong, J. H. Nordlien and K. Nisancioglu, *J. Electrochem. Soc.*, **148**, B497 (2001).
6. A. Afseth, J. H. Nordlien, G. M. Scamans, and K. Nisancioglu, *Corros. Sci.*, **43**, 2259 (2002).
7. A. Afseth, J. H. Nordlien, G. M. Scamans, and K. Nisancioglu, *Corros. Sci.*, **44**, 145 (2002).
8. Y. W. Keuong, S. Ono, J. H. Nordlien and K. Nisancioglu, *J. Electrochem. Soc.*, **150**, B547 (2003).
9. J. T. B. Gundersen, A. Aytac, J. H. Nordlien and K. Nisancioglu, *Corros. Sci.*, **46**, 265 (2004).
10. K. Nisancioglu and H. Holtan, *Corros. Sci.*, **18**, 835 (1978).
11. J. T. B. Gundersen, A. Aytac, J. H. Nordlien and K. Nisancioglu, *Corros. Sci.*, **46**, 679 (2004).
12. Ø. Sævik, Y. Yu, J. H. Nordlien and K. Nisancioglu, submitted to *J. Electrochem. Soc.* (2004).
13. H. Tsubakino, A. Nogami, A. Yamamoto, M Terasawa, T, Mitamura, T. Yamanoi, A. Kinomura and Y. Horino, *Materials Sci. Form*, **396-402**, 435 (2002).
14. R. Furneaux, G.E. Thompson and G.C. Wood, *Corros. Sci.*, **18**, 853 (1978).
15. D.J. Field, G. M. Scamans and E.P. Butler, *Metall. Tran.* **18A**, 463 (1987).
16. C. Lea and M.P. Seah, Philos. Mag., **35**, 213 (1977).

Mater. Res. Soc. Symp. Proc. Vol. 843 © 2005 Materials Research Society T3.5.

Fabrication of Surface Nitridation Mask on GaAs Substrate for Nano-Lithography

Yo Yamamoto[1], Kohji Saito[2], Toshiyuki Kondo[2], Takahiro Maruyama[1,2] and Shigeya Naritsuka[1,2]
[1]Meijo Univ. 21st Century COE Program, Nagoya, Japan
[2]Dept. of Materials Science & Engineering, Meijo University, Nagoya, Japan.

ABSTRACT

The surface nitridation of GaAs substrate under various conditions were performed and its surface morphology was studied by using atomic force microscope. The RMS roughness of the surfaces was closely related to formation of the particles which was formed by As evaporation and subsequent Ga migration, which was enhanced by higher substrate temperature.

INTRODUCTION

Recently, the hetero epitaxial growth of III-V semiconductors on Si, such as GaAs/Si, has been extensively studied for making opto-electronic integrated circuit (OEIC). In this field, the formation of dislocations caused by the lattice mismatch and the difference in thermal expansion coefficient between GaAs and Si are recognized as unavoidable problem.

As one of the effective techniques to solve these problems, epitaxial lateral overgrowth (ELO) has been proposed. It is composed by photo-lithography and selective-area growth, provides a flat epitaxial layer with very few dislocations for Si, GaAs and GaP alloy [1-6]. Also, a remarkable reduction of the dislocation density even in the hetero epitaxial system with large lattice mismatching, such as GaAs on Si, also has been reported [1-6]. By using hetero epitaxial lateral overgrowth, there is a possibility to design a novel nano-structure device with a high reliability enough for commercial applications.

However, in order to fabricate nano structure devices, finer machining technique such as electron lithography and/or scanning tunnel microscopy (STM) lithography is necessary instead of a conventional photolithography. Combination of STM lithography and selective-area growth allows us to design a nano structured optical and electronic devices. We have chosen an ultra-high-vacuumed process containing STM lithography and molecular beam epitaxy (MBE). To take full advantages of STM lithography, it is important to develop the applicable mask with good machinability and stability. One of the candidates is GaN thin layer. We studied about the formation of GaN thin layer on GaAs substrates by the nitridation. This paper especially focused on the surface morphology of the nitrided film on the GaAs substrate prepared under various conditions of nitridation and growth.

EXPERIMENTAL PROCEDURE

After chemical etching with a solution of $NH_4OH:H_2O_2:H_2O=4:1:20$, GaAs (001) substrate was heated up to 580°C in MBE chamber to remove the surface oxide layer. Then, the surface of the substrate was nitrized by RF nitrogen source, changing the RF power between 200-400W, N_2 flow rate between 1.2-0.6 SCCM, substrate temperature between 300-500°C which was set, and the nitridation time for 15-60 min. Microstructure of the specimens were studied by using atomic force microscope (AFM) and scanning electron microscope (SEM).

RESULT AND DISCUSSION

Fig.1 is typical AFM image of the nitrided surface of GaAs (001) substrate. Numerous nano particles were observed on the surfaces. The formation of nano particles on the surface, which results in the RMS roughness of the surface, is controlled by the temperature change as shown in Fig.2. For example, when the nitridation was done at 500°C for 60 min, it made a rougher surface with RMS roughness of 4.5 nm, which was brought by the formation of large particles whose diameter was as large as 300 nm. On the other hand, when the substrate was nitrided at 300°C, the surface of the substrate became flatter as the time. The RMS roughness was as small as 0.5nm for the nitridation at 300°C for 60 min.

The distribution of particle size analyzed from AFM image is shown in Fig.3. When the surfaces were nitrided at 300 °C, the particle size shows narrower distribution (i.e. homogeneous and fine microstructure consisted from fine particles) as the time, on the other hand, it shows broader distribution at 500 °C (i.e. inhomogeneous microstructure containing large particles). This tendency was confirmed also in the substrate of other orientation.

Considering that As evaporation and migration of Ga surface atoms were accelerated by high substrate temperature of 500°C, and consequently, the size of the particles was also enhanced, the formation mechanism of those particles was considered to be closely related to As evaporation and Ga migration.

Here, we discussed the cause of high RMS roughness of the surface at the substrate temperature of 500 °C. Due to high substrate temperature, As evaporation was accelerated. It generates excessive Ga atoms on the surface. By increasing the density and the migration distance of Ga, the particles were enlarged as a result, the RMS roughness became higher.

More detail information for microstructure of nitrided film, such as thickness of nitrided film and effect of each condition on it, is needed for further discussion.

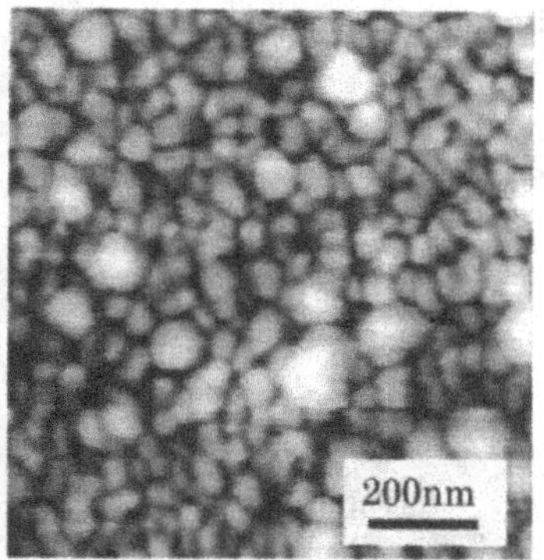

Fig. 1 Typical AFM image for the nitrided surface of GaAs (001); Sample were nitrided at 300°C for 30min.

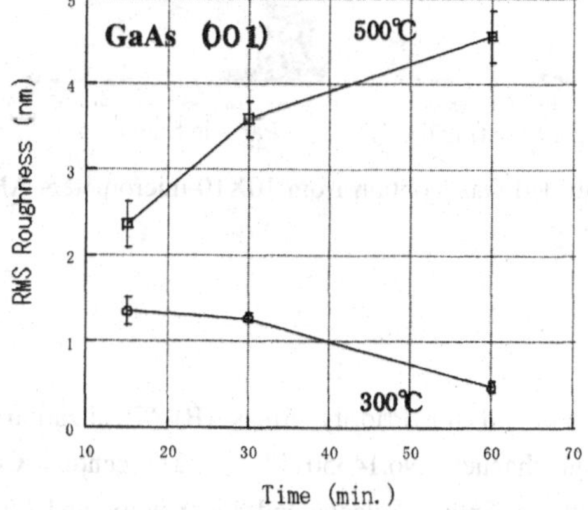

Fig. 2 The variations of RMS roughness as a function of nitridation time.

CONCLUSIONS

In order to fabricate nitride mask for a STM lithography, GaAs were nitrided with various conditions and these surface structures were studied by using AFM. The numerous particles were observed on the surfaces of nitrided GaAs, whose diameter were changed with nitridation time and substrate temperature. The formation mechanisms of particles were considered to be closely related to As evaporation and Ga migration. In case when the samples were nitrided at

300 °C, a RMS roughness of 0.5 nm was achieved after the nitridation of 60min.

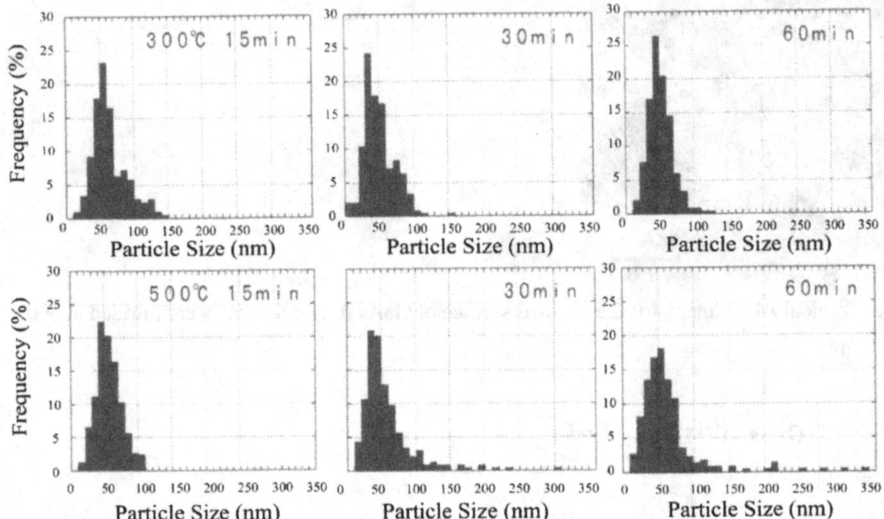

Fig. 3 The particle size distribution of the nitrided GaAs obtain from 10X10 micrometers AFM image

ACKNOWLEDGMENT

This work was partly supported by Grant-in-Aid for Priority Areas (B) "Realization of dislocation-free epitaxy with the aid of nanochannel" No.14350172, the 21st century COE program from the Ministry of Education, Culture, Sports, Science and Technology, and Meijo University Grant for the Promotion of Academic Research.

REFERENCES

[1] T. Nishinaga, T.Nakano and S. Zhang, Jpn. J. Appl. Phys. 27 (1988) L694.

[2] S. Zhang and T. Nishinaga, J. Cryst. Growth. 99 (1990) 292.

[3] Y. Suzuki and T. Nishinaga, Jpn. J. Appl. Phys. 28 (1989) 440.

[4] R. Bargman, E. Bauser and J. H. Werner, Appl. Phys. Lett. 57 (1990) 351.

[5] R. Bargman, J. Cryst. Growth 110 (1991) 823.

[6] S. Zhang and T. Nishinaga, Jpn. J. Appl. Phys. 29 (1990) 545.

AUTHOR INDEX

SUBJECT INDEX